AN
ATLAS
OF INDIA

DELHI
OXFORD UNIVERSITY PRESS
OXFORD   NEW YORK
1990

Oxford University Press, Walton Street, Oxford OX2 6DP
New York   Toronto
Delhi   Bombay   Calcutta   Madras   Karachi
Petaling Jaya   Singapore   Hong Kong   Tokyo
Nairobi   Dar es Salaam
Melbourne   Auckland
and associates in
Beirut   Berlin   Ibadan   Nicosia

EDITORIAL BOARD

SBN 0 19 562553 6

The following pertains to all maps in this atlas:

© *Government of India Copyright 1990*

Based upon Survey of India map with the permission of the Surveyor General of India.

The responsibility for the correctness of internal details rests with the publisher.

The territorial waters of India extend into the sea to a distance of twelve nautical miles measured from the appropriate base line.

The administrative headquarters of Chandigarh, Haryana and Punjab are at Chandigarh.

The boundary of Meghalaya shown on this map is as interpreted from the North-Eastern Areas (Reorganisation) Act, 1971, but has yet to be verified.

Produced in India
Designed, compiled, cartographed and printed by
TT. Maps & Publications Ltd, 328 GST Road, Chromepet, Madras 600044
and Published by S. K. Mookerjee, Oxford University Press
YMCA Library Building, Jai Singh Road, New Delhi 110001.

CONTENTS

INTRODUCTION

An Atlas of India is an exhaustive reference work on India and its states. This comprehensive volume forms a splendid companion to *A Social and Economic Atlas of India* (OUP 1987).

In this atlas, each state and union territory of India has been allotted nine maps. These have been drawn to varying scales depending upon the data available and the facts to be presented for each region. There is, however, a uniformity in the depth and detail of information provided in each map. The data presented have been indexed to the year 1989 to facilitate compatibility and analysis. Projections of data may introduce a small degree of error, but this is more than compensated for by facilitating the cross comparison of maps.

The Main Maps

The nine maps for each state or union territory deal with the following subjects: (1) physical features, (2) administration, (3) urban and rural population, (4) transport and tourist centres, (5) agriculture and irrigation, (6) industries and market potential, (7) mineral resources, (8) economic development, and (9) audio-visual mass communication.

1. The Physical map of each state depicts the main rivers and lakes, and the important peaks, hill ranges and plateaux. Regional topography is immediately apparent because of the use of a broad spectrum of colour and tones. Plains are shown in shades of green, while raised terrain appears in shades of yellow and orange. Mountainous areas are portrayed in distinctive browns and violets.

2. The Administration map is a special feature of this atlas. It provides detailed information down to the smallest administrative level, namely, the *taluk* or *tehsil*. Where *taluk*-wise information is not readily available, district-wise information has been used. The district boundaries are based on information available in 1989, but the information superimposed on each district is based on the 1981 configuration. (District boundaries for seven districts — five in U.P., and one each in Assam and Arunachal Pradesh — are not available.) Several districts formed between 1981 and 1989, were parts of bigger districts in 1981. At the time of the 1981 census there were 412 districts. In 1989 there were 449 districts, of which 442 are marked on the maps. The *taluk* boundaries, however, have been drawn in accordance with the 1981 census. In the administration maps, *taluk* headquarters have been plotted even where information about the *taluk* boundaries is not available.

3. The Population map includes every town estimated to have a population of more than 10,000 in 1989. Rural centres with populations estimated at more than 10,000 in 1989 have also been included, wherever government and census information supports the data. However, because of absence of government data and published census information, it has not been possible to plot *every* village in this category.

All population projections are estimates for 1989, based on the 1981 census figures. Estimated growth, based on the compound annual rate of growth from 1971 to 1981, has been used to arrive at the estimated populations for 1989. Only for Assam has the 1989 projection been based on the 1961-71 growth rate. For the state of Sikkim, which joined the Union of India in 1975, India's average growth rate has been used to calculate the projected population.

The colour scheme is an additional indicator of the rural or urban base of a particular state. The density of rural population by district has been estimated for 1989, again using estimated growth based on the compound annual rate of growth from 1971 to 1981. The editors have used district data for density, as statistics for all *taluks* were not available.

The Census of India classifies all towns with populations over 100,000 as Class I towns. In this atlas, the information regarding Class I towns has been further divided into four categories on the basis of population to provide a more accurate picture of the urban population in each district. All towns with populations over 250,000 are also listed.

4. The Transport and Tourism map shows all national highways, major roads, major railways, as well as all ports officially classified as major and minor. Road and railway links to important tourist destinations have also been included. A separate map of India shows the location of all civil airports.

All places of tourist interest recognized by the Government of India's Department of Tourism have been indicated in large type. Towns shown in smaller type are either road or railway junctions. Tourist centres have been broadly classified into archaeological sites, which include sites of ruins, excavations etc.; historical sites which include those of important battles and other events, old monuments, palaces and forts; religious centres which cover shrines of all faiths, ashrams etc.; hill stations, which are summer holiday resorts in the hills; holiday resorts, which include holiday facilities by the side of parks, reservoirs, hot springs etc.; wildlife sanctuaries which include national parks and tiger projects; the important beaches; and lakes or reservoirs which include waterfalls and other water bodies. However, these classifications at times overlap.

5. The Agriculture map shows the most important crop (based on annual production) grown in each district. A distinctive pattern of crop belts is obtained by the effective use of colour. Districts which contribute the most to the total production in the state of a particular crop are shown in darker shades of a colour. Districts with a lighter shade of colour contribute less to total state production of the crop.

Any other important crops grown in a particular *taluk* are shown on the map with symbols. Where *taluk*-wise information has not been available for other important crops, district-wise information is given.

This map also shows irrigated areas in each district and names the existing irrigation projects and those under construction.

6. The Industries and Markets map locates every important industrial centre in the state. These are the centres which have major industrial units with sales of over Rs 10 million each. Centres where major manufacturing projects with an investment of more than Rs 50 million were still in the process of being implemented in 1985 have also been included in this category.

The major industrial centres shown in this map are indicated by a prominent symbol. These centres have virtually *all* the categories of industry found in scattered locations elsewhere in the state.

The *Annual Survey of Industries* has listed twenty-four broad classifications of industries in the country. These classifications have been grouped into twelve categories in the industrial maps as well as in the tables at the end of the atlas. In this grouping, the following should be noted:

(i) *Beverage and tobacco products* includes alcohol and allied products, tea and coffee products, cigarettes and other tobacco products.

(ii) *Chemical products and pharmaceuticals* includes all basic chemicals and their products, petroleum and its products, all medicines and allied products, paints and inks.

(iii) *Food products* includes processed food.

(iv) *Leather, rubber and plastic products* includes all leather, rubber and plastic products.

(v) *Textile industry* includes cotton, wool, jute and synthetic textile production as well as garment manufacture using these materials.

(vi) *Electrical machinery* includes cables, wires, switches, relays, batteries, televisions, computers, calculators, radios and other communication equipment, as well as heavy electrical machinery.

(vii) *Metal products* includes all metal products and machinery made of metal.

(viii) *Non-electrical machinery* includes watches, typewriters and other light machinery which do not need electricity, and which have not been included in categories (vii) and (viii).

(ix) *Non-metallic products* includes cement, other allied products, pearls and precious stones, porcelain, sanitaryware, glass etc.

(x) *Paper products* includes all types of paper and its products, stationery items, printing and publishing.

(xi) *Transport equipment* includes all transport vehicles (including tractors).

(xii) *Wood products* includes all furniture etc.

Since much of Indian industry is urban-based, and urban areas are usually the better markets for consumer products, the district-wise market potential shown on this map is a reflection of urban buying potential. Hindustan Thompson Associates have arrived at separate market potential indices for rural and urban areas, and R.K. Swamy Advertising Associates for a combined district market, which emphasizes rural market indicators. Based on the indicators used by both these estimates and additional indicators considered valid by the editors, a district-wise market profile for 1989 has been developed for these maps. This is perhaps a truer indication of each district's potential. The indicators taken into account, using projected data for 1989, are as follows:

Area, number of households, population, population growth rate, sex ratio, density of population, male literacy, female literacy, urbanisation, workforce, number of traders, bank officers per 100,000 population, per capita bank credit to agriculture, per capita bank credit to industries, total cropped area, production of major crops, number of motor vehicles, post offices per 100,000 population, circulation of newspapers, cinema houses per 100,000 population, and number of radio and television sets.

7. In the Minerals map, mineral production in each state is plotted by *taluk*. Only mining centres which are nationally important have been named, although production in other areas may be locally significant.

8. The Economic Development map estimates the economic development of each district for 1989 based on projections of the growth rate over the previous ten years of the following indicators:

Per capita value of output of major crops, per capita bank credit to agriculture, number of mining and factory workers per 100,000 population, number of households of factory workers per 100,000 population, per capita bank credit for manufacturing sector, per capita bank deposits, per capita bank credit to services, literacy and urbanization.

The map uses a graded colour scheme by which economically developed areas are shown in shades of green, and economically underdeveloped areas in shades ranging from yellow to red. This colour scheme enables the development in each state to be understood at a glance. Information is also provided about districts which have been identified by the National Committee on the Development of Backward Areas as *Backward Districts*.

9. The Mass Communication map provides information on radio and television coverage in each state. For television coverage, the *area* reached by each television centre is indicated by the diameter of the circle centered on a particular station, and the *population covered* is indicated by the different shades used for each circle. For radio coverage, the *range* of each station in the four cardinal directions is shown by arcs. The population covered by these radio stations may be estimated by studying the extent of these arcs in conjunction with the Population map of each state. Where the scales are different, some calculation will be necessary to arrive at the population figure.

The map also includes a list of all cities where 100 or more periodicals are published.

Maps of India

In addition to state-wise information, this atlas contains eighteen introductory maps of India and two pages of diagrams. These provide additional information about some of the subjects shown in detail for each state.

For instance, the Climate and Soils maps, taken together, supplement, state-wise, the information given in the Agriculture map. The Power Generation map includes *all* existing power projects as well as those under construction. This map could be used to supplement the information in the Agriculture or Industries map, which in turn would provide further indicators to the map on Economic Development. The maps on Religion, Literacy and Workforce are district-wise, and supplement the information provided on the Population map of each state or union territory. Infrastructure Development is mapped at the district level, and gives information about the basic infrastructural amenities in the villages, the index being arrived at on the basis of the following parameters:

Railway station within 5 km; road and bus stop within 2 km; primary and middle schools within 2 km; high school within 5 km; dispensaries and health centres within 2 km; hospital within 5 km; fertilizer depots, seed stores, pump repair shops, warehousing and veterinary dispensaries within 5 km; credit cooperatives and banks within 5 km; post offices within 2 km; electricity and water supply within the village.

Of particular interest in this section of the atlas are two pages of diagrams which provide a comparison of the area, population and income of the different states and union territories in India, and of the population and area of selected countries in the rest of the world.

Tables and Index

The five tables at the end of the atlas supplement the information found on the maps. The tables provide information about the average rainfall and temperature in 212 centres; male and female as well as rural and urban populations; the area and production of 14 major crops; the annual production of all minerals found in India; and all towns which have one or more of the 12 industries classified. This information is based on the 1981 district boundaries in all cases. If a district does not appear in the Agriculture or Minerals table, it is because the output is either negligible or non-existent, or because the data was not available.

The atlas concludes with a comprehensive index of every town, village and physical feature marked on the maps and includes over 12,000 entries.

We thank the following organizations whose material formed our primary sources:

Census of India
Centre for Monitoring Indian Economy
Department of Agriculture and Cooperation
Department of Statistics, Ministry of Planning
Department of Tourism
Meteorological Department
Ministry of Information and Broadcasting
Ministry of Industry and Company Affairs
National Atlas and Thematic Mapping Organization
Survey of India.

The spellings in this atlas are in accordance with those used by the Survey of India.

Madras/Delhi
September 1990

India

This section comprises 18 maps and five diagrams. They present India at a glance, providing an objective view of the country's topography, federal structure, human habitation and infrastructural development, climate, religion, literacy, workforce and communication routes. The content is arranged on these maps in such a manner as to provide the user with ready information on the country.

Included in this section are five diagrams that have been drawn so as to show information in proportion to the actual data. Two diagrams are comparative statements of the Area and Population of India and the countries of the world. The other three diagrams are comparative statements of the Area, Population and Income of the states of India.

Note:

The maps of India in this section are on a Lambert azimuthal equal area projection and are in three scales: the full page map, 1:15,000,000 (1 cm = 150 kms); the half page map, 1:21,000,000 (1 cm = 210 kms); and the inset map, 1:350,000,000 (1 cm = 350 kms). The Survey of India approved latitudes and longitudes appear only on the Political map in this section.

INDIA

PHYSICAL

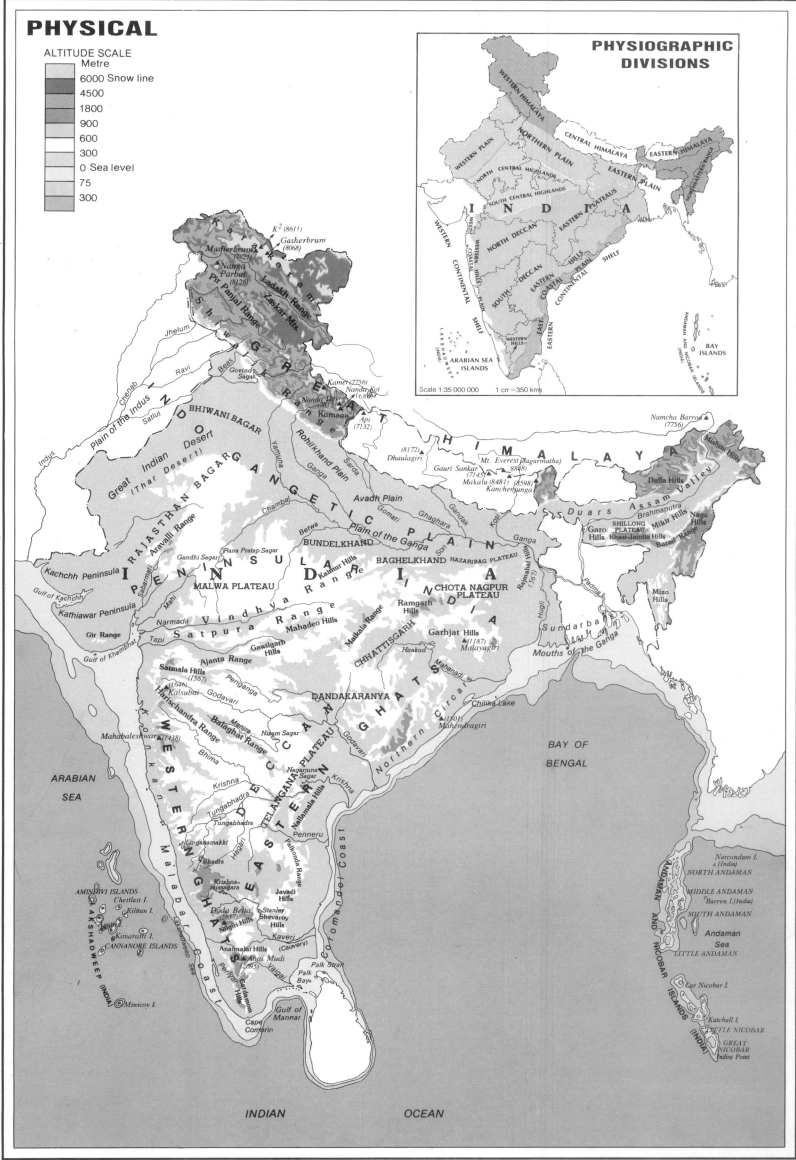

ALTITUDE SCALE
Metre

6000 Snow line
4500
1800
900
600
300
0 Sea level
75
300

PHYSIOGRAPHIC DIVISIONS

WESTERN HIMALAYA
NORTHERN PLAIN
CENTRAL HIMALAYA
EASTERN HIMALAYA
WESTERN PLAIN
NORTH CENTRAL HIGHLANDS
EASTERN PLAIN
NORTHEASTERN RANGES
I N D I A
SOUTH CENTRAL HIGHLANDS
EASTERN PLATEAUS
WESTERN COASTAL HILLS
NORTH DECCAN
EASTERN HILLS
WESTERN CONTINENTAL SHELF
SOUTH DECCAN
EASTERN COASTAL PLAIN
CONTINENTAL SHELF
WESTERN HILLS
EASTERN
ANDAMAN AND NICOBAR ISLANDS
BAY ISLANDS
LAKSHADWEEP (INDIA)
ARABIAN SEA ISLANDS

Scale 1:35 000 000   1 cm = 350 kms

K² (8611)
Gasherbrum (8068)
Masherbrum (7821)
Karakoram
Nanga Parbat (8126)
Pir Panjal Range
Ladakh Range
Zaskar Mts.
Shiwalik Range
GREAT Range
Jhelum
Beas
Govind Sagar
Chenab
Ravi
Satluj
Indus
Plain of the Indus
Kamet (7756)
Nanda Kot (6804)
Nanda Devi (7817)
Kumaon
Api (7132)
BHIWANI BAGAR
INDO GANGETIC PLAIN
Great Indian Desert (Thar Desert)
RAJASTHAN BAGAR
Rohilkhand Plain
Yamuna
Ganga
Sarda
Ghaghara
Gandak
Kosi
Ganga
(8172)
Dhaulagiri
Gauri Sankar (7145)
Mt. Everest (Sagarmatha) (8848)
Makalu (8481) (8598)
Kanchenjunga
Avadh Plain
Gomati
HIMALAYA
Namcha Barwa (7756)
Mishmi Hills
Dafla Hills
Duars
Assam Valley
Brahmaputra
Garo Hills
SHILLONG PLATEAU
Khasi-Jaintia Hills
Mikir Hills
Naga Hills
Barail Range
Padma
Hugli
Mizo Hills
ARAVALLI Range
Aravalli Range
Chambal
Betwa
Son
Plain of the Ganga
BUNDELKHAND
BAGHELKHAND
HAZARIBAG PLATEAU
Kaimur Hills
Kaimur Range
Rajmahal Hills (567)
CHOTA NAGPUR PLATEAU
Sundarbans
Mouths of the Ganga
Kachchh Peninsula
Gulf of Kachchh
Kathiawar Peninsula
PENINSULAR INDIA
Sabarmati
Gandhi Sagar
Rana Pratap Sagar
MALWA PLATEAU
Mahi
Narmada
Vindhya Range
Satpura Range
Mahadeo Hills
Ramgarh Hills
Malkala Range
Garhjat Hills
(1187)
Malayagiri
Hirakud
CHHATTISGARH
Gir Range
Tapi
Gulf of Khambhat
Gawilgarh Hills
Ajanta Range
DANDAKARANYA
Mahanadi
Chilika Lake
(1501)
Mahendragiri
Satmala Hills (1567)
(1646)
Kalsubai
Godavari
Penganga
Harischandra Range
Balaghat Range
Manjra
Nizam Sagar
NORTHERN CIRCARS
Mahabaleshwar (1438)
Bhima
DECCAN PLATEAU
TELANGANA PLATEAU
Godavari
EASTERN GHATS
WESTERN GHATS
Krishna
Nagarjuna Sagar
Krishna
BAY OF BENGAL
ARABIAN SEA
Tungabhadra
Tungabhadra
Penneru
Nallamala Hills
Palkonda Range
Lingamakki
Hagari
Krishna-rajasagara
Bhadra
Javadi Hills
Coromandel Coast
Doda Betta (2637)
Stanley Shevaroy Hills
Nilgiri Hills
Kaveri
AMINDIVI ISLANDS
Chettlatt I.
Kiltan I.
Kavaratti I.
CANNANORE ISLANDS
LAKSHADWEEP (INDIA)
Minicoy I.
Konkan Coast
Malabar Coast
Lakshadweep Sea
Anaimalai Hills
Anai Mudi (2695)
Yaigai
Cardamom Hills
Periyar
Palk Strait
Palk Bay
Gulf of Mannar
Cape Comorin
Narcondam I. (India)
NORTH ANDAMAN
MIDDLE ANDAMAN
Barren I. (India)
SOUTH ANDAMAN
Andaman Sea
LITTLE ANDAMAN
ANDAMAN AND NICOBAR ISLANDS (INDIA)
Car Nicobar I.
Katchall I.
LITTLE NICOBAR
GREAT NICOBAR
Indira Point

INDIAN OCEAN

Scale 1:15 000 000   1 Cm = 150 Kms

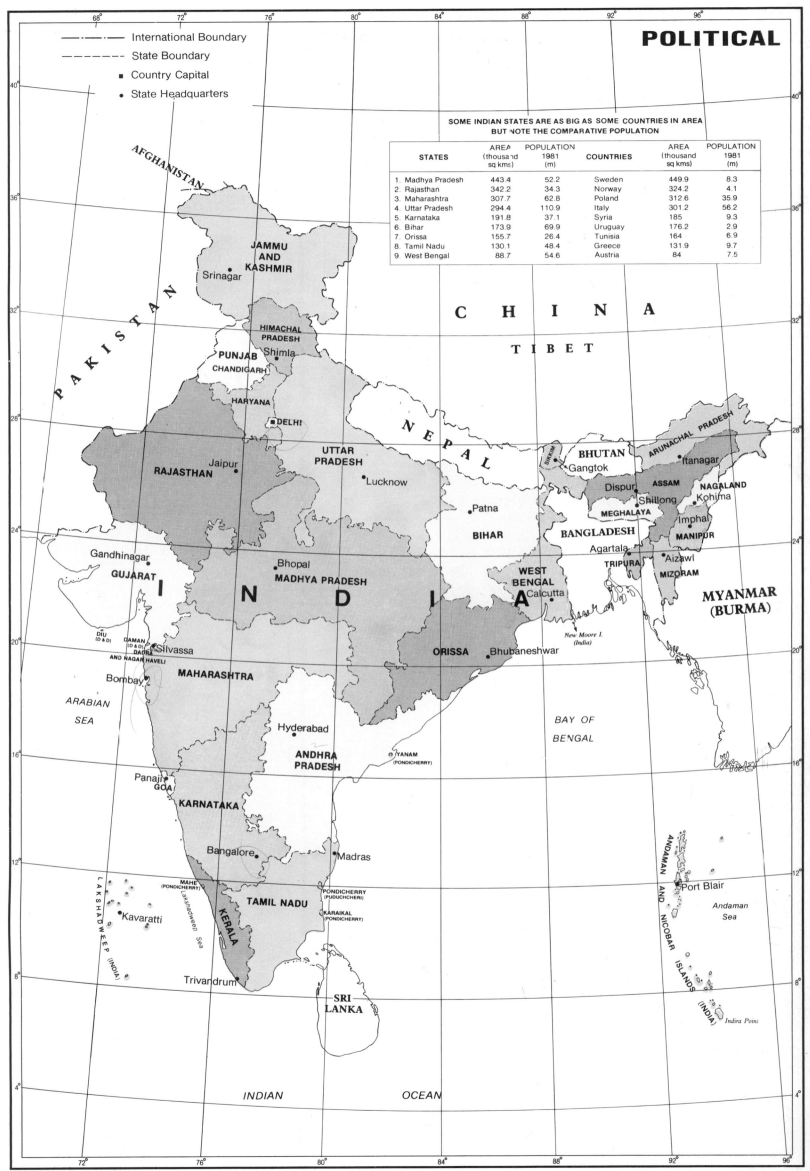

POLITICAL

SOME INDIAN STATES ARE AS BIG AS SOME COUNTRIES IN AREA
BUT NOTE THE COMPARATIVE POPULATION

STATES	AREA (thousand sq kms)	POPULATION 1981 (m)	COUNTRIES	AREA (thousand sq kms)	POPULATION 1981 (m)
1. Madhya Pradesh	443.4	52.2	Sweden	449.9	8.3
2. Rajasthan	342.2	34.3	Norway	324.2	4.1
3. Maharashtra	307.7	62.8	Poland	312.6	35.9
4. Uttar Pradesh	294.4	110.9	Italy	301.2	56.2
5. Karnataka	191.8	37.1	Syria	185	9.3
6. Bihar	173.9	69.9	Uruguay	176.2	2.9
7. Orissa	155.7	26.4	Tunisia	164	6.9
8. Tamil Nadu	130.1	48.4	Greece	131.9	9.7
9. West Bengal	88.7	54.6	Austria	84	7.5

International Boundary
State Boundary
■ Country Capital
• State Headquarters

AFGHANISTAN

PAKISTAN

CHINA

TIBET

NEPAL

JAMMU AND KASHMIR
Srinagar

HIMACHAL PRADESH
Shimla

PUNJAB
CHANDIGARH

HARYANA

■ DELHI

RAJASTHAN
Jaipur

UTTAR PRADESH
Lucknow

Patna

BIHAR

SIKKIM
Gangtok

BHUTAN

ARUNACHAL PRADESH
Itanagar

Dispur
ASSAM
Shillong
MEGHALAYA

NAGALAND
Kohima

Imphal
MANIPUR

BANGLADESH

Agartala
TRIPURA

Aizawl
MIZORAM

MYANMAR (BURMA)

Gandhinagar

GUJARAT

Bhopal
MADHYA PRADESH

INDIA

WEST BENGAL
Calcutta

New Moore I. (India)

DIU (D & D)
DAMAN (D & D)
DADRA AND NAGAR HAVELI
Silvassa

ORISSA
Bhubaneshwar

BAY OF BENGAL

Bombay

MAHARASHTRA

ARABIAN SEA

Hyderabad

ANDHRA PRADESH

YANAM (PONDICHERRY)

Panaji
GOA

KARNATAKA

Bangalore

Madras

MAHE (PONDICHERRY)

PONDICHERRY (PUDUCHERI)

ANDAMAN AND NICOBAR ISLANDS

Port Blair

Andaman Sea

LAKSHADWEEP (INDIA)

Kavaratti

Lakshadweep Sea

KARAIKAL (PONDICHERRY)

TAMIL NADU

KERALA

Trivandrum

SRI LANKA

(INDIA)
Indira Point

INDIAN   OCEAN

Scale 1:15 000 000   1 Cm = 150 Kms

3

INDIA AND THE WORLD — AREA

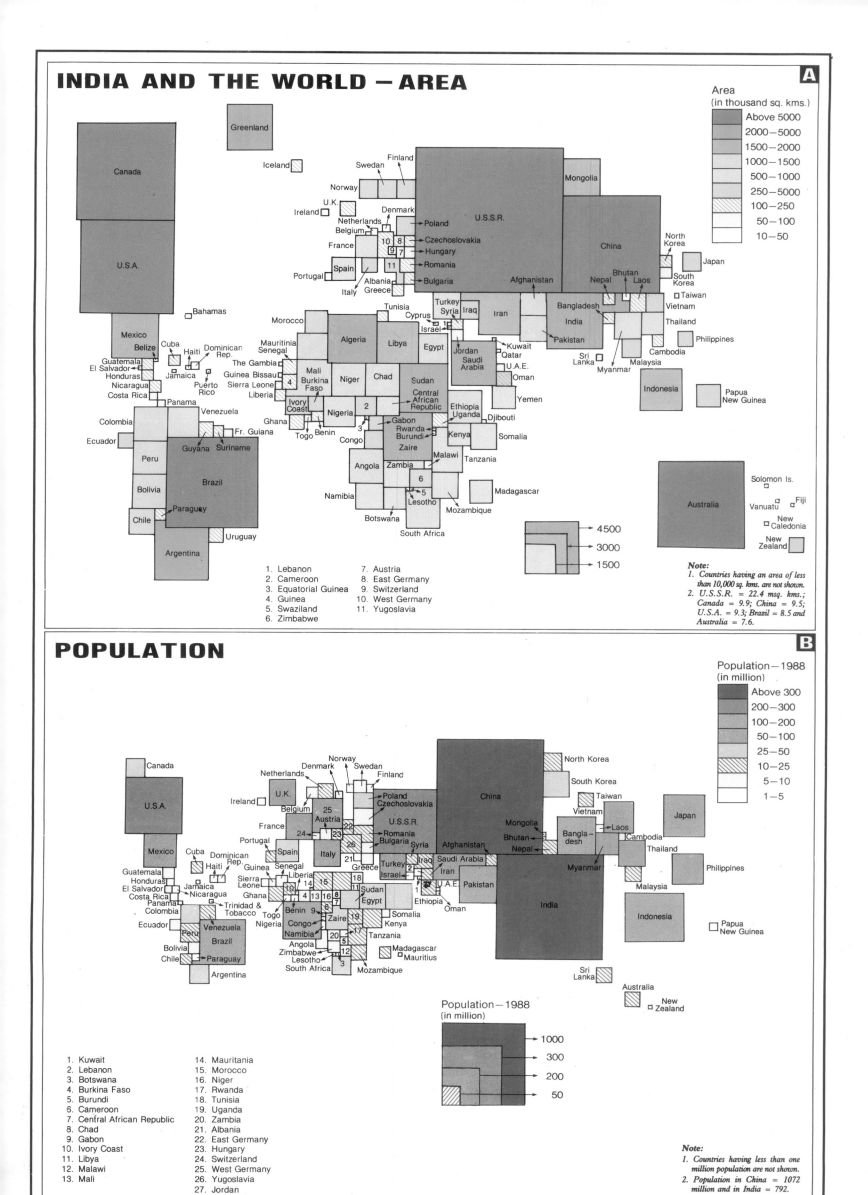

A

Area
(in thousand sq. kms.)

- Above 5000
- 2000 — 5000
- 1500 — 2000
- 1000 — 1500
- 500 — 1000
- 250 — 5000
- 100 — 250
- 50 — 100
- 10 — 50

1. Lebanon
2. Cameroon
3. Equatorial Guinea
4. Guinea
5. Swaziland
6. Zimbabwe
7. Austria
8. East Germany
9. Switzerland
10. West Germany
11. Yugoslavia

Note:
1. Countries having an area of less than 10,000 sq. kms. are not shown.
2. U.S.S.R. = 22.4 msq. kms.; Canada = 9.9; China = 9.5; U.S.A. = 9.3; Brazil = 8.5 and Australia = 7.6.

POPULATION

B

Population — 1988
(in million)

- Above 300
- 200 — 300
- 100 — 200
- 50 — 100
- 25 — 50
- 10 — 25
- 5 — 10
- 1 — 5

1. Kuwait
2. Lebanon
3. Botswana
4. Burkina Faso
5. Burundi
6. Cameroon
7. Central African Republic
8. Chad
9. Gabon
10. Ivory Coast
11. Libya
12. Malawi
13. Mali
14. Mauritania
15. Morocco
16. Niger
17. Rwanda
18. Tunisia
19. Uganda
20. Zambia
21. Albania
22. East Germany
23. Hungary
24. Switzerland
25. West Germany
26. Yugoslavia
27. Jordan

Population — 1988
(in million)

Note:
1. Countries having less than one million population are not shown.
2. Population in China = 1072 million and in India = 792.

4

A ANNUAL RAINFALL

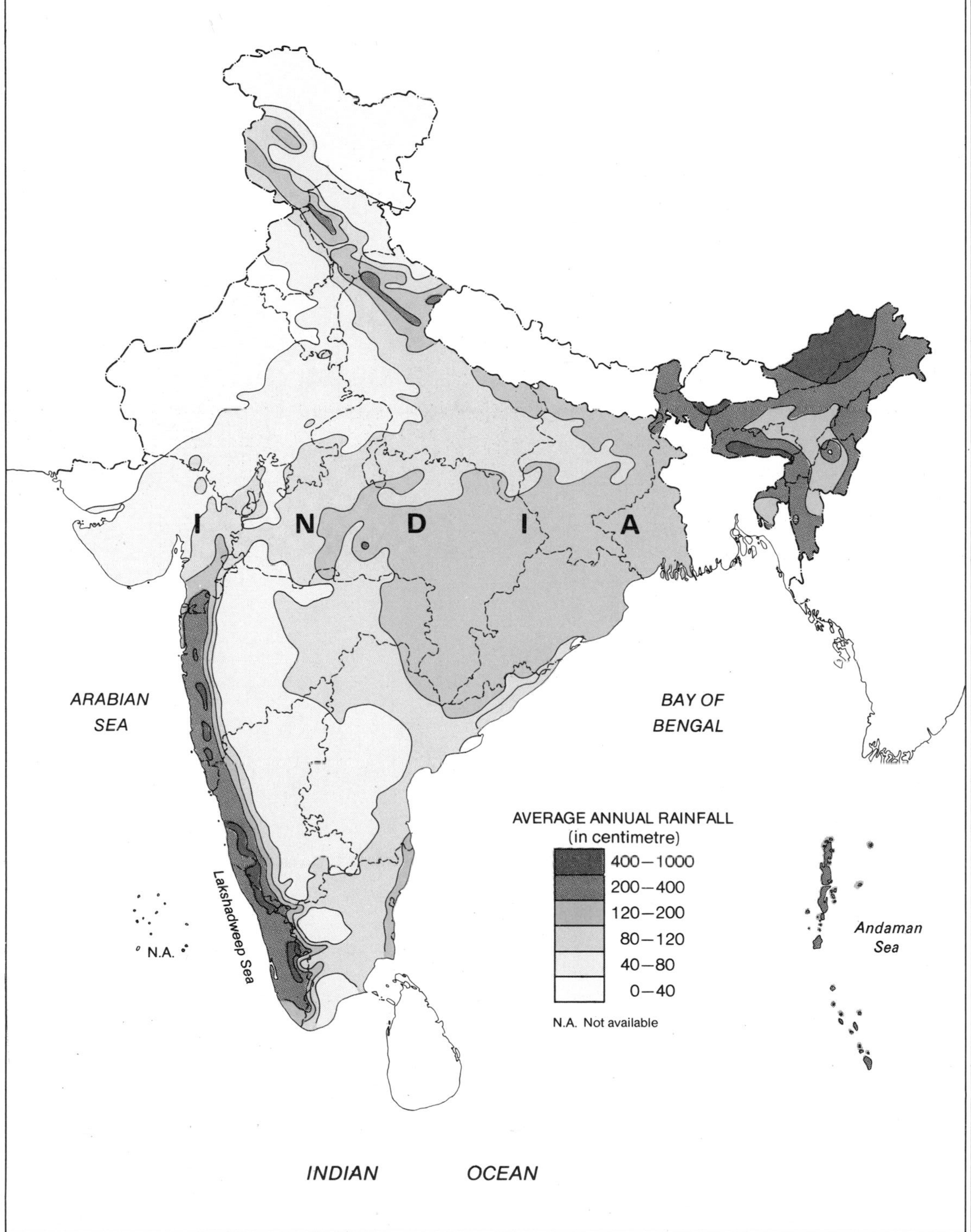
I N D I A
ARABIAN SEA
BAY OF BENGAL
Lakshadweep Sea
N.A.
Andaman Sea
INDIAN OCEAN
AVERAGE ANNUAL RAINFALL
(in centimetre)
400–1000
200–400
120–200
80–120
40–80
0–40
N.A. Not available

B ANNUAL TEMPERATURE

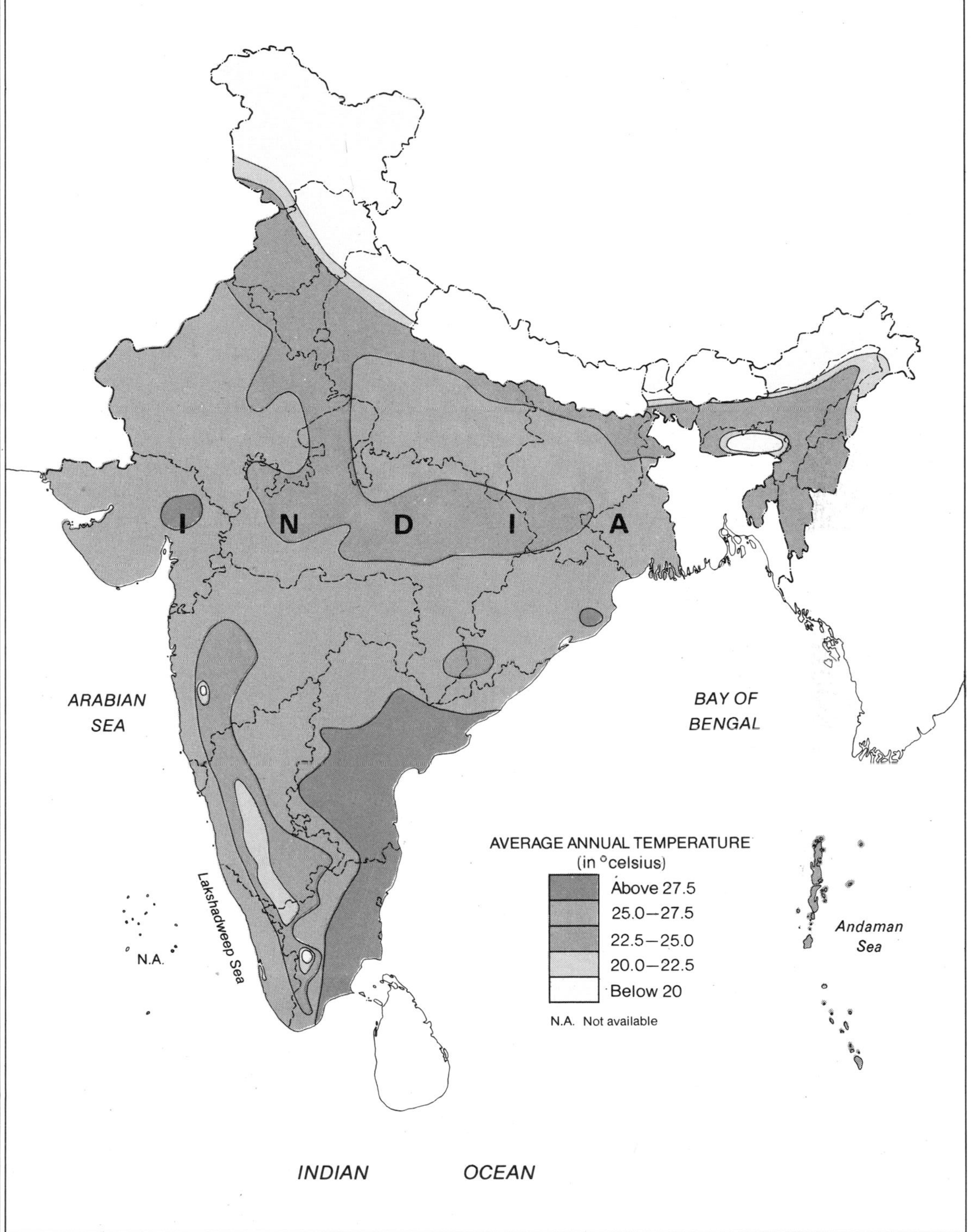
I N D I A
ARABIAN SEA
BAY OF BENGAL
Lakshadweep Sea
N.A.
Andaman Sea
INDIAN OCEAN
AVERAGE ANNUAL TEMPERATURE
(in °celsius)
Above 27.5
25.0–27.5
22.5–25.0
20.0–22.5
Below 20
N.A. Not available

Scale 1:21 000 000   1 Cm = 210 Kms

SOILS

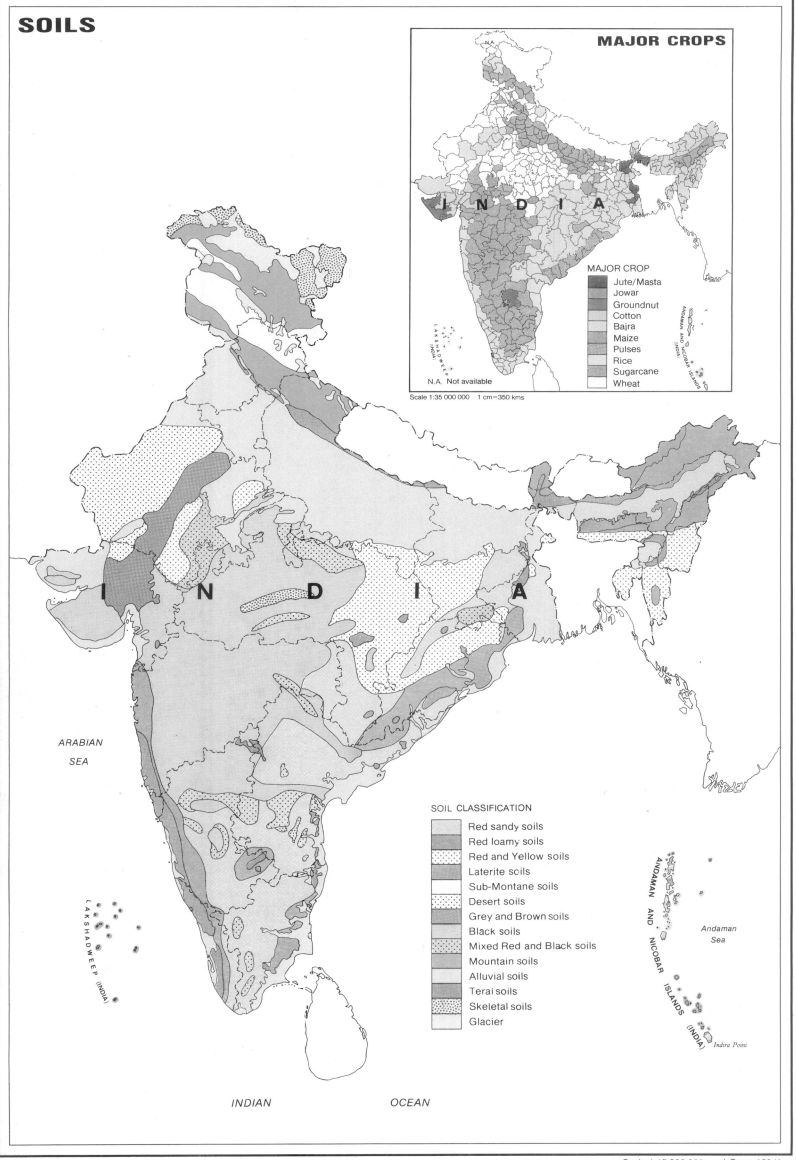

MAJOR CROPS

MAJOR CROP

- Jute/Masta
- Jowar
- Groundnut
- Cotton
- Bajra
- Maize
- Pulses
- Rice
- Sugarcane
- Wheat

N.A. Not available

Scale 1:35 000 000   1 cm=350 kms

INDIA

ARABIAN
SEA

LAKSHADWEEP (INDIA)

ANDAMAN AND NICOBAR ISLANDS (INDIA)

Andaman
Sea

Indira Point

SOIL CLASSIFICATION

- Red sandy soils
- Red loamy soils
- Red and Yellow soils
- Laterite soils
- Sub-Montane soils
- Desert soils
- Grey and Brown soils
- Black soils
- Mixed Red and Black soils
- Mountain soils
- Alluvial soils
- Terai soils
- Skeletal soils
- Glacier

INDIAN   OCEAN

Scale 1:15 000 000   1 Cm = 150 Kms

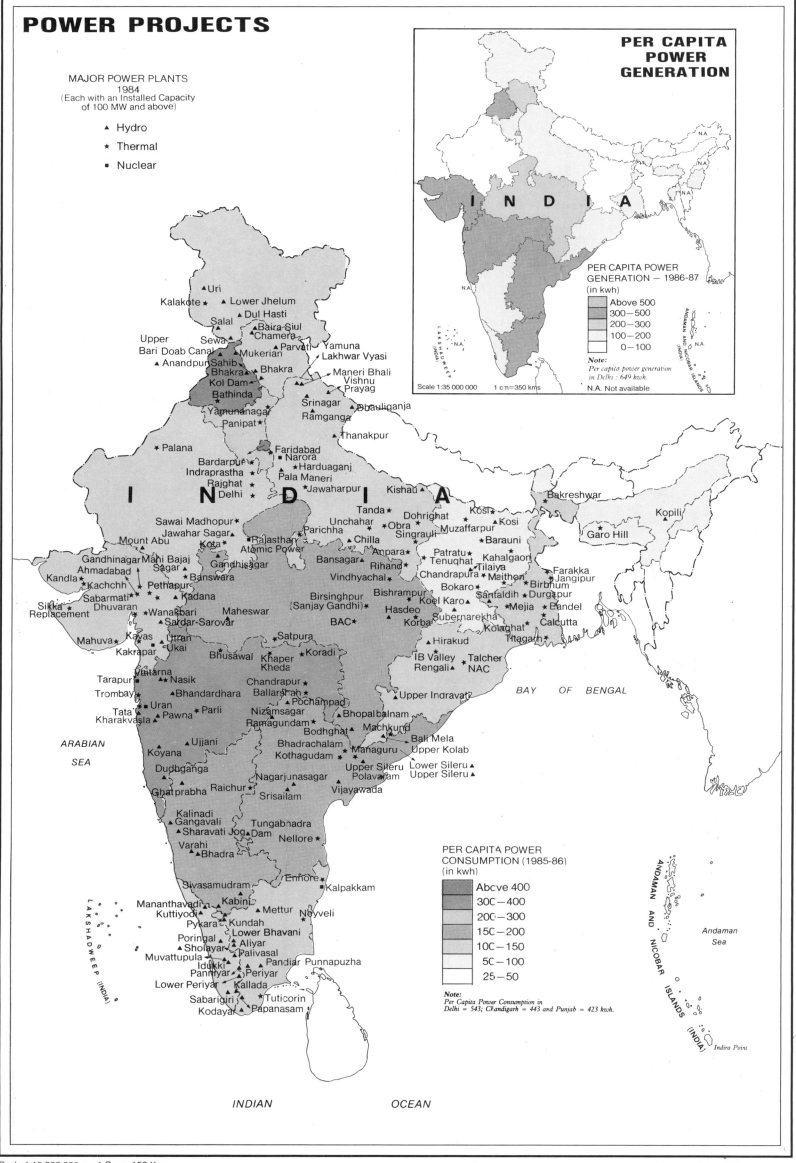

POWER PROJECTS

MAJOR POWER PLANTS
1984
(Each with an Installed Capacity
of 100 MW and above)

▲ Hydro
★ Thermal
■ Nuclear

PER CAPITA POWER GENERATION

INDIA

PER CAPITA POWER
GENERATION — 1986-87
(in kwh)

- Above 500
- 300 — 500
- 200 — 300
- 100 — 200
- 0 — 100

Note:
*Per capita power generation
in Delhi : 649 kwh.*

Scale 1:35 000 000   1 cm = 350 kms   N.A. Not available

Uri
Kalakote
Lower Jhelum
Salal
Dul Hasti
Baira Siul
Sewa
Chamera
Upper
Bari Doab Canal
Mukerian
Parvati
Yamuna
Lakhwar Vyasi
Anandpur Sahib
Bhakra
Bhakra
Maneri Bhali
Vishnu
Prayag
Kol Dam
Bathinda
Srinagar
Dhauliganja
Yamunanagar
Ramganga
Panipat
Thanakpur
Palana
Faridabad
Bardarpur
Narora
Indraprastha
Harduaganj
Rajghat
Pala Maneri
Delhi
Jawaharpur
Kishau
Bakreshwar
Kopili
Tanda
Dohrighat
Kosi
Sawai Madhopur
Unchahar
Obra
Kosi
Garo Hill
Jawahar Sagar
Parichha
Chilla
Singrauli
Muzaffarpur
Mount Abu
Kota
Rajasthan
Atomic Power
Ampara
Patratu
Barauni
Gandhinagar
Mahi Bajaj
Bansagar
Rihand
Tenughat
Kahalgaon
Ahmadabad
Sagar
Gandhisagar
Vindhyachal
Chandrapura
Tilaiya
Farakka
Kandla
Banswara
Bishrampur
Bokaro
Maithon
Jangipur
Kachchh
Pethapur
Birsinghpur
Koel Karo
Santaldih
Birbhum
Sabarmati
Kadana
(Sanjay Gandhi)
Hasdeo
Subernarekha
Mejia
Durgapur
Sikka
Dhuvaran
Wanakbari
Maheswar
BAC
Korba
Kolaghat
Bandel
Replacement
Sardar-Sarovar
Bishrampur
Titagarh
Calcutta
Mahuva
Kavas
Uttan
Satpura
Hirakud
Kakrapar
Ukai
Bhusawal
Koradi
IB Valley
Talcher
Tarapur
Vaitarna
Khaper
Kheda
Rengali
NAC
Trombay
Nasik
Chandrapur
Upper Indravati
BAY   OF   BENGAL
Tata
Bhandardhara
Ballarshah
Pochampad
Kharakvasla
Uran
Parli
Nizamsagar
Bhopalpalnam
Pawna
Ramagundam
Machkund
Ujjani
Bodhghat
Bali Mela
Koyana
Bhadrachalam
Managuru
Upper Kolab
Dudhganga
Kothagudam
Upper Sileru
Lower Sileru
Ghatprabha
Raichur
Nagarjunasagar
Polavaram
Upper Sileru
Srisailam
Vijayawada
Kalinadi
Gangavali
Tungabhadra
Sharavati Jog Dam
Nellore
Varahi
Bhadra
Enhore
Sivasamudram
Kalpakkam
Mananthavadi
Kabini
Mettur
Kuttiyodi
Neyveli
Pykara
Kundah
Poringal
Lower Bhavani
Sholayar
Aliyar
Muvattupula
Palivasal
Pandiar Punnapuzha
Idukki
Panniyar
Periyar
Lower Periyar
Kallada
Sabarigiri
Tuticorin
Kodayar
Papanasam

ARABIAN
SEA

LAKSHADWEEP
(INDIA)

ANDAMAN
AND
NICOBAR
ISLANDS
(INDIA)

Andaman
Sea

Indira Point

PER CAPITA POWER CONSUMPTION (1985-86)
(in kwh)

- Above 400
- 300 — 400
- 200 — 300
- 150 — 200
- 100 — 150
- 50 — 100
- 25 — 50

Note:
*Per Capita Power Consumption in
Delhi = 543; Chandigarh = 443 and Punjab = 423 kwh.*

INDIAN   OCEAN

POWER GENERATION

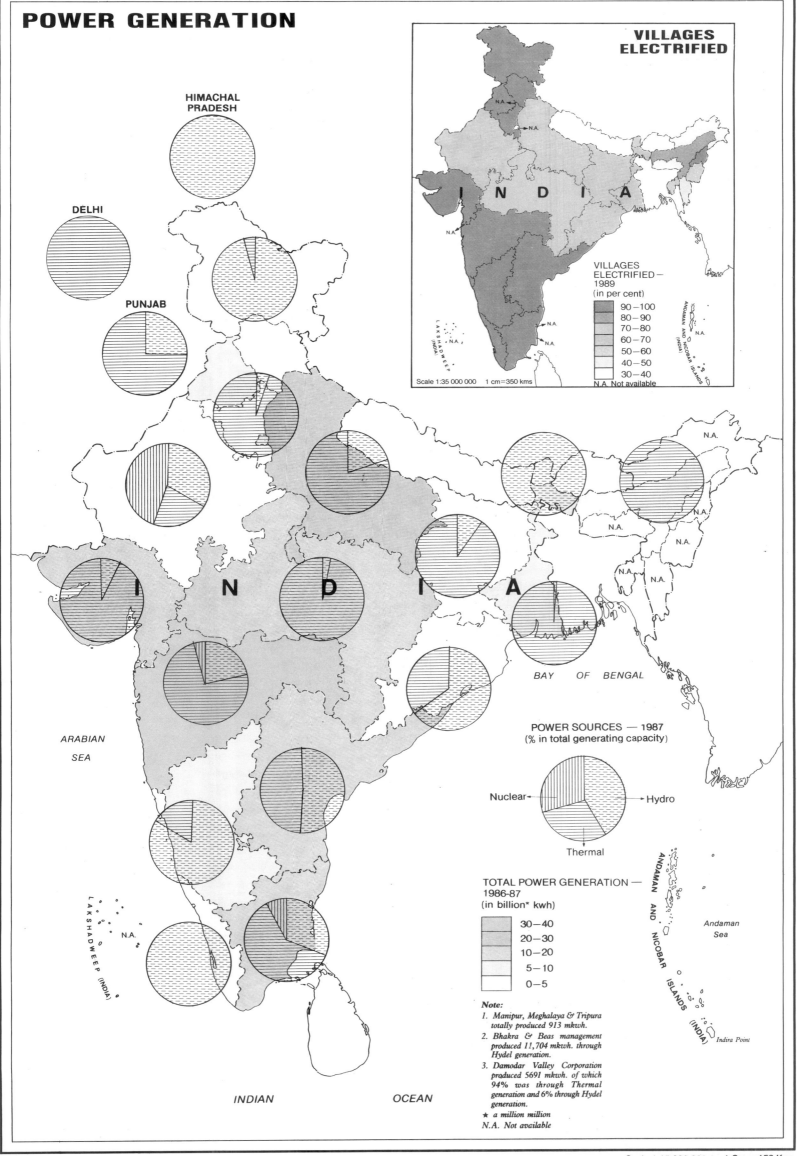

VILLAGES ELECTRIFIED

INDIA

VILLAGES
ELECTRIFIED —
1989
(in per cent)

90 — 100
80 — 90
70 — 80
60 — 70
50 — 60
40 — 50
30 — 40
N.A. Not available

Scale 1:35 000 000   1 cm=350 kms

HIMACHAL
PRADESH

DELHI

PUNJAB

ARABIAN
SEA

BAY   OF   BENGAL

POWER SOURCES — 1987
(% in total generating capacity)

Nuclear ← → Hydro

Thermal

TOTAL POWER GENERATION —
1986-87
(in billion* kwh)

30 — 40
20 — 30
10 — 20
5 — 10
0 — 5

Note:
1. *Manipur, Meghalaya & Tripura
    totally produced 913 mkwh.*
2. *Bhakra & Beas management
    produced 11,704 mkwh. through
    Hydel generation.*
3. *Damodar Valley Corporation
    produced 5691 mkwh. of which
    94% was through Thermal
    generation and 6% through Hydel
    generation.*
★ *a million million*
N.A. *Not available*

ARABIAN SEA

LAKSHADWEEP (INDIA)

N.A.

ANDAMAN   AND   NICOBAR   ISLANDS   (INDIA)

Andaman
Sea

Indira Point

INDIAN   OCEAN

Scale 1:15 000 000   1 Cm = 150 Kms

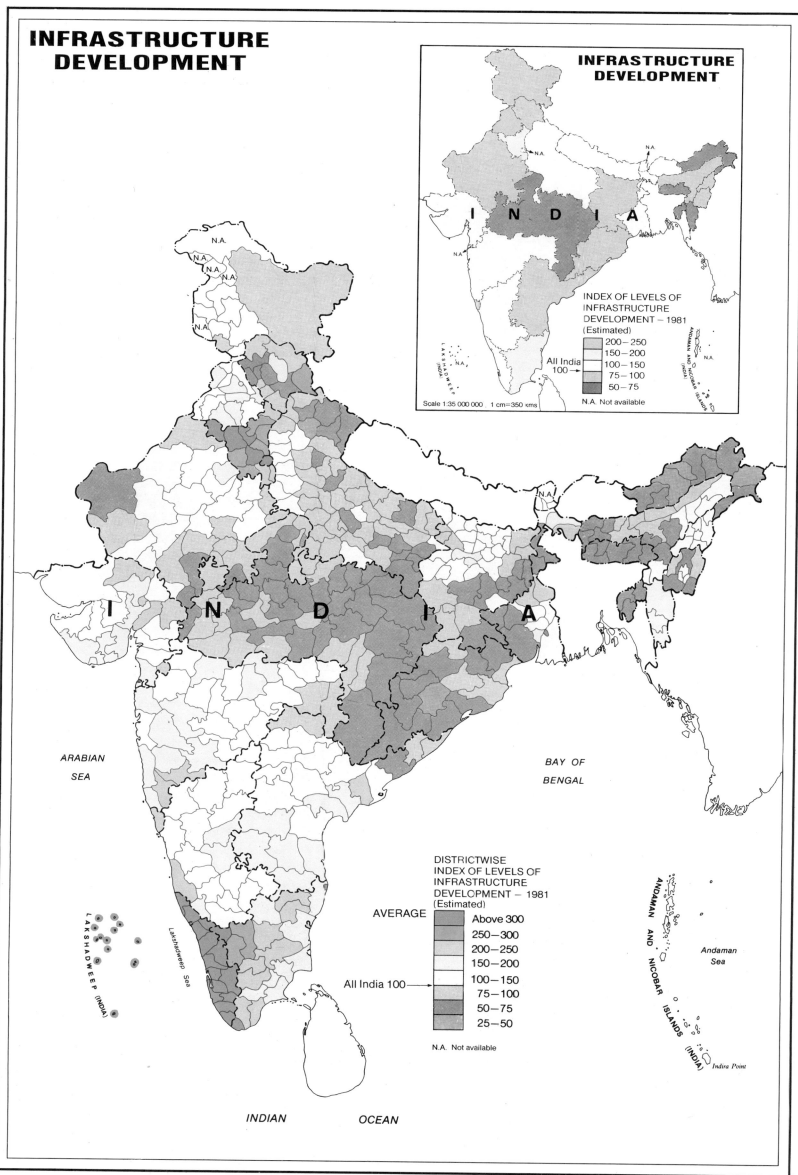

INFRASTRUCTURE DEVELOPMENT

INFRASTRUCTURE
DEVELOPMENT

INDEX OF LEVELS OF
INFRASTRUCTURE
DEVELOPMENT – 1981
(Estimated)

200 – 250
150 – 200
100 – 150
All India
100 →
75 – 100
50 – 75

N.A. Not available

Scale 1:35 000 000 : 1 cm=350 ᴋms

ARABIAN
SEA

BAY OF
BENGAL

LAKSHADWEEP (INDIA)

Lakshadweep Sea

ANDAMAN AND NICOBAR ISLANDS (INDIA)

Andaman
Sea

DISTRICTWISE
INDEX OF LEVELS OF
INFRASTRUCTURE
DEVELOPMENT – 1981
(Estimated)

AVERAGE

Above 300
250 – 300
200 – 250
150 – 200
100 – 150
All India 100 →
75 – 100
50 – 75
25 – 50

N.A. Not available

Indira Point

INDIAN   OCEAN

RELIGION

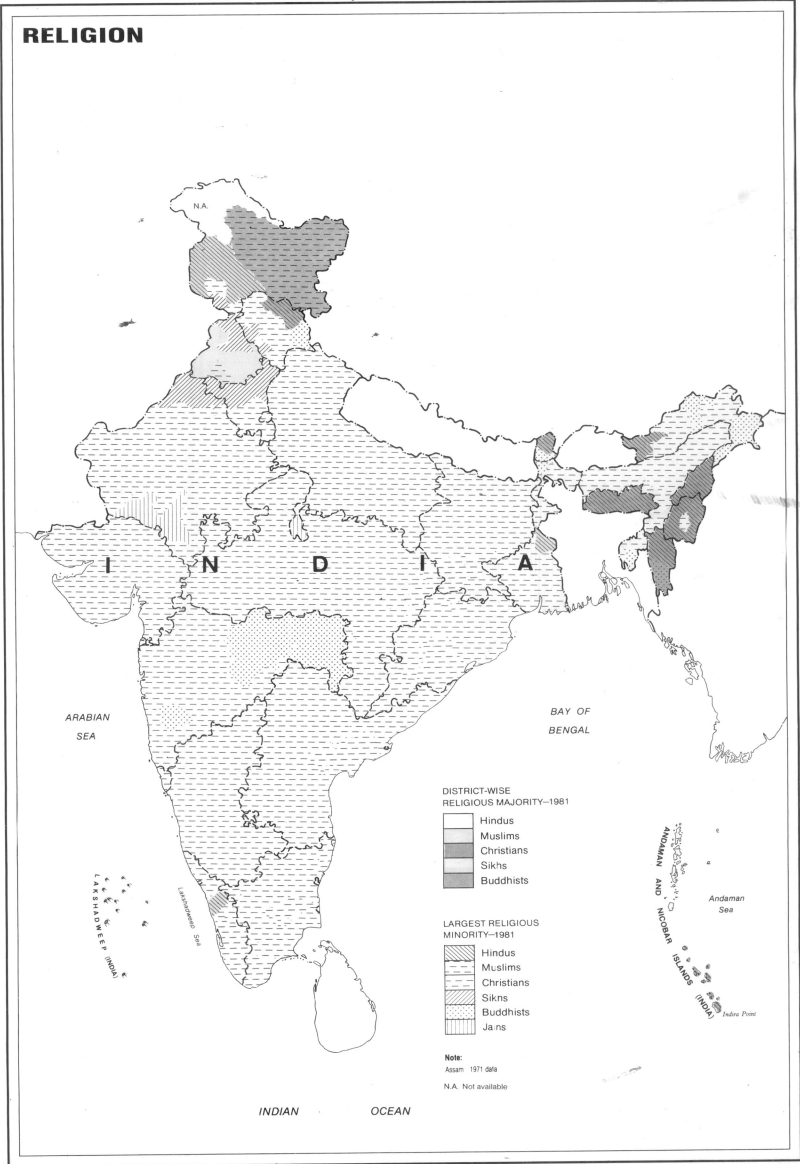

N.A.

ARABIAN
SEA

I N D I A

BAY OF
BENGAL

LAKSHADWEEP (INDIA)

Lakshadweep Sea

ANDAMAN AND NICOBAR ISLANDS (INDIA)

Andaman
Sea

Indira Point

**DISTRICT-WISE
RELIGIOUS MAJORITY—1981**

- Hindus
- Muslims
- Christians
- Sikhs
- Buddhists

**LARGEST RELIGIOUS
MINORITY—1981**

- Hindus
- Muslims
- Christians
- Sikns
- Buddhists
- Jains

Note:
Assam 1971 data

N.A. Not available

INDIAN   OCEAN

Scale 1:15 000 000   1 Cm = 150 Kms

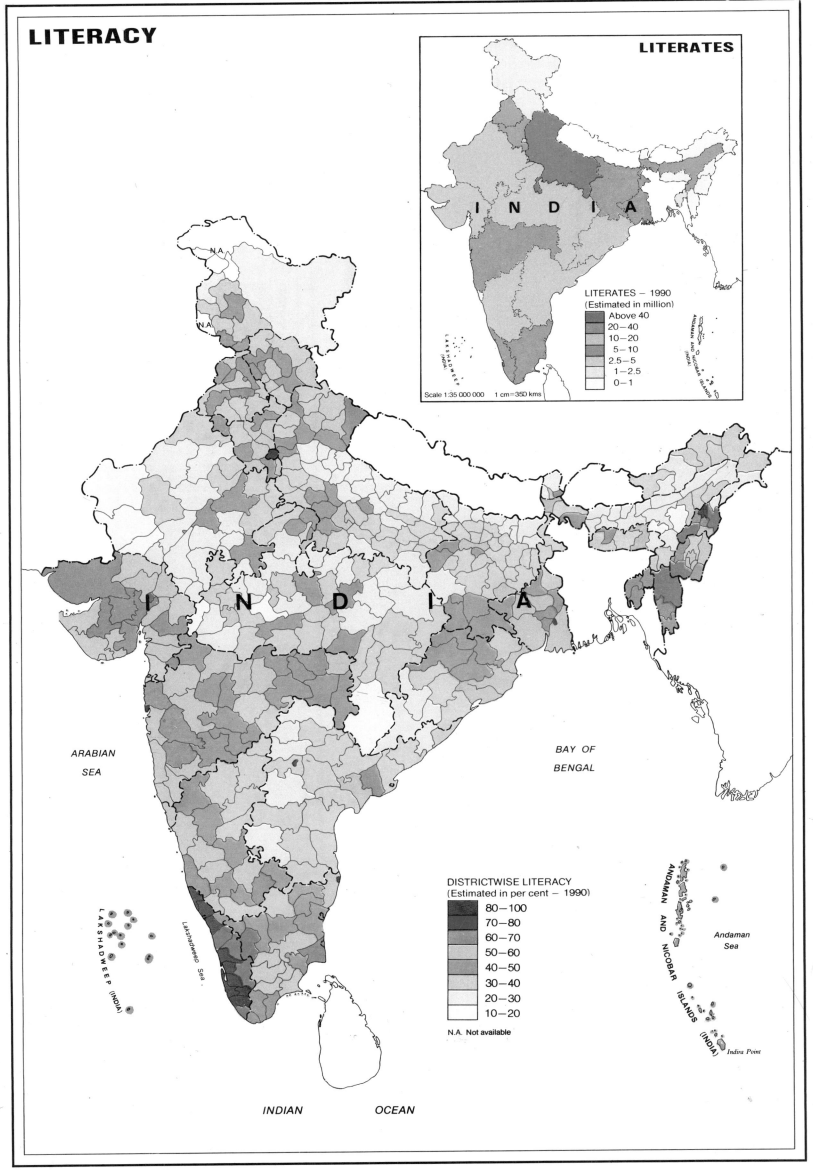

LITERACY

LITERATES

LITERATES — 1990
(Estimated in million)

	Above 40
	20—40
	10—20
	5—10
	2.5—5
	1—2.5
	0—1

Scale 1:35 000 000   1 cm=350 kms

N.A.

N.A.

I N D I A

ARABIAN

SEA

BAY OF

BENGAL

L A K S H A D W E E P   (INDIA)

Lakshadweep Sea

ANDAMAN AND NICOBAR ISLANDS (INDIA)

Andaman
Sea

DISTRICTWISE LITERACY
(Estimated in per cent — 1990)

	80—100
	70—80
	60—70
	50—60
	40—50
	30—40
	20—30
	10—20

N.A. Not available

Indira Point

INDIAN   OCEAN

Scale 1:15 000 000   1 Cm = 150 Kms

WORKFORCE

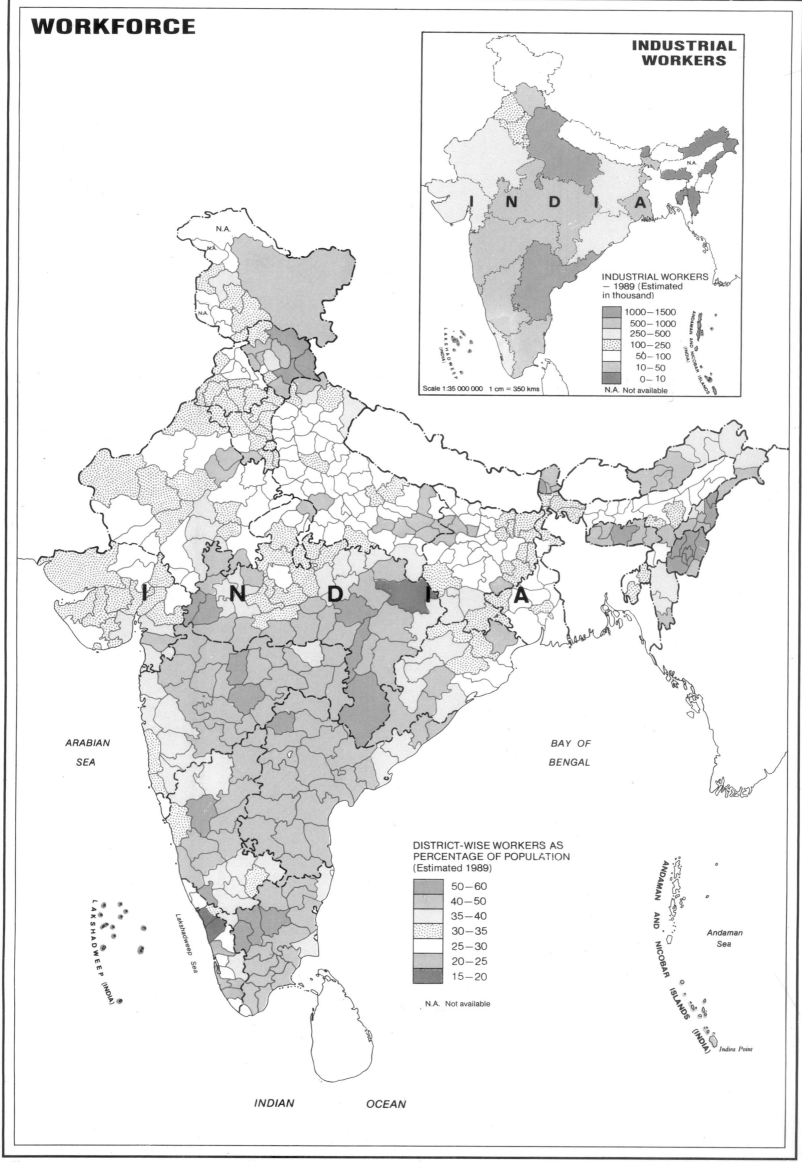

INDUSTRIAL WORKERS

I N D I A

INDUSTRIAL WORKERS
— 1989 (Estimated
in thousand)

	1000—1500
	500—1000
	250—500
	100—250
	50—100
	10—50
	0— 10
	N.A. Not available

Scale 1:35 000 000   1 cm = 350 kms

L A K S H A D W E E P (INDIA)

A N D A M A N   A N D   N I C O B A R   I S L A N D S (INDIA)

N.A.

I N D I A

ARABIAN SEA

BAY OF BENGAL

L A K S H A D W E E P (INDIA)

Lakshadweep Sea

DISTRICT-WISE WORKERS AS
PERCENTAGE OF POPULATION
(Estimated 1989)

	50—60
	40—50
	35—40
	30—35
	25—30
	20—25
	15—20

N.A. Not available

A N D A M A N   A N D   N I C O B A R   I S L A N D S (INDIA)

Andaman Sea

Indira Point

INDIAN   OCEAN

Scale 1:15 000 000   1 Cm = 150 Kms

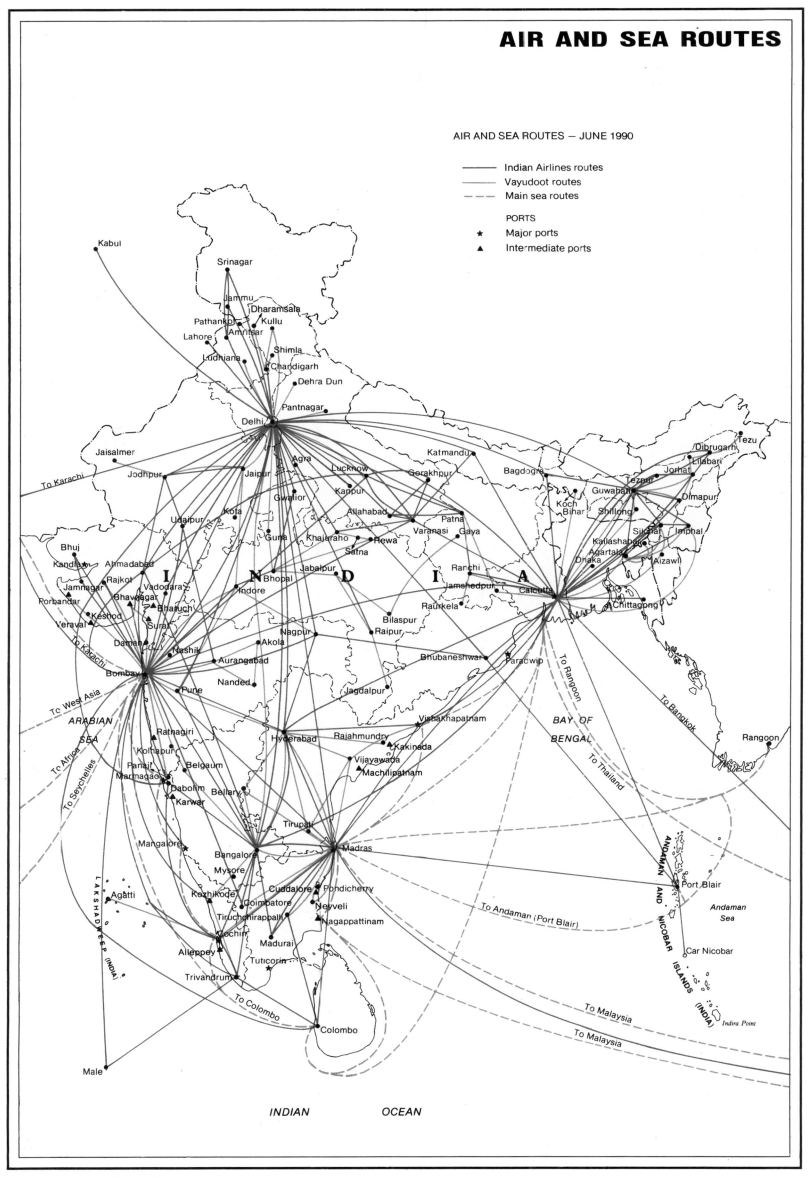

AIR AND SEA ROUTES

AIR AND SEA ROUTES — JUNE 1990

——————— Indian Airlines routes
——————— Vayudoot routes
— — — — Main sea routes

PORTS
★ Major ports
▲ Intermediate ports

Kabul

Srinagar
Jammu
Dharamsala
Pathankot
Kullu
Lahore
Amritsar
Ludhiana
Shimla
Chandigarh
Dehra Dun
Pantnagar
Delhi

Jaisalmer
To Karachi
Jodhpur
Agra
Lucknow
Gorakhpur
Katmandu
Bagdogra
Tezu
Dibrugarh
Lilabari
Jorhat
Jaipur
Tezpur
Guwahati
Dimapur
Gwalior
Kanpur
Koch Bihar
Shillong
Kota
Allahabad
Patna
Silchar
Imphal
Udaipur
Khajuraho
Rewa
Varanasi
Gaya
Kailashahar
Agartala
Aizawl
Bhuj
Guna
Satna
Ranchi
Dhaka
Kandla
Ahmadabad
Jabalpur
Bhopal
Indore
Jamshedpur
Calcutta
Chittagong
Rajkot
Vadodara
Jamnagar
Bhavnagar
Bharuch
Raurkela
Porbandar
Keshod
Surat
Bilaspur
Nagpur
Raipur
Veraval
Daman
Nashik
Akola
Bhubaneshwar
Paradwip
To Rangoon
Bombay
Aurangabad
Pune
Nanded
Jagdalpur
ARABIAN SEA
To West Asia
Ratnagiri
Vishakhapatnam
BAY OF BENGAL
To Bangkok
To Africa
Kolhapur
Hyderabad
Rajahmundry
Kakinada
Rangoon
To Seychelles
Panaji
Marmagao
Belgaum
Vijayawada
Machilipatnam
To Thailand
Dabolim
Bellary
Karwar
Tirupati
Mangalore
Bangalore
Madras
Port Blair
Mysore
ANDAMAN AND NICOBAR ISLANDS (INDIA)
Andaman Sea
Agatti
Kozhikode
Cuddalore
Pondicherry
Coimbatore
Neyveli
To Andaman (Port Blair)
Tiruchchirappalli
Nagappattinam
LAKSHADWEEP (INDIA)
Kochin
Car Nicobar
Madurai
Alleppey
Tuticorin
Trivandrum
To Colombo
To Malaysia
Indira Point
Colombo
To Malaysia

Male

INDIA

INDIAN   OCEAN

Scale 1:15 000 000   1 Cm = 150 Kms

13

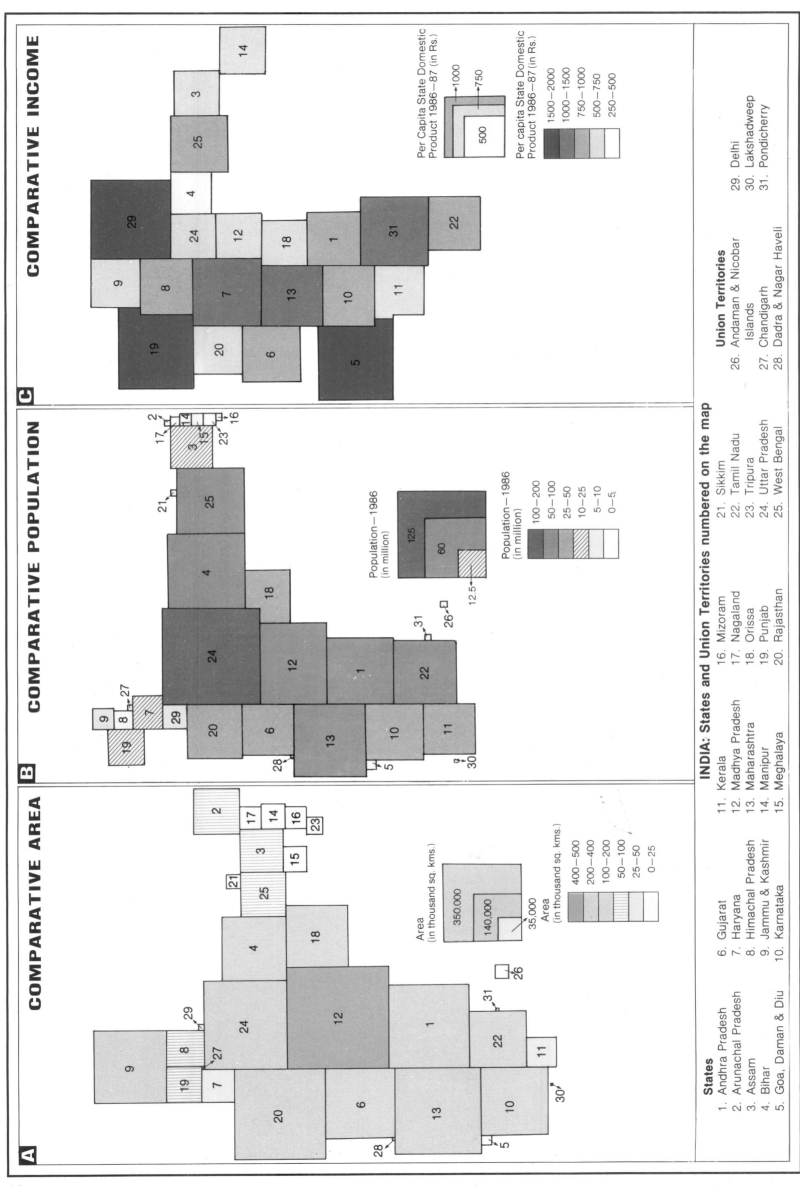

A COMPARATIVE AREA

B COMPARATIVE POPULATION

C COMPARATIVE INCOME

Area
(in thousand sq. kms.)

Area
(in thousand sq. kms.)
- 400–500
- 200–400
- 100–200
- 50–100
- 25–50
- 0–25

Population—1986
(in million)

Population—1986
(in million)
- 100–200
- 50–100
- 25–50
- 10–25
- 5–10
- 0–5

Per Capita State Domestic
Product 1986–87 (in Rs.)

Per capita State Domestic
Product 1986–87 (in Rs.)
- 1500–2000
- 1000–1500
- 750–1000
- 500–750
- 250–500

INDIA: States and Union Territories numbered on the map

States
1. Andhra Pradesh
2. Arunachal Pradesh
3. Assam
4. Bihar
5. Goa, Daman & Diu
6. Gujarat
7. Haryana
8. Himachal Pradesh
9. Jammu & Kashmir
10. Karnataka
11. Kerala
12. Madhya Pradesh
13. Maharashtra
14. Manipur
15. Meghalaya
16. Mizoram
17. Nagaland
18. Orissa
19. Punjab
20. Rajasthan
21. Sikkim
22. Tamil Nadu
23. Tripura
24. Uttar Pradesh
25. West Bengal

Union Territories
26. Andaman & Nicobar Islands
27. Chandigarh
28. Dadra & Nagar Haveli
29. Delhi
30. Lakshadweep
31. Pondicherry

14

States and Union Territories

This section consists of a series of 230 thematic maps providing a detailed picture of the 25 states and seven union territories in India. The maps for each state provide information on nine subjects, namely, Physical Features, Administration, Population, Transport & Tourism, Agriculture, Industries & Markets, Minerals, Economic Development and Mass Communication.

The maps are arranged in the same order for each state. All locations marked on the maps in this section have also been indexed with a cross-grid number for easy reference. The maps are up to date and give the user access to information at a glance.

Note:

In this section, most of the data used is based on Census 1981 when there were 412 districts. The maps, however, show the districts as they existed at the end of 1989, when there were 449 districts (442 districts for which boundaries were available are shown on the maps; district boundaries for five districts in Uttar Pradesh and one each in Arunachal Pradesh and Assam were not available at the time of printing). All information published in this section is in accordance with data published by official sources and which was available with us as on 1.1.1990. Where no data was available, it has been indicated by N.A. (Data Not Available).

The maps in this section have been drawn to different scales and these scales are indicated on the respective maps. The projection used for all these maps is International (Polyconic).

The Survey of India approved latitudes and longitudes appear on all maps in this section for which permission has been granted.

ANDHRA PRADESH

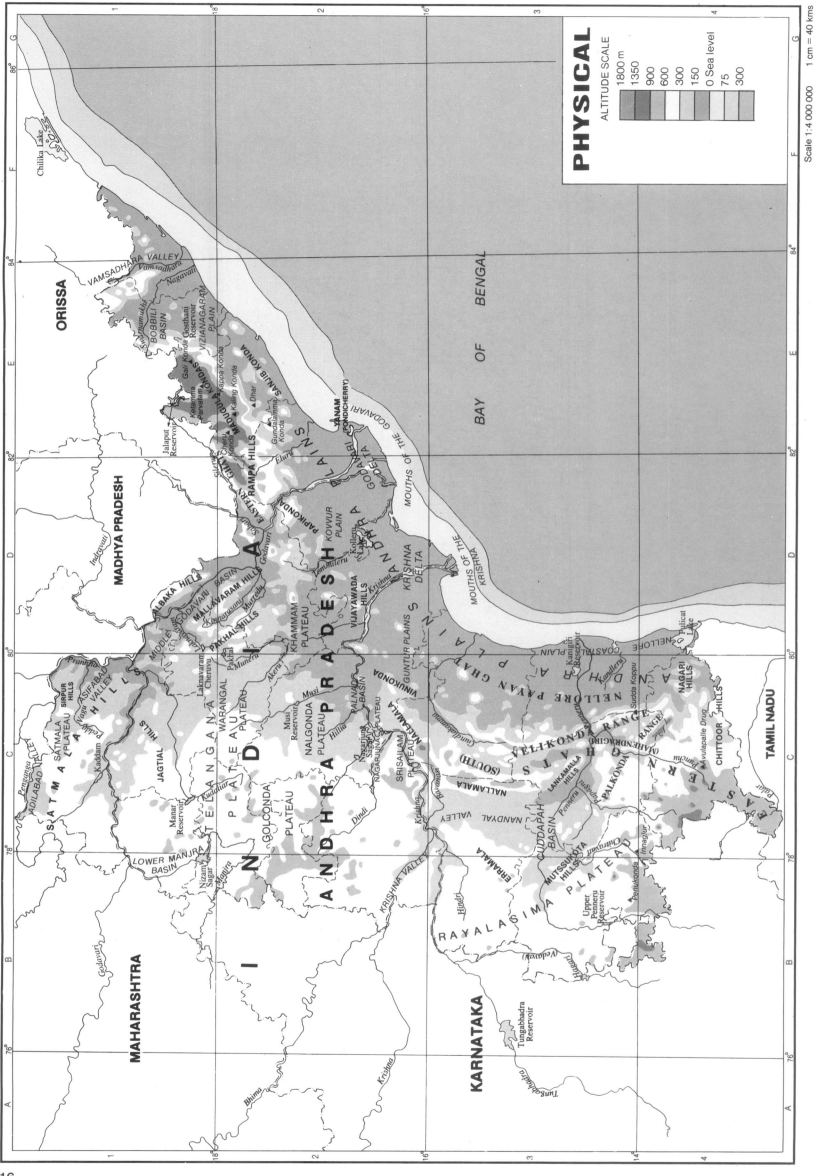

PHYSICAL
ALTITUDE SCALE

1800 m
1350
900
600
300
150
0 Sea level
75
300

Scale 1:4 000 000   1 cm = 40 kms

ORISSA

MADHYA PRADESH

MAHARASHTRA

KARNATAKA

TAMIL NADU

BAY OF BENGAL

RAYALASIMA

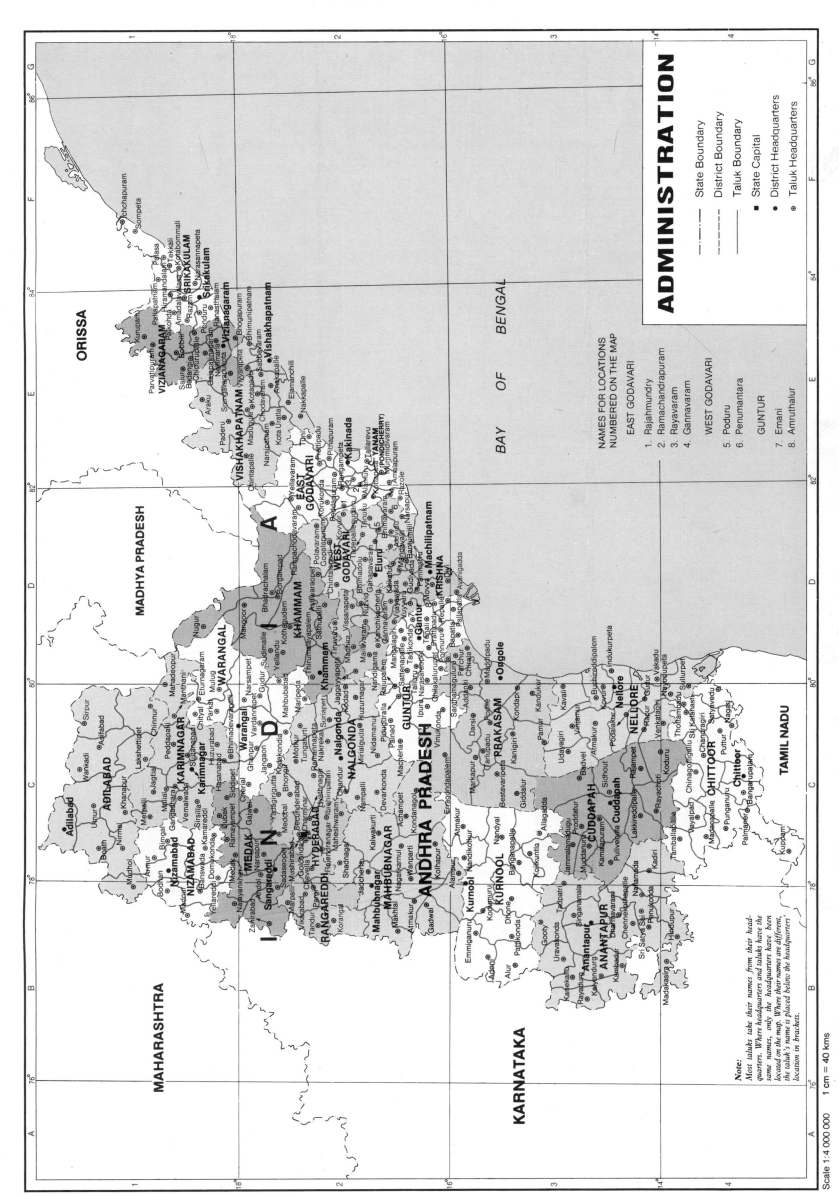

ADMINISTRATION

NAMES FOR LOCATIONS
NUMBERED ON THE MAP

EAST GODAVARI
1. Rajahmundry
2. Ramachandrapuram
3. Rayavaram
4. Gannavaram

WEST GODAVARI
5. Poduru
6. Penumantara

GUNTUR
7. Emani
8. Amruthalur

State Boundary
District Boundary
Taluk Boundary
■ State Capital
● District Headquarters
◉ Taluk Headquarters

Note:
Most taluks take their names from their head-quarters. Where headquarters and taluks have the same names, only the headquarters have been located on the map. Where their names are different, the taluk's name is placed below the headquarters' location in brackets.

Scale 1:4 000 000   1 cm = 40 kms

MAHARASHTRA

ORISSA

MADHYA PRADESH

BAY   OF   BENGAL

KARNATAKA

TAMIL NADU

ANDHRA PRADESH

17

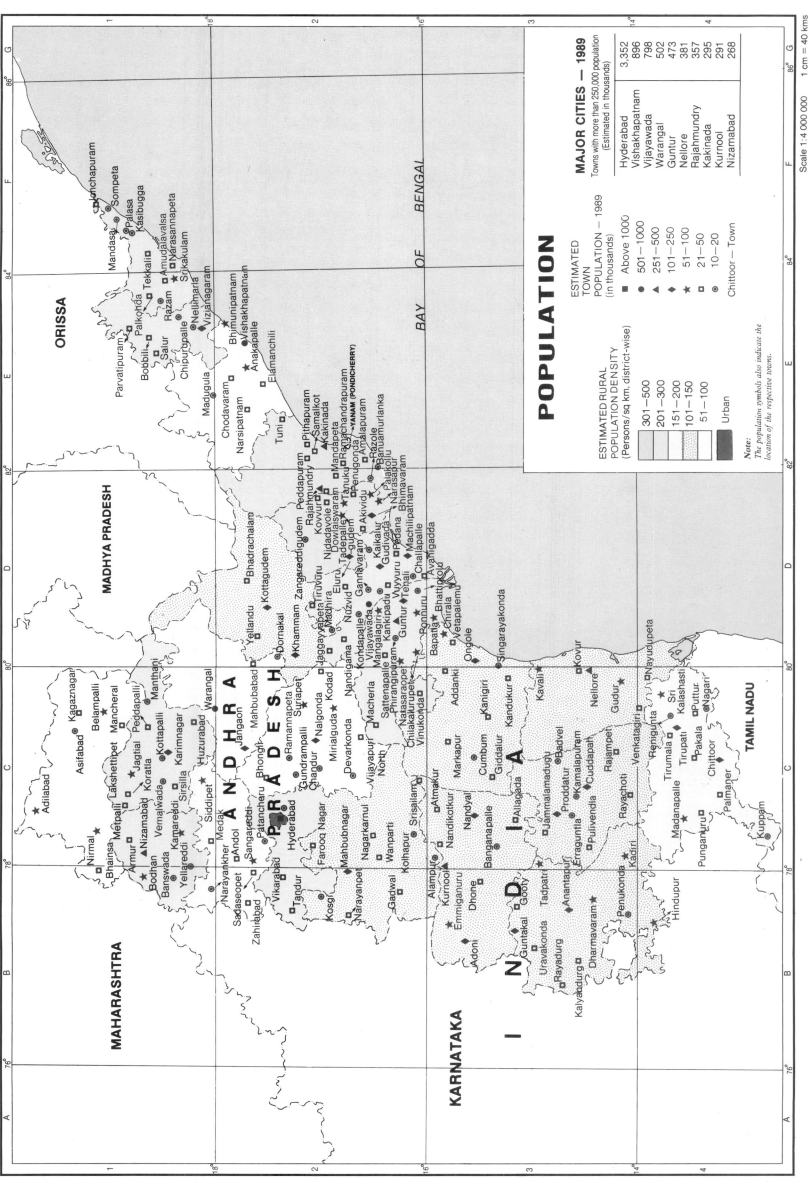

POPULATION

MAJOR CITIES — 1989

Towns with more than 250,000 population
(Estimated in thousands)

Hyderabad	3,352
Vishakhapatnam	896
Vijayawada	798
Warangal	502
Guntur	473
Nellore	381
Rajahmundry	357
Kakinada	295
Kurnool	291
Nizamabad	268

Scale 1:4 000 000   1 cm = 40 kms

ESTIMATED TOWN POPULATION — 1989
(in thousands)

- ■ Above 1000
- ● 501—1000
- ▲ 251—500
- ◆ 101—250
- ★ 51—100
- □ 21—50
- ◎ 10—20

Chittoor — Town

ESTIMATED RURAL POPULATION DENSITY
(Persons/sq km, district-wise)

- 301—500
- 201—300
- 151—200
- 101—150
- 51—100

Urban

Note:
The population symbols also indicate the location of the respective towns.

MAHARASHTRA

ORISSA

MADHYA PRADESH

KARNATAKA

TAMIL NADU

ANDHRA PRADESH

INDIA

BAY OF BENGAL

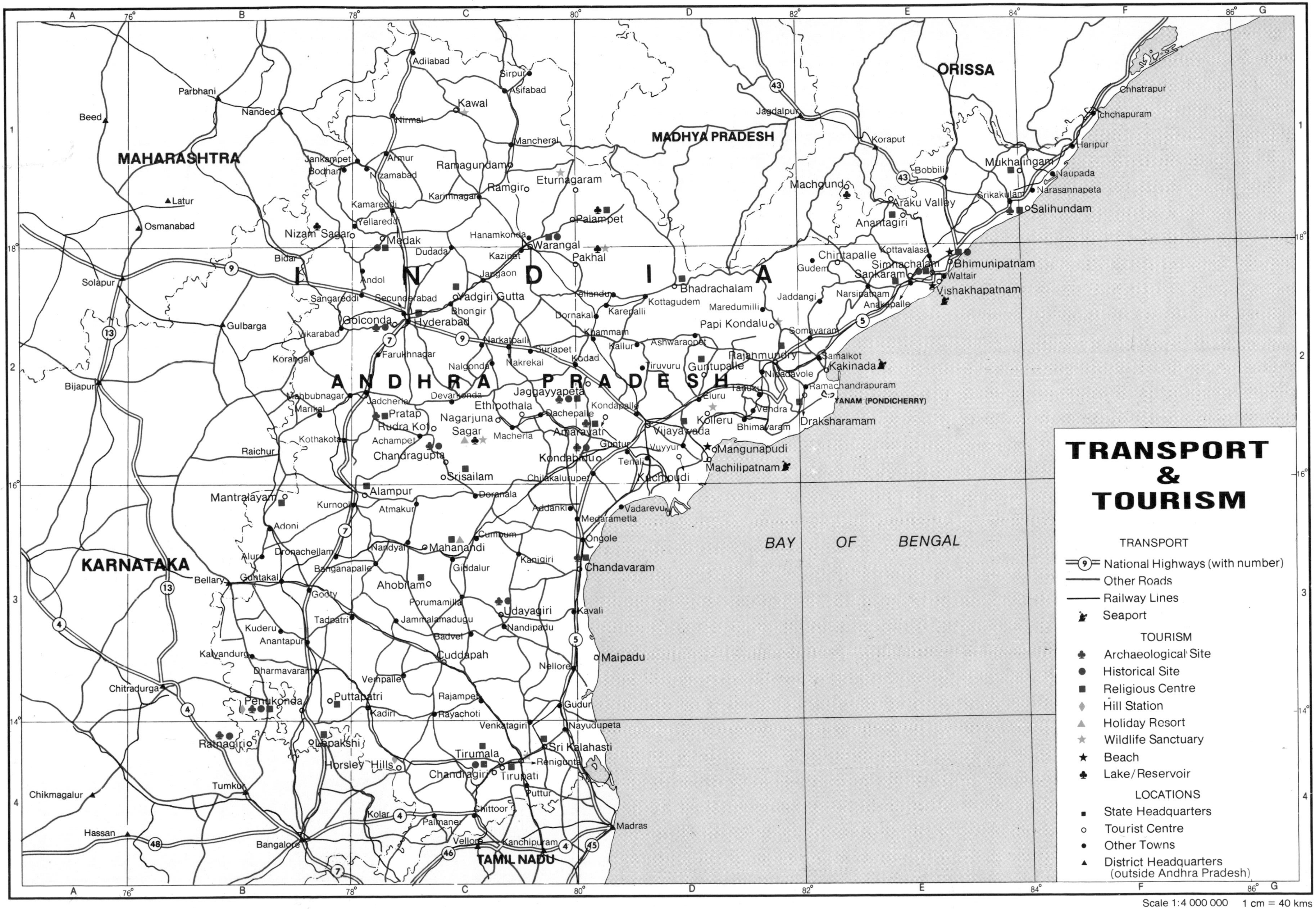

TRANSPORT & TOURISM

TRANSPORT

- ⑨ National Highways (with number)
- Other Roads
- Railway Lines
- Seaport

TOURISM

- ♣ Archaeological Site
- ● Historical Site
- ■ Religious Centre
- ◆ Hill Station
- ◀ Holiday Resort
- ★ Wildlife Sanctuary
- ★ Beach
- ♣ Lake/Reservoir

LOCATIONS

- ■ State Headquarters
- ○ Tourist Centre
- ● Other Towns
- ◀ District Headquarters (outside Andhra Pradesh)

Scale 1:4 000 000   1 cm = 40 kms

ORISSA

MADHYA PRADESH

MAHARASHTRA

KARNATAKA

TAMIL NADU

ANDHRA PRADESH

INDIA

BAY OF BENGAL

19

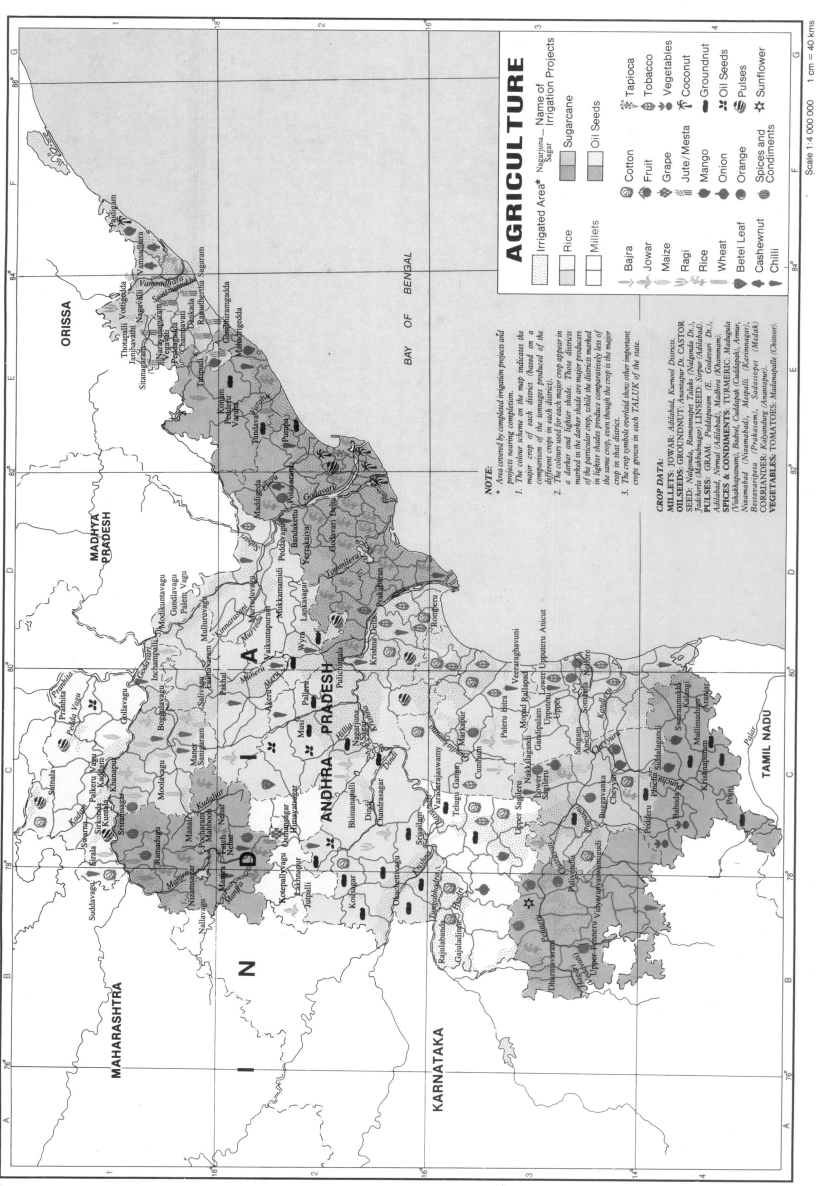

AGRICULTURE

Irrigated Area Nagarjuna — Name of
Sagar ★   Irrigation Projects

Rice	Sugarcane	Cotton	Tapioca
Millets	Oil Seeds	Fruit	Tobacco

Sugarcane
Oil Seeds

Tapioca
Tobacco
Vegetables
Coconut
Groundnut
Oil Seeds
Pulses
Sunflower

Cotton
Fruit
Grape
Jute/Mesta
Mango
Onion
Orange
Spices and
Condiments

Bajra
Jowar
Maize
Ragi
Rice
Wheat
Betel Leaf
Cashewnut
Chilli

Scale 1:4 000 000   1 cm = 40 kms

ORISSA

MADHYA PRADESH

MAHARASHTRA

KARNATAKA

TAMIL NADU

BAY OF BENGAL

ANDHRA PRADESH

INDIA

NOTE:
★ Area covered by completed irrigation projects and
   projects nearing completion.

1. The colour scheme on the map indicates the
   major crop of each district (based on a
   comparison of the tonnages produced of the
   different crops in each district).
2. The colours used for each major crop appear in
   a darker and lighter shade. Those districts
   marked in the darker shade are major producers
   of the particular crop, while the districts marked
   in lighter shades produce comparatively less of
   the same crop, even though the crop is the major
   crop in that district.
3. The crop symbols overlaid show other important
   crops grown in each TALUK of the state.

CROP DATA:
MILLETS: JOWAR: Adilabad, Kurnool Districts.
OILSEEDS: GROUNDNUT: Anantapur Dt. CASTOR
SEED: Nalgonda, Ramannapet Taluks (Nalgonda Dt.),
Jadcherla (Mahbubnagar) LINSEED: Sirpur (Adilabad).
PULSES: GRAM: Peddapuram (E. Godavari Dt.),
Adilabad, Nirmal (Adilabad), Madhira (Khammam).
SPICES & CONDIMENTS: TURMERIC: Maduguia
(Vishakhapatnam), Badvel, Cuddapah (Cuddapah), Armur,
Nizamabad (Nizamabad), Metpalli (Karimnagar),
Bestavaripeta (Prakasam), Sadasivpet (Medak).
CORRIANDER: Kalyandurg (Anantapur).
VEGETABLES: TOMATOES: Madanapalle (Chittoor).

20

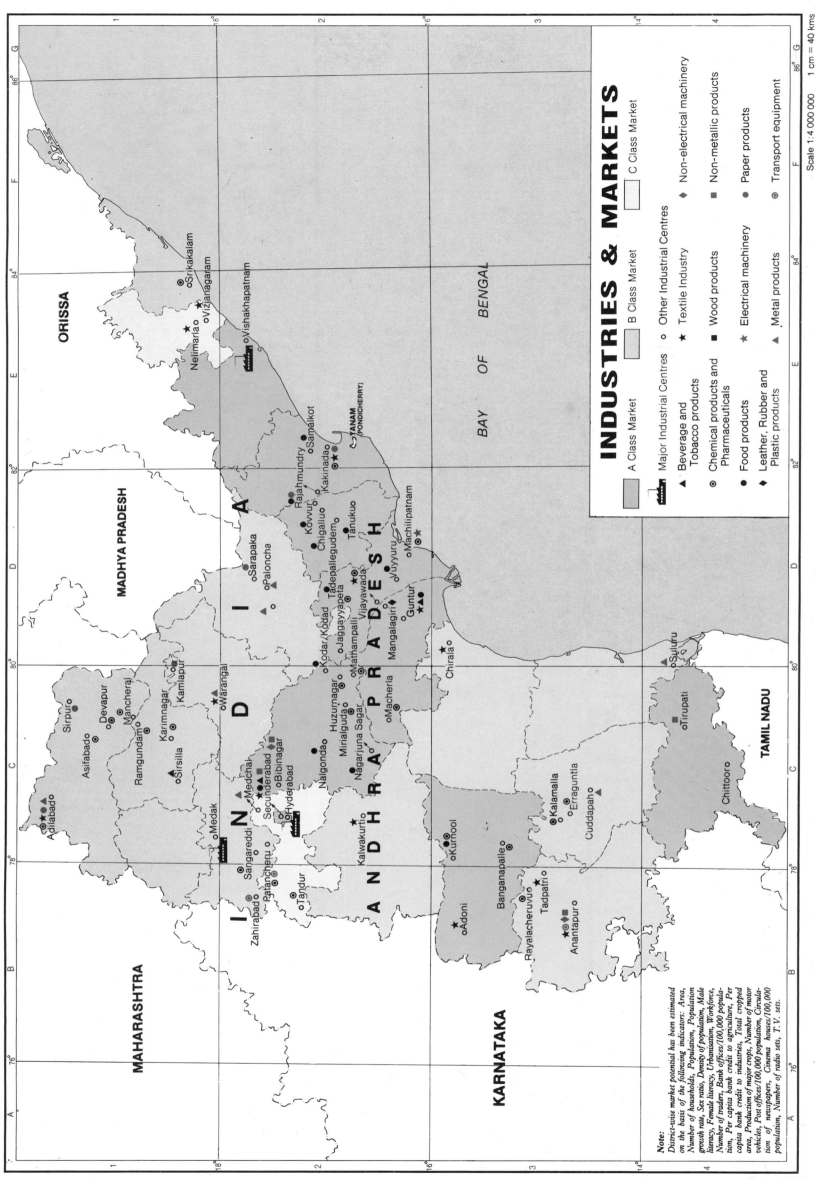

INDUSTRIES & MARKETS

ORISSA

MADHYA PRADESH

MAHARASHTRA

KARNATAKA

TAMIL NADU

A N D H R A   P R A D E S H

I N D I A

BAY OF BENGAL

Srikakalam
Vizianagaram
Nelimarla
Vishakhapatnam

Samalkot
Rajahmundry
Kovvur
Chigallu
Kakinada
Tadepallegudem
Tanuku
TANAM (PONDICHERRY)
Sarapaka
Paloncha
Jaggayyapeta
Vijayawada
Machilipatnam
Uyyuru
Mathampalli
Kodar/Kodad
Mangalagiri
Guntur
Macherla
Chirala
Sujuru

Warangal
Kamlapur
Karimnagar
Mancheral
Devapur
Sirpur
Asifabad
Ramgundam
Sirsilla
Medak
Medchal
Secunderabad
Bibinagar
Hyderabad
Nalgonda
Huzurnagar
Miriraiguda
Nagarjuna Sagar
Kalwakurtio
Sangareddi
Patancheru
Zahirabad
Tandur
Adilabad

Kurnool
Adoni
Banganapalle
Tadpatri
Rayalacheruvu
Anantapur
Kalamalla
Erraguntla
Cuddapah
Chittoor
Tirupati

Scale 1:4 000 000   1 cm = 40 kms

Legend

	A Class Market		Major Industrial Centres		Other Industrial Centres
	B Class Market				Non-electrical machinery
	C Class Market				Textile Industry		Non-metallic products

- Beverage and Tobacco products
- Chemical products and Pharmaceuticals
- Food products
- Leather, Rubber and Plastic products
- Wood products
- Electrical machinery
- Metal products
- Non-metallic products
- Paper products
- Transport equipment

Note:
District-wise market potential has been estimated on the basis of the following indicators: Area, Number of households, Population, Population growth rate, Sex ratio, Density of population, Male literacy, Female literacy, Urbanisation, Workforce, Number of traders, Bank offices/100,000 population, Per capita bank credit to agriculture, Per capita bank credit to industries, Total cropped area, Production of major crops, Number of motor vehicles, Post offices/100,000 population, Circulation of newspapers, Cinema houses/100,000 population, Number of radio sets, T.V. sets.

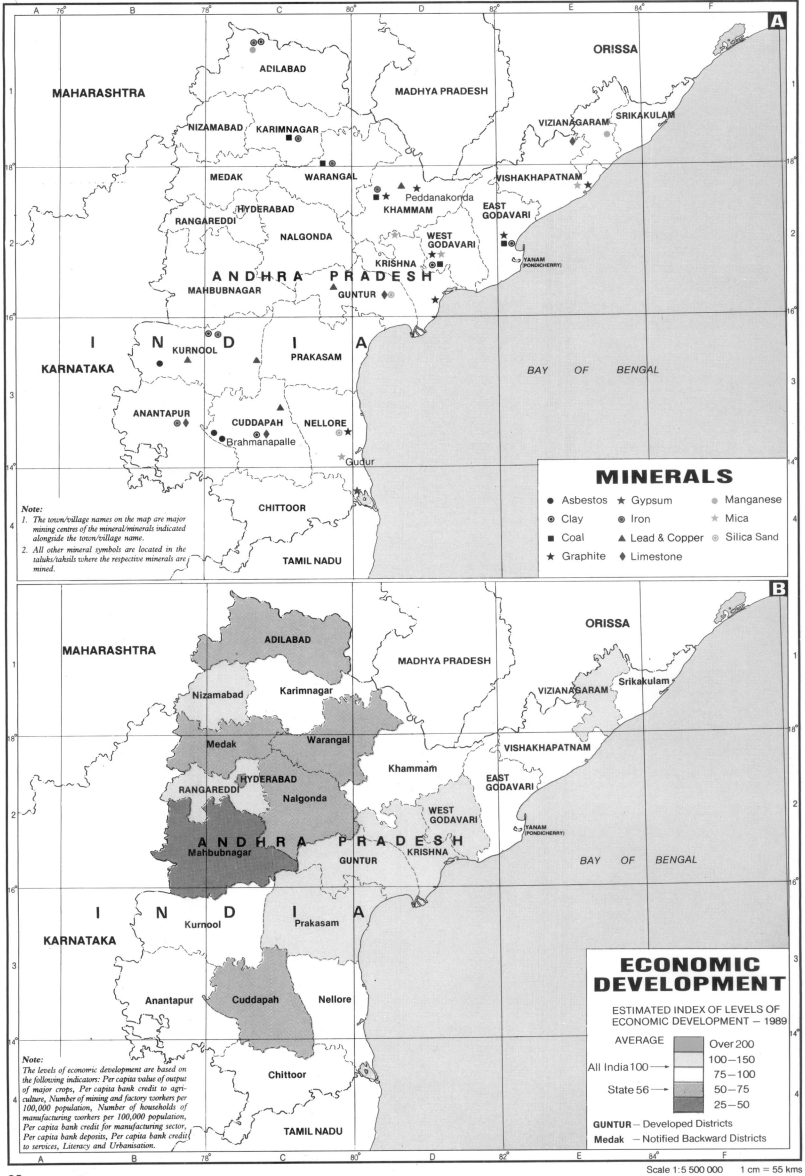

A

MAHARASHTRA

ORISSA

ADILABAD

MADHYA PRADESH

VIZIANAGARAM SRIKAKULAM

NIZAMABAD KARIMNAGAR

MEDAK WARANGAL

VISHAKHAPATNAM

HYDERABAD

RANGAREDDI KHAMMAM

Peddanakonda EAST
GODAVARI

NALGONDA

WEST
GODAVARI

A N D H R A   P R A D E S H

KRISHNA

MAHBUBNAGAR GUNTUR

YANAM
(PONDICHERRY)

I N D I A

KURNOOL PRAKASAM

KARNATAKA

BAY   OF   BENGAL

ANANTAPUR

CUDDAPAH NELLORE

Brahmanapalle

Gudur

CHITTOOR

Note:
*1. The town/village names on the map are major
mining centres of the mineral/minerals indicated
alongside the town/village name.*

*2. All other mineral symbols are located in the
taluks/tahsils where the respective minerals are
mined.*

TAMIL NADU

MINERALS

● Asbestos	★ Gypsum	● Manganese
◎ Clay	◉ Iron	★ Mica
■ Coal	▲ Lead & Copper	◉ Silica Sand
★ Graphite	◆ Limestone	

B

MAHARASHTRA

ORISSA

ADILABAD

MADHYA PRADESH

VIZIANAGARAM Srikakulam

Nizamabad Karimnagar

Medak Warangal

VISHAKHAPATNAM

HYDERABAD

RANGAREDDI

Khammam

EAST
GODAVARI

Nalgonda

WEST
GODAVARI

A N D H R A   P R A D E S H

KRISHNA

Mahbubnagar

YANAM
(PONDICHERRY)

GUNTUR

BAY   OF   BENGAL

I N D I A

Kurnool Prakasam

KARNATAKA

Anantapur Cuddapah Nellore

Chittoor

Note:
*The levels of economic development are based on
the following indicators: Per capita value of output
of major crops, Per capita bank credit to agriculture, Number of mining and factory workers per
100,000 population, Number of households of
manufacturing workers per 100,000 population,
Per capita bank credit for manufacturing sector,
Per capita bank deposits, Per capita bank credit
to services, Literacy and Urbanisation.*

TAMIL NADU

ECONOMIC DEVELOPMENT

ESTIMATED INDEX OF LEVELS OF
ECONOMIC DEVELOPMENT — 1989

AVERAGE

	Over 200
All India 100 →	100—150
	75—100
State 56 →	50—75
	25—50

GUNTUR — Developed Districts

Medak — Notified Backward Districts

Scale 1:5 500 000   1 cm = 55 kms

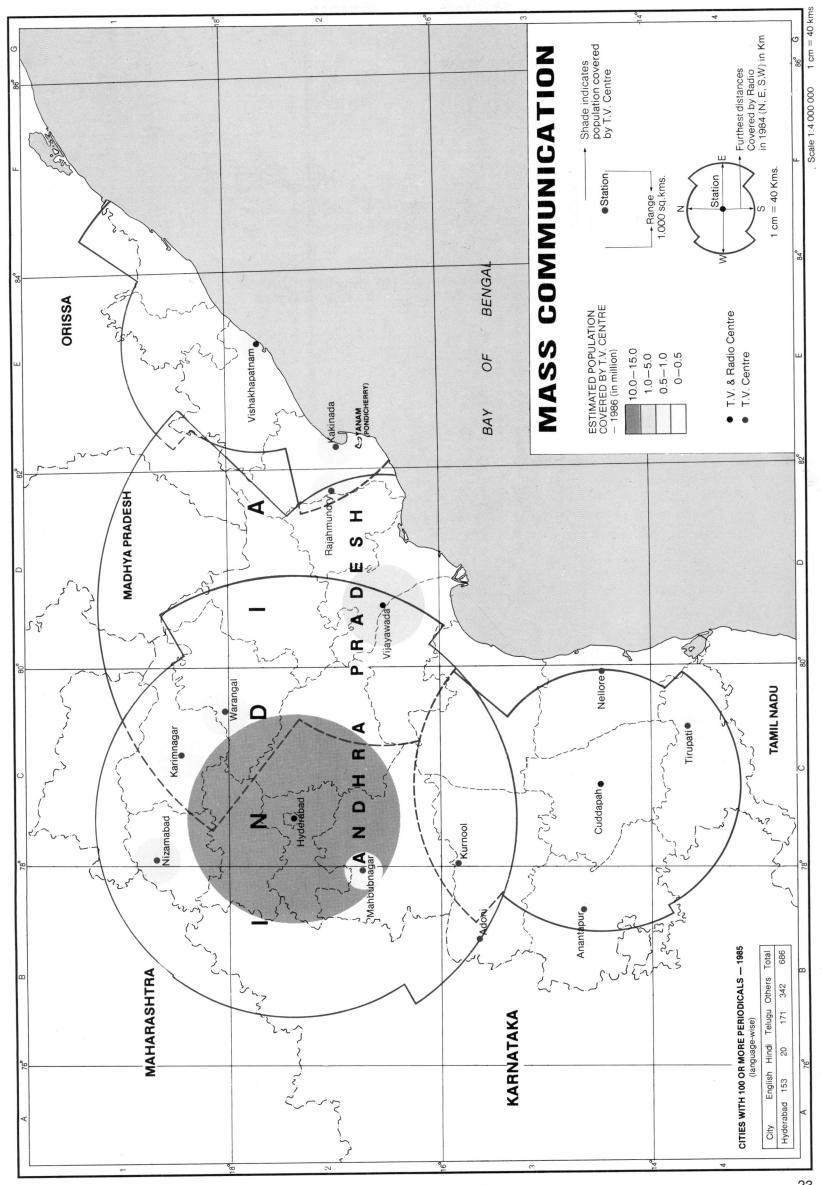

MASS COMMUNICATION

ORISSA

MADHYA PRADESH

MAHARASHTRA

KARNATAKA

TAMIL NADU

I N D I A

A N D H R A   P R A D E S H

BAY   OF   BENGAL

Vishakhapatnam
Kakinada
YANAM (PONDICHERRY)
Rajahmundry
Vijayawada
Warangal
Karimnagar
Nizamabad
Hyderabad
Mahbubnagar
Kurnool
Adoni
Anantapur
Cuddapah
Nellore
Tirupati

CITIES WITH 100 OR MORE PERIODICALS — 1985
(language-wise)

City	English	Hindi	Telugu	Others	Total
Hyderabad	153	20	171	342	686

ESTIMATED POPULATION
COVERED BY T.V. CENTRE
— 1986 (in million)

- 10.0 – 15.0
- 1.0 – 5.0
- 0.5 – 1.0
- 0 – 0.5

● T.V. & Radio Centre
○ T.V. Centre

→ Shade indicates
population covered
by T.V. Centre

● Station

Range
1.000 sq.kms.

Station
N E W S

Furthest distances
Covered by Radio
in 1984 (N, E, S,W) in Km

1 cm = 40 Kms.

Scale 1:4 000 000   1 cm = 40 kms

23

ARUNACHAL PRADESH

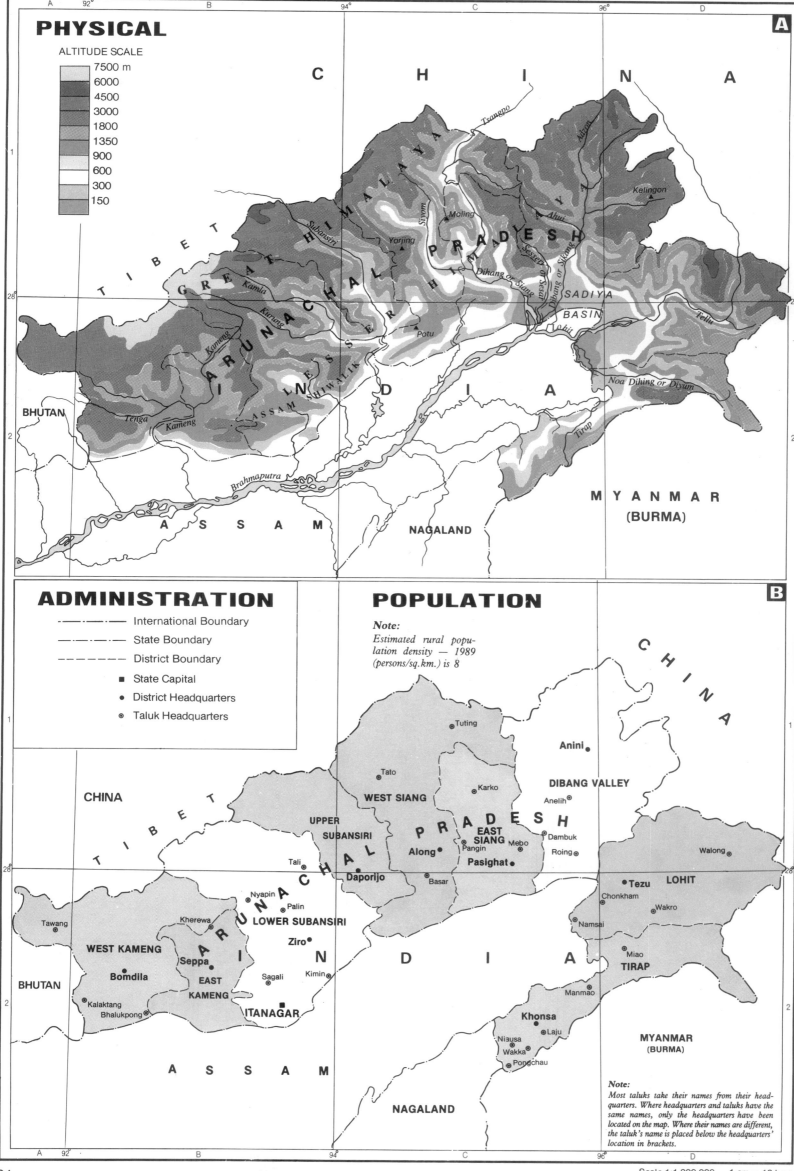

PHYSICAL [A]

ALTITUDE SCALE

	7500 m
	6000
	4500
	3000
	1800
	1350
	900
	600
	300
	150

CHINA

TIBET

GREAT HIMALAYA

ARUNACHAL PRADESH

LESSER HIMALAYA

SHIWALIK

ASSAM

Subansiri

Kamla

Kurung

Kameng

Tenga

Kameng

Yorjing

Siyom

Moling

Potu

Dihang or Siang

Sessen or Sikang

SADIYA

BASIN

Lohit

Ahui

Kelingon

Tellu

Tsangpo

Adzon

Noa Dihing or Diyum

Tirap

Brahmaputra

BHUTAN

INDIA

ASSAM

NAGALAND

MYANMAR
(BURMA)

ADMINISTRATION

———————	International Boundary
—·—·—·—	State Boundary
———————	District Boundary
■	State Capital
●	District Headquarters
⊙	Taluk Headquarters

POPULATION [B]

Note:
Estimated rural population density — 1989 (persons/sq.km.) is 8

CHINA

TIBET

BHUTAN

ASSAM

NAGALAND

INDIA

MYANMAR
(BURMA)

CHINA

ARUNACHAL PRADESH

Tawang

WEST KAMENG

Bomdila

Kalaktang

Bhalukpong

EAST KAMENG

Seppa

Kherewa

Nyapin

Palin

LOWER SUBANSIRI

Ziro

Sagali

Kimin

ITANAGAR

UPPER SUBANSIRI

Tali

Daporijo

Tuting

Tato

WEST SIANG

Karko

Along

Basar

Pangin

EAST SIANG

Pasighat

Mebo

Roing

Dambuk

Anini

DIBANG VALLEY

Anelih

Walong

LOHIT

Tezu

Chonkham

Wakro

Namsai

Miao

TIRAP

Manmao

Khonsa

Laju

Niausa

Wakka

Pongchau

Note:
Most taluks take their names from their headquarters. Where headquarters and taluks have the same names, only the headquarters have been located on the map. Where their names are different, the taluk's name is placed below the headquarters' location in brackets.

Scale 1:1 200 000   1 cm = 12 kms

A TRANSPORT & TOURISM

TRANSPORT
- 38 National Highways (with number)
- Other Roads
- Railway Lines

TOURISM
- ★ Historical Site
- ● Religious Centre
- ★ Wildlife Sanctuary

LOCATIONS
- ■ State Headquarters
- ○ Tourist Centre
- ● Other Towns
- ▲ District Headquarters (outside Arunachal Pradesh)

CHINA
TIBET
CHINA
TIBET

Tawang
Dirang
Bomdila
Rizon
Riang
Bhalukpong
Sonai Rupai
Tezpur
Itanagar
Hapoli
Ziro
Hach
Gochang
Dolungmukh
Daporijo
Along
Pangin
Bomdo
Jidu
Bruini
Minutang
Parasuram Kund
Bhismaknagar
Pasighat
Ledum
Raiing
Amilli
Tezu
Brahmakund
Tawaji
Trapun
Vijayanagar
Malinthan
Dibrugarh
North Lakhimpur
Sibsagar
Jorhat
Tirap
Khonsa
Miauba
Lakang

ARUNACHAL PRADESH
INDIA
ASSAM
NAGALAND
BHUTAN
MYANMAR (BURMA)

37
38

B INDUSTRIES & MARKETS

- ▨ C Class Market
- ○ Other Industrial Centres
- ■ Wood products
- ★ Electrical machinery
- ■ Non-metallic products

MINERALS
- □ Asbestos
- ● Coal
- ▲ Dolomite
- ⊠ Lime Stone

Note:
District-wise market potential has been estimated on the basis of the following indicators; Area, Number of households, Population, Population growth rate, Sex ratio, Density of population, Male literacy, Female literacy, Urbanisation, Workforce, Number of traders, Bank offices/100,000 population, Per capita bank credit to agriculture, Per capita bank credit to industries, Total cropped area, Production of major crops, Number of motor vehicles, Post offices/100,000 population, Circulation of newspapers, Cinema houses/100,000 population, Number of radio sets, T.V. sets.

CHINA
TIBET
CHINA
TIBET

DIBANG VALLEY
LOHIT
Tezu
EAST SIANG
WEST SIANG
UPPER SUBANSIRI
LOWER SUBANSIRI
EAST KAMENG
WEST KAMENG
Rupa
Itanagar
Pasighat
TIRAP

ARUNACHAL PRADESH
INDIA
ASSAM
NAGALAND
BHUTAN
MYANMAR (BURMA)

C ECONOMIC DEVELOPMENT

ESTIMATED INDEX OF LEVELS OF ECONOMIC DEVELOPMENT – 1989

AVERAGE
- All India 100
- State 12

- 25–50
- 0–25

Tirap – Notified Backward Districts

Note:
The levels of economic development are based on the following indicators; Per capita value of output of major crops, Per capita bank credit to agriculture, Number of mining and factory workers per 100,000 population, Number of households of manufacturing workers per 100,000 population, Per capita bank credit for manufacturing sector, Per capita bank deposits, Per capita bank credit to services, Literacy and Urbanisation.

CHINA
TIBET
CHINA
TIBET

Dibang Valley
Lohit
East Siang
West Siang
Upper Subansiri
Lower Subansiri
East Kameng
West Kameng
Tirap

ARUNACHAL PRADESH
INDIA
ASSAM
NAGALAND
BHUTAN
MYANMAR (BURMA)

D MASS COMMUNICATION

ESTIMATED POPULATION COVERED BY T.V. CENTRE – 1986 (in million)
- 0–0.5

- ● T.V. Centre
- ★ Radio Centre

→ Shade indicates population covered by T.V. Centre

Station → Range 1000 sq. kms

Furthest distance covered by Radio in 1984 (N.E.W.S.) in km

1 cm. = 27 kms.

N E S W Station

CHINA
TIBET
CHINA
TIBET

Tezu
Pasighat
Itanagar
Tawang

ARUNACHAL PRADESH
INDIA
ASSAM
NAGALAND
BHUTAN
MYANMAR (BURMA)

Scale 1:2 700 000   1 cm = 27 kms

ASSAM

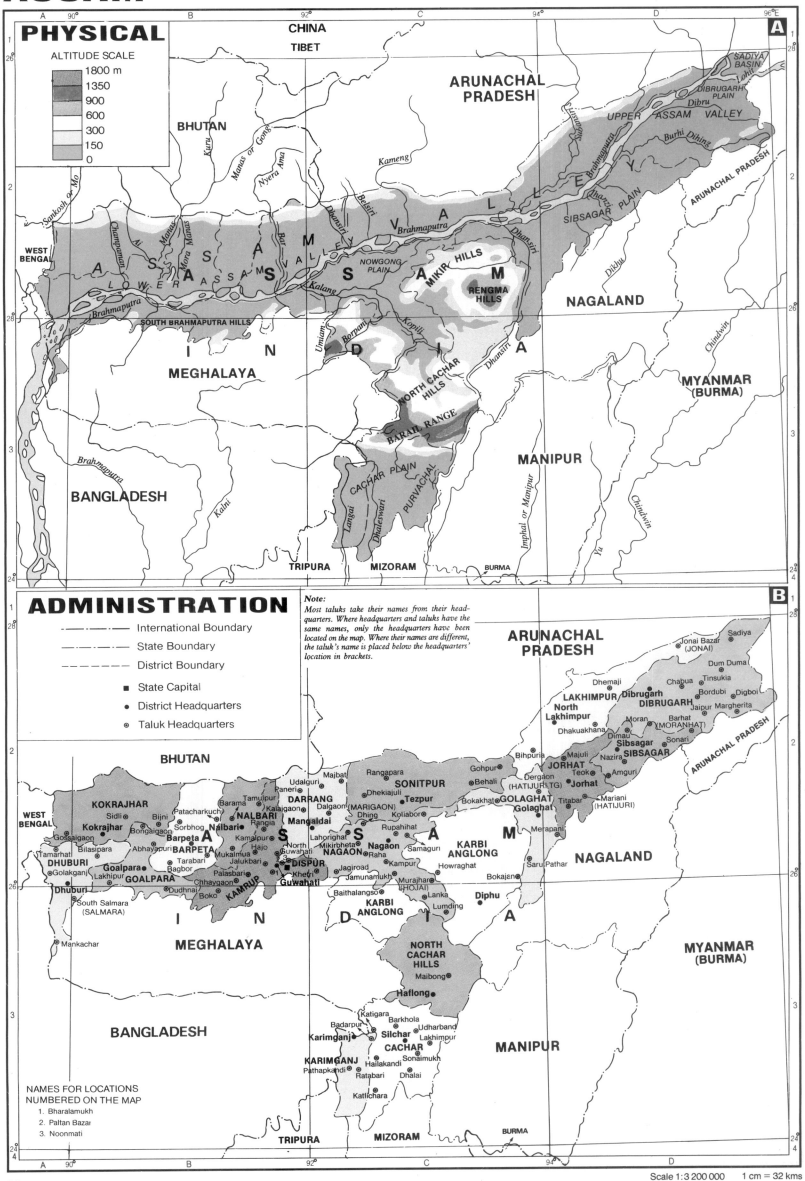

PHYSICAL

ALTITUDE SCALE

	1800 m
	1350
	900
	600
	300
	150
	0

CHINA
TIBET

BHUTAN

ARUNACHAL PRADESH

WEST BENGAL

BANGLADESH

MEGHALAYA

NAGALAND

MANIPUR

MYANMAR (BURMA)

TRIPURA   MIZORAM   BURMA

ASSAM VALLEY

LOWER ASSAM

UPPER ASSAM VALLEY

SADIYA BASIN

DIBRUGARH PLAIN

SIBSAGAR PLAIN

NOWGONG PLAIN

MIKIR HILLS

RENGMA HILLS

SOUTH BRAHMAPUTRA HILLS

NORTH CACHAR HILLS

BARAIL RANGE

CACHAR PLAIN

PURVACHAL

I N D I A

ADMINISTRATION

—————————	International Boundary
—·—·—·—·—	State Boundary
— — — — —	District Boundary
■	State Capital
●	District Headquarters
⊙	Taluk Headquarters

Note:
Most taluks take their names from their head-quarters. Where headquarters and taluks have the same names, only the headquarters have been located on the map. Where their names are different, the taluk's name is placed below the headquarters' location in brackets.

ARUNACHAL PRADESH

BHUTAN

WEST BENGAL

KOKRAJHAR

DHUBURI

GOALPARA

KAMRUP

NALBARI

BARPETA

DARRANG

SONITPUR

NAGAON

KARBI ANGLONG

LAKHIMPUR

DIBRUGARH

SIBSAGAR

JORHAT

GOLAGHAT

NORTH CACHAR HILLS

CACHAR

KARIMGANJ

MEGHALAYA

BANGLADESH

NAGALAND

MANIPUR

MYANMAR (BURMA)

TRIPURA   MIZORAM   BURMA

I N D I A

NAMES FOR LOCATIONS
NUMBERED ON THE MAP
1. Bharalamukh
2. Paltan Bazar
3. Noonmati

Scale 1:3 200 000   1 cm = 32 kms

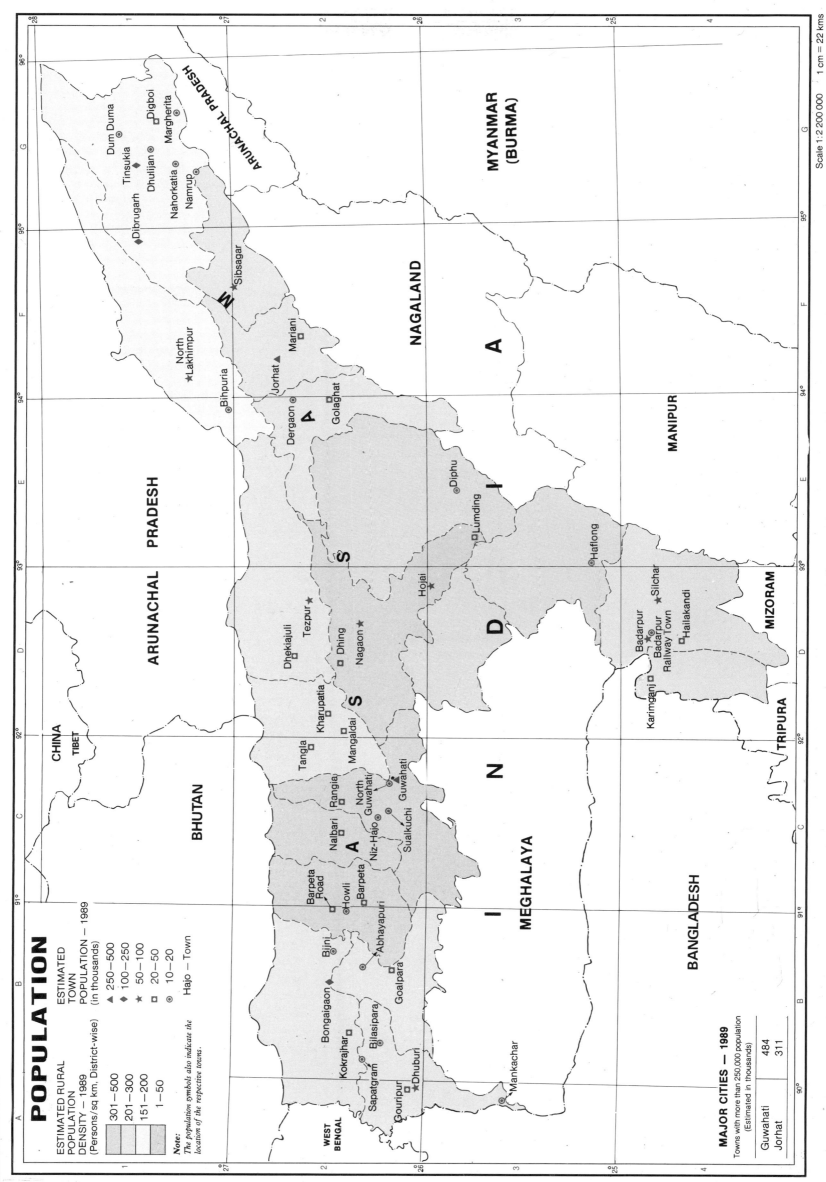

POPULATION

ESTIMATED RURAL
POPULATION
DENSITY – 1989
(Persons/sq km, District-wise)

301–500	
201–300	
151–200	
1–50	

ESTIMATED
TOWN
POPULATION – 1989
(in thousands)

▲ 250–500
◆ 100–250
★ 50–100
▢ 20–50
◉ 10–20

Hajo – Town

Note:
The population symbols also indicate the location of the respective towns.

CHINA
TIBET

BHUTAN

ARUNACHAL PRADESH

WEST
BENGAL

BANGLADESH

MEGHALAYA

NAGALAND

MANIPUR

MIZORAM

TRIPURA

MYANMAR
(BURMA)

A S S A M

N A G A O N

A S S A M

Dum Duma ◉
Tinsukia ◆
Dibrugarh ◆
Dhulijan ◉
Nahorkatia ◉
Digboi ▢
Margherita ◉
Namrup ◉

North
★Lakhimpur
Bihpuria ◉

★ Sibsagar

Mariani ▢
Jorhat ▲
Dergaon ◉
Golaghat ▢

Diphu ◉

Lumding ▢

Haflong ◉

Dhekiajuli ◉
Tezpur ★
Dhing ▢
Nagaon ★
Hojai ★

Kharupatia ◉

Tangla ◉
Mangaldai ▢

Rangia ◉
North ◉
Guwahati
Nalbari ◉
Niz-Hajo ◉
Sualkuchi ◉
Guwahati ◉

Barpeta
Road ▢
Howli ◉ Barpeta ▢
Abhayapuri ▲
Bijni ◉
Goalpara ▢

Bongaigaon ◉
Kokrajhar ▢
Bilasipara ◉
Sapatgram ◉
Gouripur ◉
Dhuburi ★

Mankachar ◉

Karimganj ▢
Badarpur ◉
Badarpur
Railway Town ★
Silchar ★
Hailakandi ▢

MAJOR CITIES — 1989

Towns with more than 250,000 population
(Estimated in thousands)

Guwahati	484
Jorhat	311

Scale 1:2 200 000   1 cm = 22 kms

3

27

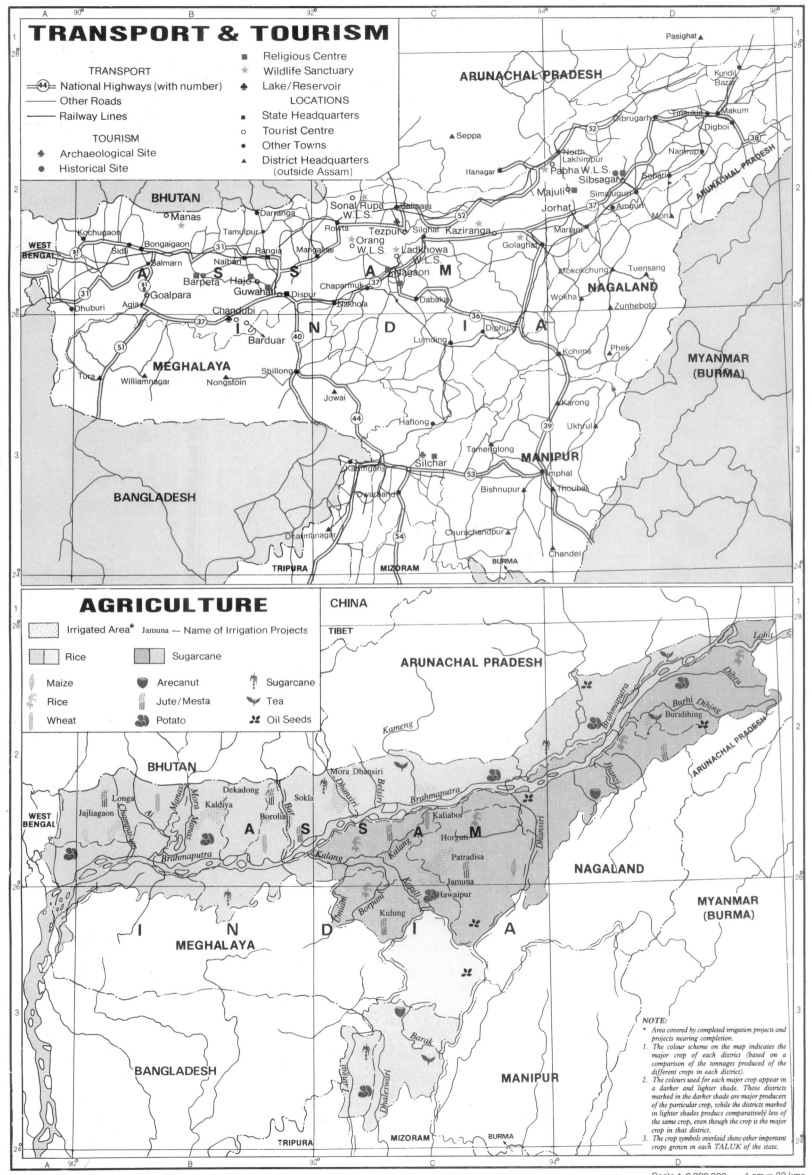

TRANSPORT & TOURISM

TRANSPORT
- 44 National Highways (with number)
- Other Roads
- Railway Lines

TOURISM
- Archaeological Site
- Historical Site

- Religious Centre
- Wildlife Sanctuary
- Lake/Reservoir

LOCATIONS
- State Headquarters
- Tourist Centre
- Other Towns
- District Headquarters (outside Assam)

ARUNACHAL PRADESH

Pasighat
Kundil Bazar
Makum
Digboi
Namrup
Tinsukia
Dibrugarh
52
Sepa
Sibsagar
North Lakhimpur
Pabha W.L.S.
Majuli
Simaluguri
Itanagar
Jorhat
37
Amguri
Mariani
Mon
BHUTAN
Manas
Darranga
Sonai Rupa W.L.S.
Balipara
Kaziranga
Golaghat
Tamulpur
Rowta
Tezpur
Silghat
Kochugaon
Bongaigaon
Rangia
Mangaldai
Orang W.L.S.
Ladkhowa W.L.S.
Mokokchung
Tuensang
Sidli
Nalbari
Salmari
ASSAM
Wokha
Zunheboto
WEST BENGAL
31
Barpeta
Hajo
Guwahati
Dispur
Nagaon
37
NAGALAND
Goalpara
Chaparmuk
Dabaka
Dhuburi
Agia
Chandubi
Nakhola
31
37
INDIA
Lumding
36
Diphu
Barduar
40
Kohima
Phek
MYANMAR (BURMA)
51
Tura
Shillong
Williamnagar
Nongstoin
Karong
Ukhrul
Jowai
39
MEGHALAYA
44
Haflong
Tamenglong
MANIPUR
Silchar
Imphal
Karimganj
53
Bishnupur
Thoubal
BANGLADESH
Dwarband
Churachandpur
Dharmanagar
54
Chandel
TRIPURA
MIZORAM
BURMA

AGRICULTURE

- Irrigated Area* Jamuna — Name of Irrigation Projects
- Rice
- Sugarcane
- Maize
- Arecanut
- Sugarcane
- Rice
- Jute/Mesta
- Tea
- Wheat
- Potato
- Oil Seeds

CHINA
TIBET
ARUNACHAL PRADESH
Lohit
Kameng
Brahmaputra
Dibru
Burhi Dihing
Buridihing
BHUTAN
Mora Dhansiri
Longa
Dekadong
Sokla
Dhansiri
Belsiri
Jhanzi
Jajliagaon
Kaldiya
Borolia
Brahmaputra
Champamati
Ai
Manas
Buri
Kaliabor
ASSAM
Kalang
Horguti
Mora Manas
Brahmaputra
Kalang
Patradisa
Dhansiri
NAGALAND
Kalang
Jamuna
WEST BENGAL
Kapili
Hawaipur
MYANMAR (BURMA)
INDIA
Dimani
Borpani
Kulung
MEGHALAYA
Barak
Langai
Dhaleswari
BANGLADESH
MANIPUR
MIZORAM
BURMA
TRIPURA

NOTE:
* Area covered by completed irrigation projects and projects nearing completion.
1. The colour scheme on the map indicates the major crop of each district (based on a comparison of the tonnages produced of the different crops in each district).
2. The colours used for each major crop appear in a darker and lighter shade. Those districts marked in the darker shade are major producers of the particular crop, while the districts marked in lighter shades produce comparatively less of the same crop, even though the crop is the major crop in that district.
3. The crop symbols overlaid show other important crops grown in each TALUK of the state.

Scale 1:3 200 000   1 cm = 32 kms

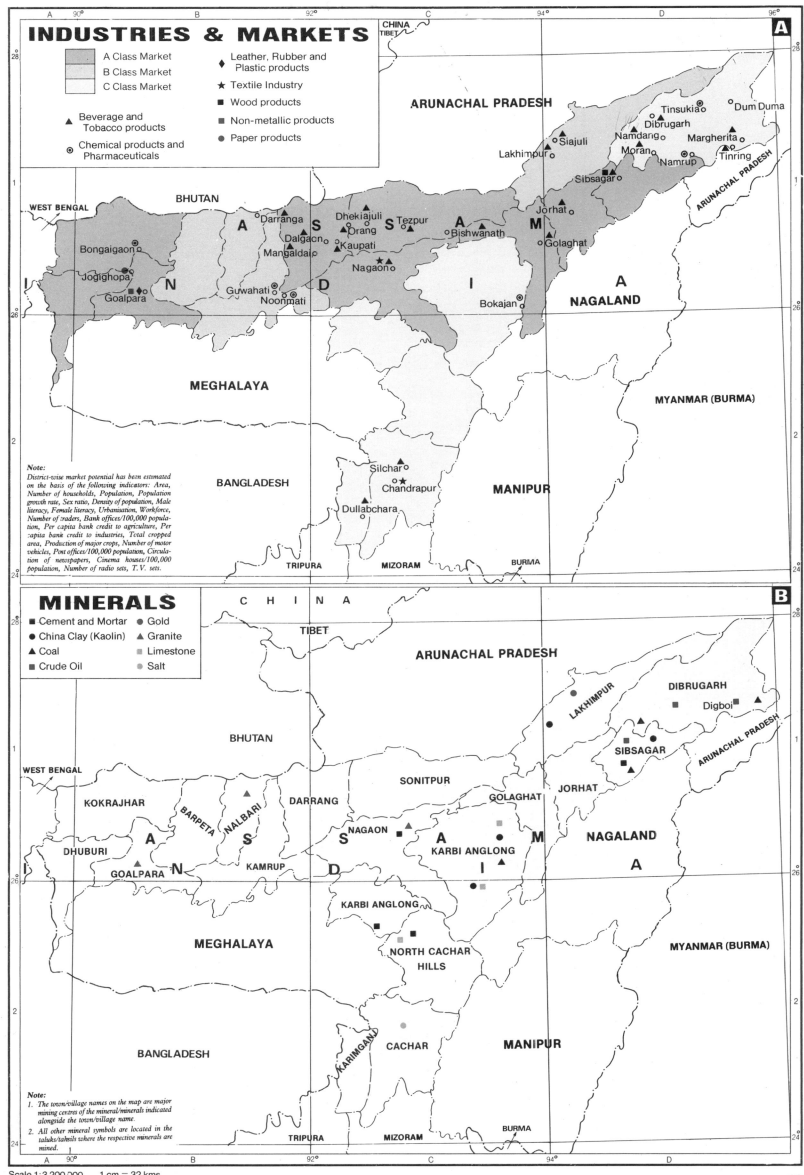

INDUSTRIES & MARKETS

A Class Market
B Class Market
C Class Market

▲ Beverage and Tobacco products
◎ Chemical products and Pharmaceuticals
◆ Leather, Rubber and Plastic products
★ Textile Industry
■ Wood products
■ Non-metallic products
● Paper products

CHINA
TIBET

ARUNACHAL PRADESH

BHUTAN

WEST BENGAL

Tinsukia   Dum Duma
Dibrugarh
Namdang   Margherita
Siajuli   Moran   Tinring
Lakhimpur   Namrup
Sibsagar

A   S   S   A   M
Darranga   Dhekiajuli   Jorhat
Orang   Tezpur   Golaghat
Dalgaon   Bishwanath
Bongaigaon   Mangaldai   Kaupati
Jogighopa   Nagaon
Goalpara   I   N   D   I   A

I   N   D   Guwahati   NAGALAND
Noonmati   Bokajan

MEGHALAYA   MYANMAR (BURMA)

Silchar
Chandrapur   MANIPUR
Dullabchara

BANGLADESH

Note:
District-wise market potential has been estimated on the basis of the following indicators: Area, Number of households, Population, Population growth rate, Sex ratio, Density of population, Male literacy, Female literacy, Urbanisation, Workforce, Number of traders, Bank offices/100,000 population, Per capita bank credit to agriculture, Per capita bank credit to industries, Total cropped area, Production of major crops, Number of motor vehicles, Post offices/100,000 population, Circulation of newspapers, Cinema houses/100,000 population, Number of radio sets, T.V. sets.

TRIPURA   MIZORAM   BURMA

MINERALS

■ Cement and Mortar   ● Gold
● China Clay (Kaolin)   ▲ Granite
▲ Coal   ■ Limestone
■ Crude Oil   ● Salt

CHINA
TIBET

ARUNACHAL PRADESH

BHUTAN

WEST BENGAL

LAKHIMPUR   DIBRUGARH
Digboi
SIBSAGAR

KOKRAJHAR   SONITPUR   JORHAT
BARPETA   NALBARI   DARRANG   GOLAGHAT
A   S   S   NAGAON   A   M   NAGALAND
DHUBURI   KAMRUP   KARBI ANGLONG
GOALPARA   N   D   I   A

KARBI ANGLONG

MEGHALAYA   NORTH CACHAR   MYANMAR (BURMA)
HILLS

KARIMGANJ   CACHAR   MANIPUR

BANGLADESH

Note:
1. *The town/village names on the map are major mining centres of the mineral/minerals indicated alongside the town/village name.*
2. *All other mineral symbols are located in the taluks/tahsils where the respective minerals are mined.*

TRIPURA   MIZORAM   BURMA

Scale 1:3 200 000   1 cm = 32 kms

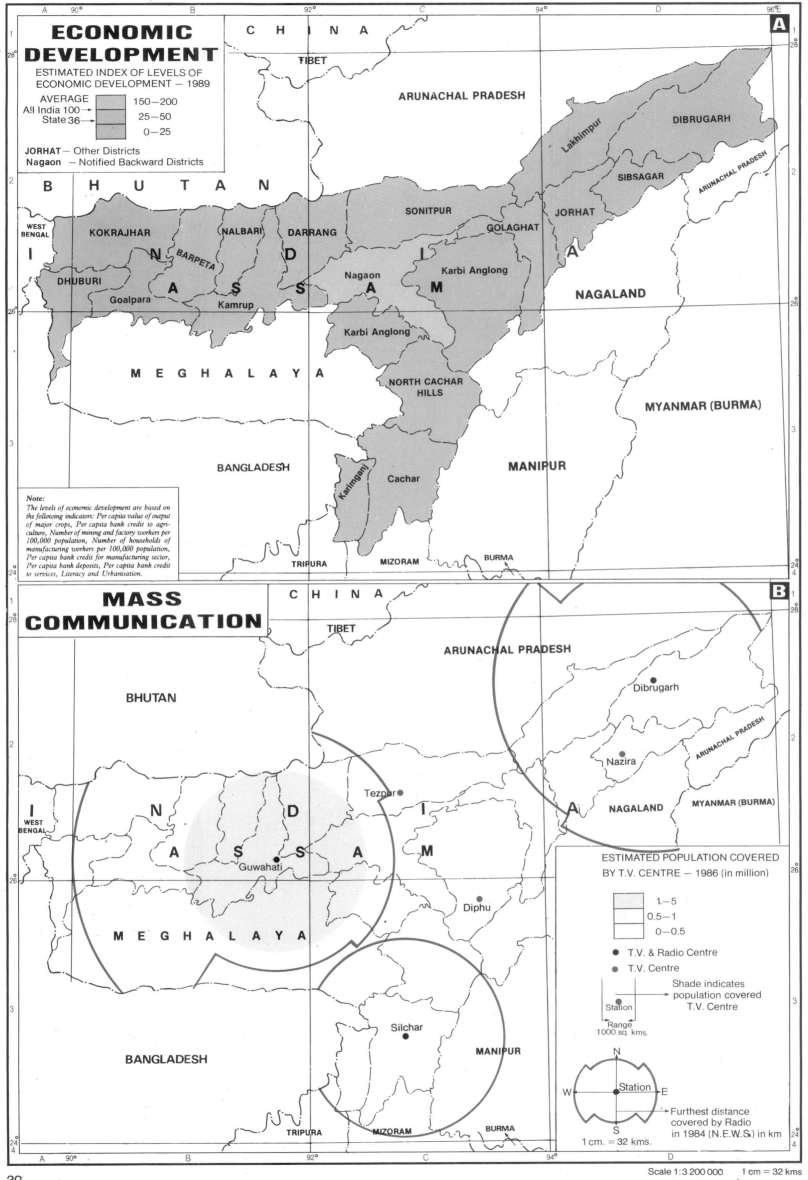

ECONOMIC DEVELOPMENT

ESTIMATED INDEX OF LEVELS OF ECONOMIC DEVELOPMENT — 1989

AVERAGE
All India 100 →
State 36 →

150—200
25—50
0—25

JORHAT — Other Districts
Nagaon — Notified Backward Districts

Note:
The levels of economic development are based on the following indicators: Per capita value of output of major crops, Per capita bank credit to agriculture, Number of mining and factory workers per 100,000 population, Number of households of manufacturing workers per 100,000 population, Per capita bank credit for manufacturing sector, Per capita bank deposits, Per capita bank credit to services, Literacy and Urbanisation.

CHINA
TIBET
ARUNACHAL PRADESH
BHUTAN
WEST BENGAL
DHUBURI
KOKRAJHAR
Goalpara
BARPETA
NALBARI
Kamrup
DARRANG
Nagaon
SONITPUR
Karbi Anglong
GOLAGHAT
JORHAT
SIBSAGAR
Lakhimpur
DIBRUGARH
ARUNACHAL PRADESH
NAGALAND
Karbi Anglong
NORTH CACHAR HILLS
MYANMAR (BURMA)
MEGHALAYA
BANGLADESH
Karimganj
Cachar
MANIPUR
TRIPURA
MIZORAM
BURMA
I N A S S A M

A

MASS COMMUNICATION

CHINA
TIBET
ARUNACHAL PRADESH
BHUTAN
WEST BENGAL
I N A S S A M
Dibrugarh
Nazira
Tezpur
NAGALAND
MYANMAR (BURMA)
Guwahati
Diphu
MEGHALAYA
Silchar
BANGLADESH
MANIPUR
TRIPURA
MIZORAM
BURMA

B

ESTIMATED POPULATION COVERED
BY T.V. CENTRE — 1986 (in million)

1—5
0.5—1
0—0.5

● T.V. & Radio Centre
● T.V. Centre

Station

Range
1000 sq. kms.

Shade indicates
population covered
T.V. Centre

N
W — Station — E
S

1 cm. = 32 kms.

Furthest distance
covered by Radio
in 1984 (N.E.W.S.) in km

30

Scale 1:3 200 000   1 cm = 32 kms

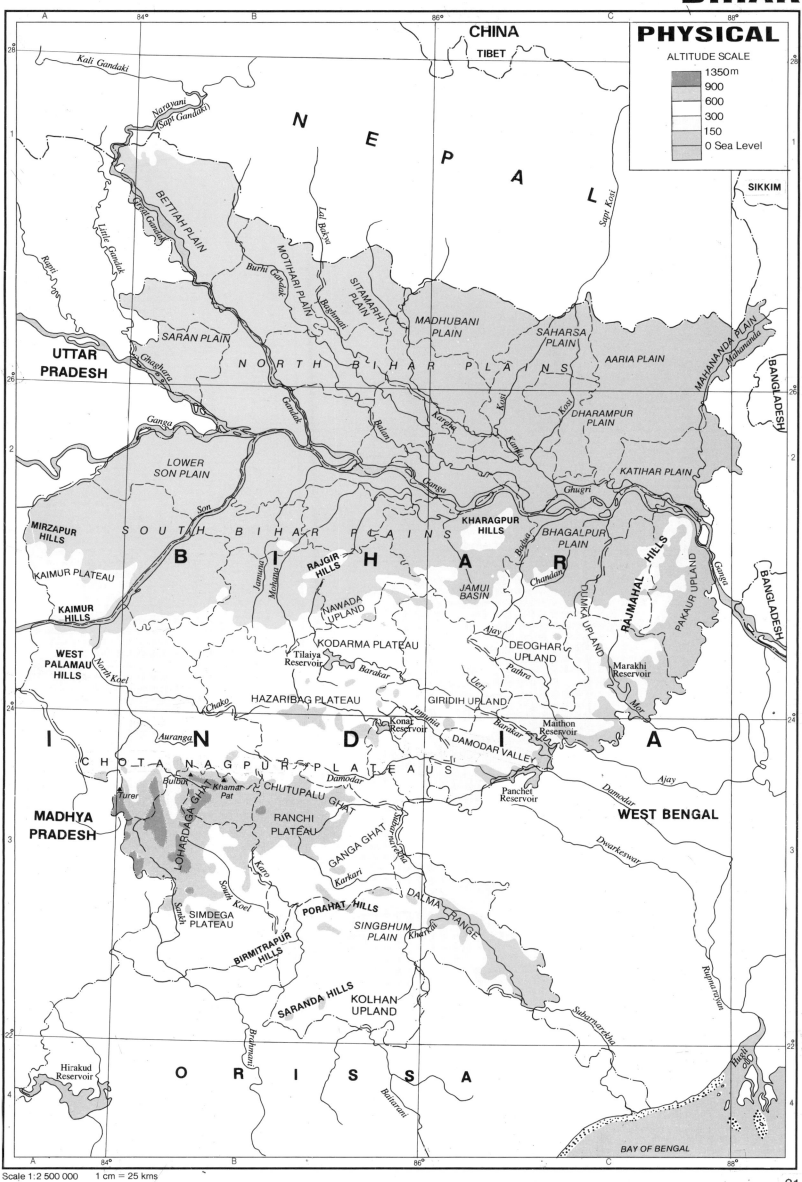

BIHAR

ALTITUDE SCALE

1350m
900
600
300
150
0 Sea Level

CHINA

TIBET

N E P A L

SIKKIM

Kali Gandaki

Narayani
(Sapt Gandaki)

Little Gandak

Rupni

Lal Bakya

Sapt Kosi

BETTIAH PLAIN

Great Gandak

MOTIHARI PLAIN

Burhi Gandak

SITAMARHI PLAIN

Baghmati

MADHUBANI PLAIN

SARAN PLAIN

SAHARSA PLAIN

AARIA PLAIN

MAHANANDA PLAIN

Mahananda

BANGLADESH

UTTAR PRADESH

N O R T H   B I H A R   P L A I N S

Ghaghara

Ganga

Gandak

Balan

Kareha

Kosi

Kosi

Kamla

DHARAMPUR PLAIN

LOWER SON PLAIN

KATIHAR PLAIN

Ganga

Ghugri

Son

MIRZAPUR HILLS

S O U T H   B I H A R   P L A I N S

KHARAGPUR HILLS

BHAGALPUR PLAIN

Ganga

KAIMUR PLATEAU

B

Jamuna

Mohana

I

RAJGIR HILLS

H

A

R

Badua

Chandan

DUMKA UPLAND

RAJMAHAL HILLS

PAKAUR UPLAND

KAIMUR HILLS

NAWADA UPLAND

JAMUI BASIN

WEST PALAMAU HILLS

North Koel

KODARMA PLATEAU

Tilaiya Reservoir

Barakar

Ajay

DEOGHAR UPLAND

Pathra

Marakhi Reservoir

Mor

Chako

HAZARIBAG PLATEAU

GIRIDIH UPLAND

Jamunia

Ueri

Barakar

Maithon Reservoir

Ganga

Auranga

I

N

Konar Reservoir

D

DAMODAR VALLEY

I

A

Ajay

CHOTA NAGPUR PLATEAUS

Damodar

Panchet Reservoir

Damodar

WEST BENGAL

MADHYA PRADESH

Turer

Bulbut

LOHARDAGA GHAT

Khamar Pat

CHUTUPALU GHAT

RANCHI PLATEAU

Subarnarekha

Dwarkeswar

Karo

Ganga Ghat

Karkari

South Koel

Sankh

SIMDEGA PLATEAU

PORAHAT HILLS

DALMA RANGE

SINGBHUM PLAIN

Kharkai

Rupnarayan

BIRMITRAPUR HILLS

Subarnarekha

SARANDA HILLS

KOLHAN UPLAND

Brahmani

Hirakud Reservoir

O   R   I   S   S   A

Baitarani

Hugli

BAY OF BENGAL

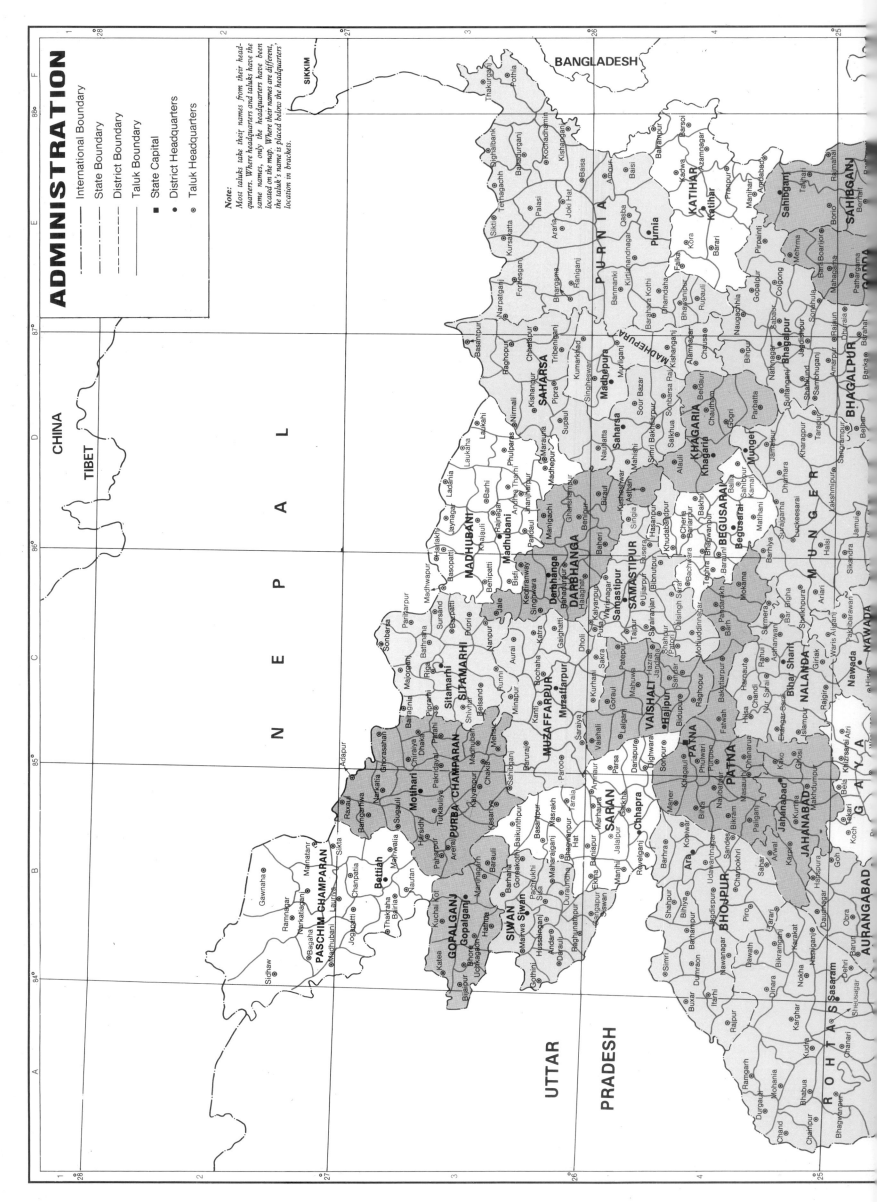

ADMINISTRATION

Note:
Most taluks take their names from their head-quarters. Where headquarters and taluks have the same names, only the headquarters have been located on the map. Where their names are different, the taluk's name is placed below the headquarters' location in brackets.

32

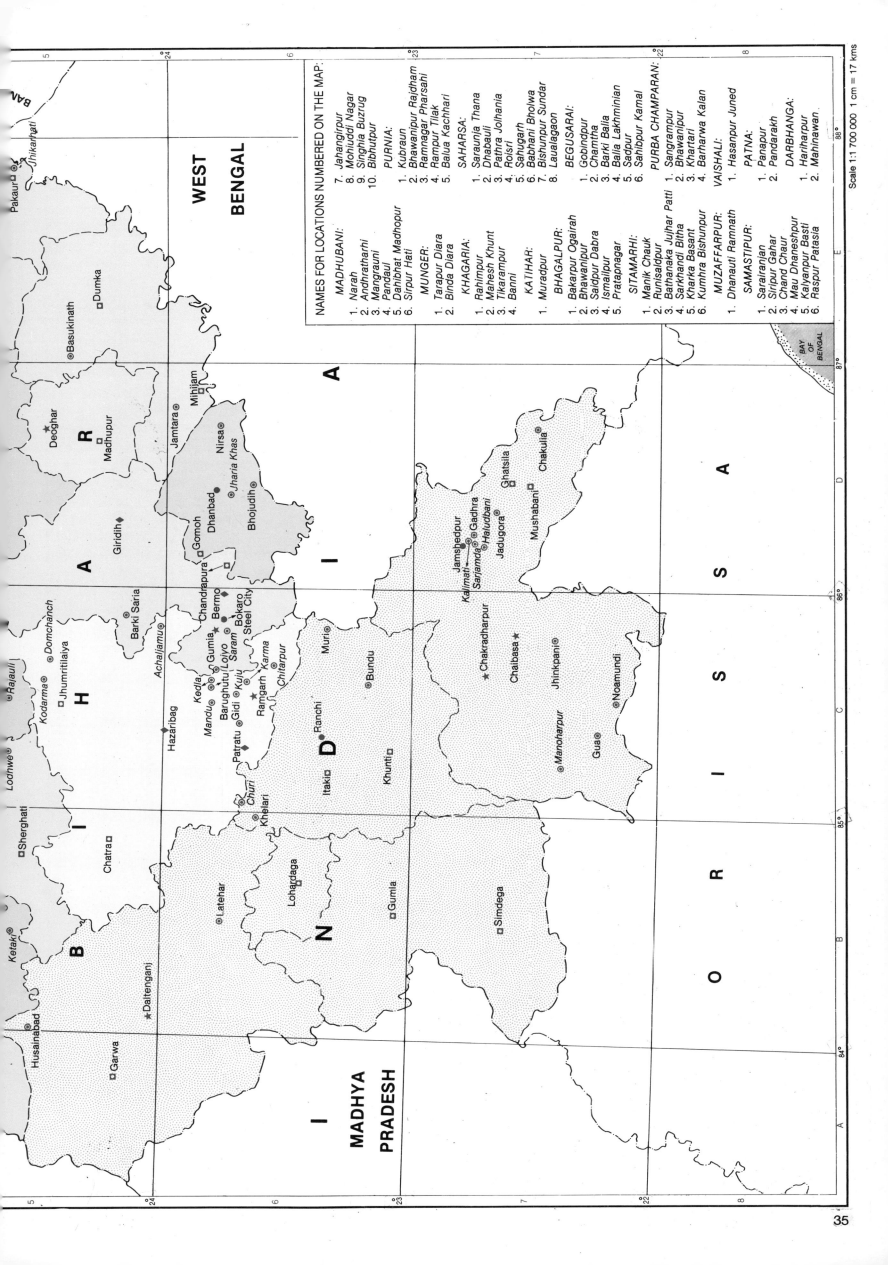

NAMES FOR LOCATIONS NUMBERED ON THE MAP:

MADHUBANI:
1. Narah
2. Andhratharhi
3. Mangrauni
4. Pandaul
5. Dahibhat Madhopur
6. Sirpur Hati
7. Jahangirpur
8. Mohiuddi Nagar
9. Singhia Buzrug
10. Bibhutpur

PURNIA:
1. Kubraun
2. Bhawanipur Rajdham
3. Ramnagar Pharsahi
4. Rampur Tilak
5. Balua Kachhari

MUNGER:
1. Tarapur Diara
2. Binda Diara

KHAGARIA:
1. Rahimpur
2. Mahesh Khunt
3. Tikarampur
4. Banni

SAHARSA:
1. Saraunja Thana
2. Dhabauli
3. Pathra Jolhania
4. Roisri
5. Sahugarh
6. Babhani Bholwa
7. Bishunpur Sundar
8. Laualagaon

KATIHAR:
1. Muradpur

BHAGALPUR:
1. Bakarpur Ogairah
2. Bhawanipur
3. Saidpur Dabra
4. Ismailpur
5. Pratapnagar

BEGUSARAI:
1. Gobindpur
2. Chamtha
3. Barki Balla
4. Balia Lakhminian
5. Sadpur
6. Sahibpur Kamal

SITAMARHI:
1. Manik Chauk
2. Runisaldpur
3. Bathanaka Jujhar Patti
4. Sarkhandi Bitha
5. Kharka Basant
6. Kumhra Bishunpur

PURBA CHAMPARAN:
1. Sangrampur
2. Bhawanipur
3. Khartari
4. Barharwa Kalan

MUZAFFARPUR:
1. Dhanauti Ramnath

VAISHALI:
1. Hasanpur Juned

SAMASTIPUR:
1. Sarairanjan
2. Siripur Gahar
3. Chand Chaur
4. Mau Dhaneshpur
5. Kalyanpur Basti
6. Raspur Patasia

PATNA:
1. Panapur
2. Pandarakh

DARBHANGA:
1. Hariharpur
2. Mahinawan.

Scale 1:1 700 000 1 cm = 17 kms

35

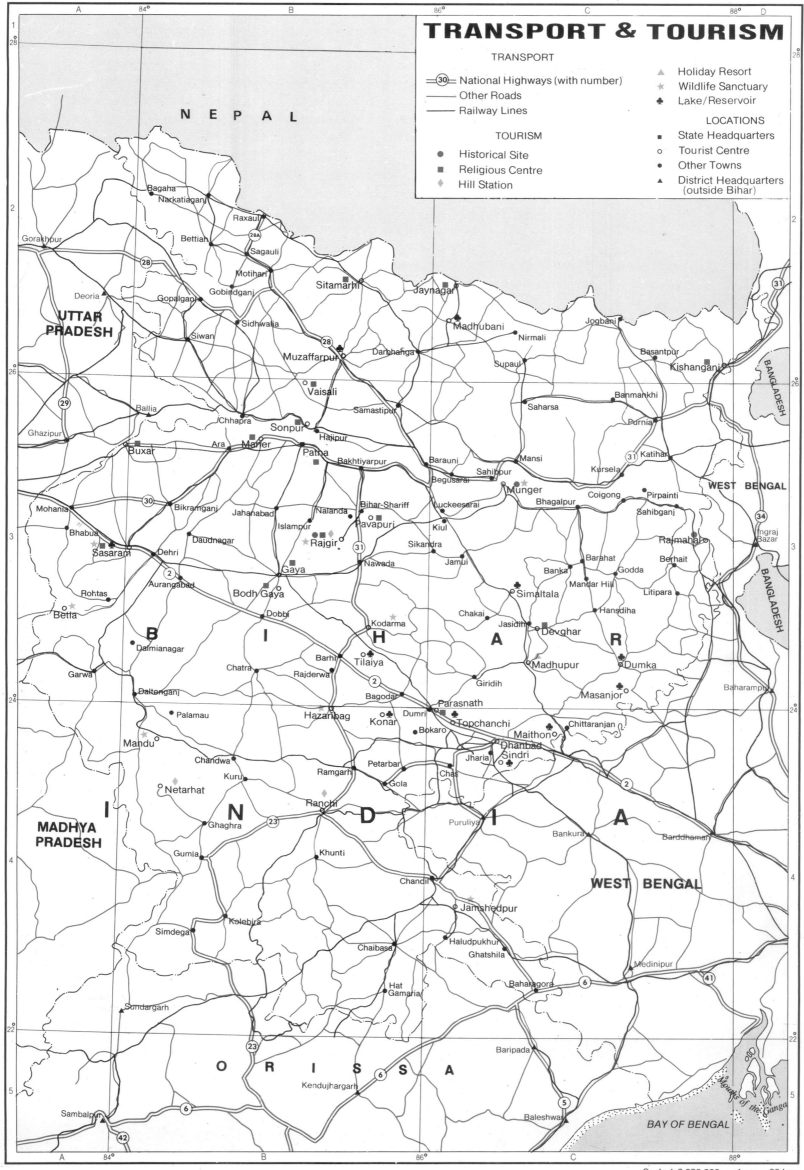

TRANSPORT & TOURISM

TRANSPORT

⊟30⊟ National Highways (with number)
—— Other Roads
—— Railway Lines

TOURISM

● Historical Site
■ Religious Centre
◆ Hill Station

▲ Holiday Resort
★ Wildlife Sanctuary
♣ Lake/Reservoir

LOCATIONS

■ State Headquarters
○ Tourist Centre
● Other Towns
▲ District Headquarters (outside Bihar)

NEPAL

UTTAR PRADESH

Gorakhpur
Bagaha
Narkatiaganj
Raxaul
Bettiah
Sagauli
Deoria
Motihari
Gopalganj
Gobindganj
Sidhwalia
Sitamarhi
Jaynagar
Siwan
Madhubani
Nirmali
Jogbani
Basantpur
Muzaffarpur
Darbhanga
Kishanganj
Vaisali
Samastipur
Supaul
Ballia
Saharsa
Banmankhi
Ghazipur
Chhapra
Sonpur
Hajipur
Purnia
Buxar
Ara
Maner
Patna
Bakhtiyarpur
Barauni
Sahibpur
Mansi
Kursela
Katihar
Mohania
Bikramganj
Jahanabad
Nalanda
Bihar-Shariff
Luckeesarai
Begusarai
Munger
Coigong
Pirpainti
Sahibganj
WEST BENGAL
Bhabua
Islampur
Pavapuri
Bhagalpur
Rajmahal
Ingraj Bazar
Sasaram
Dehri
Daudnagar
Rajgir
Kiul
Barahat
Berhait
Rohtas
Aurangabad
Gaya
Nawada
Sikandra
Banka
Godda
Betla
Bodh Gaya
Jamui
Simaltala
Mandar Hill
Litipara
Dobbi
Kodarma
Jasidih
Hansdiha
Dalmianagar
Chakai
Devghar
Chatra
Barhi
Tilaiya
Madhupur
Dumka
Garwa
Rajderwa
Giridih
Masanjor
Baharampur
Daltenganj
Bagodar
Parasnath
Palamau
Hazaribag
Dumri
Topchanchi
Chittaranjan
Mandu
Konar
Bokaro
Maithon
Chandwa
Petarbar
Chas
Jharia
Dhanbad
Sindri
Netarhat
Kuru
Ramgarh
Gola
Ranchi
Chandra
Ghaghra
Puruliya
Bankura
Barddhaman
Gumla
Khunti
MADHYA PRADESH
Chandil
WEST BENGAL
Simdega
Kolebira
Jamshedpur
Chaibasa
Haludpukhur
Ghatshila
Medinipur
Hat Gamaria
Bahragora
Sundargarh
ORISSA
Baripada
Kendujhargarh
Baleshwar
Sambalpur
BAY OF BENGAL
Mouth of the Ganga

B I H A R

I N D I A

BANGLADESH

Scale 1:2 500 000   1 cm = 25 kms

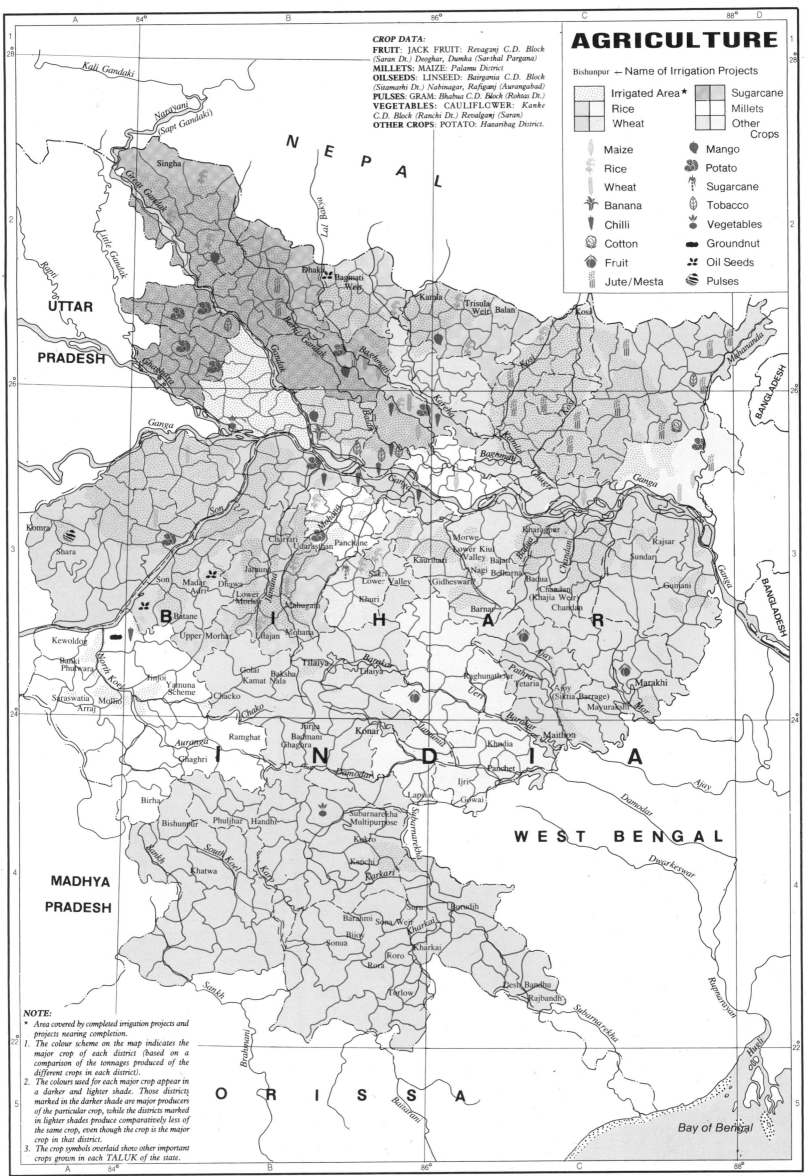

AGRICULTURE

Bishunpur ← Name of Irrigation Projects

CROP DATA:
FRUIT: JACK FRUIT: *Revaganj C.D. Block (Saran Dt.) Deoghar, Dumka (Santhal Pargana)*
MILLETS: MAIZE: *Palamu District*
OILSEEDS: LINSEED: *Bairgania C.D. Block (Sitamarhi Dt.) Nabinagar, Rafiganj (Aurangabad)*
PULSES: GRAM: *Bhabua C.D. Block (Rohtas Dt.)*
VEGETABLES: CAULIFLOWER: *Kanke C.D. Block (Ranchi Dt.) Revalganj (Saran)*
OTHER CROPS: POTATO: *Hazaribag District.*

Irrigated Area★   Sugarcane
Rice   Millets
Wheat   Other Crops

Maize   Mango
Rice   Potato
Wheat   Sugarcane
Banana   Tobacco
Chilli   Vegetables
Cotton   Groundnut
Fruit   Oil Seeds
Jute/Mesta   Pulses

NEPAL

UTTAR PRADESH

B I H A R

INDIA

MADHYA PRADESH

WEST BENGAL

BANGLADESH

ORISSA

Bay of Bengal

NOTE:
★ *Area covered by completed irrigation projects and projects nearing completion.*
1. *The colour scheme on the map indicates the major crop of each district (based on a comparison of the tonnages produced of the different crops in each district).*
2. *The colours used for each major crop appear in a darker and lighter shade. Those districts marked in the darker shade are major producers of the particular crop, while the districts marked in lighter shades produce comparatively less of the same crop, even though the crop is the major crop in that district.*
3. *The crop symbols overlaid show other important crops grown in each TALUK of the state.*

Scale 1:2 500 000   1 cm = 25 kms

37

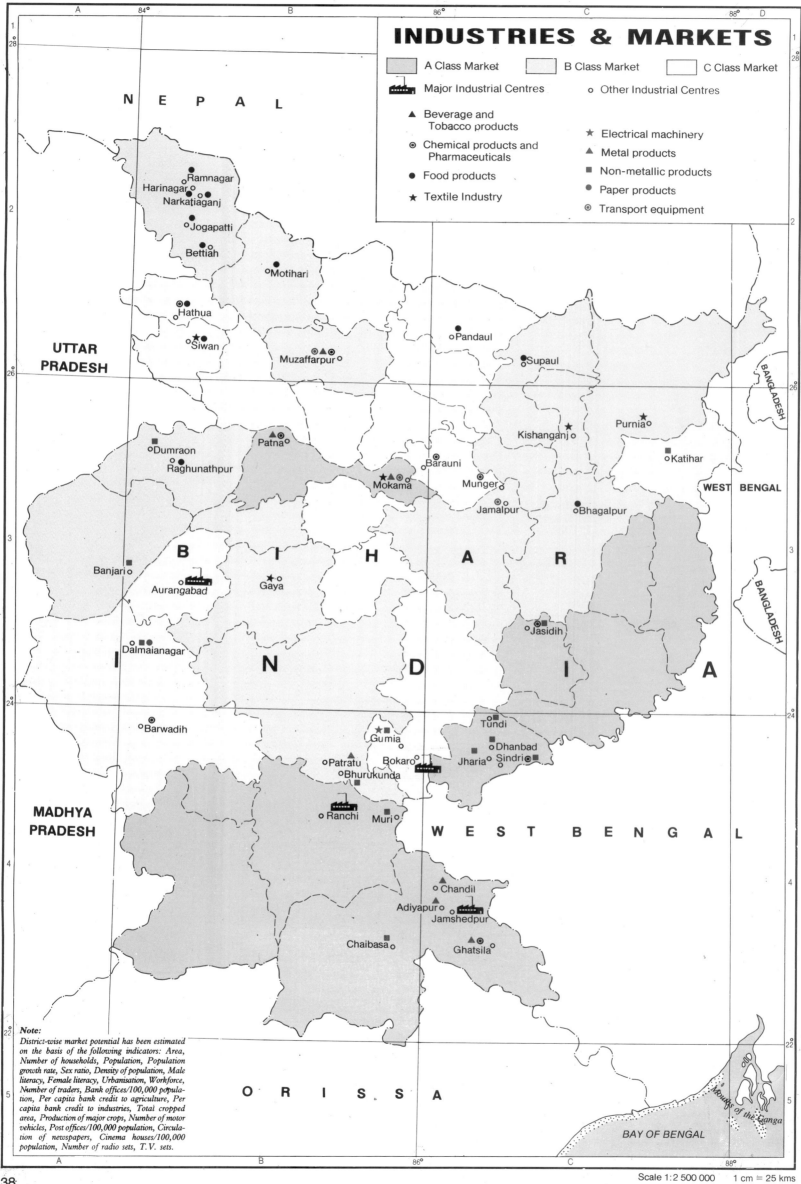

INDUSTRIES & MARKETS

| | A Class Market | | B Class Market | | C Class Market |
| | Major Industrial Centres | | ○ Other Industrial Centres | | |

▲ Beverage and Tobacco products
⊙ Chemical products and Pharmaceuticals
● Food products
★ Textile Industry

★ Electrical machinery
▲ Metal products
■ Non-metallic products
● Paper products
⊙ Transport equipment

N E P A L

UTTAR PRADESH

Ramnagar
Harinagar
Narkatiaganj
Jogapatti
Bettiah
Motihari
Hathua
Siwan
Muzaffarpur
Pandaul
Supaul
Kishanganj
Purnia
Katihar

WEST BENGAL

Dumraon
Patna
Raghunathpur
Barauni
Mokama
Munger
Jamalpur
Bhagalpur

B I H A R

Banjari
Aurangabad
Gaya
Jasidih

I N D I A

Dalmaianagar

Barwadih

Tundi
Gumia
Dhanbad
Patratu
Bokaro
Jharia
Sindri
Bhurukunda

MADHYA PRADESH

Ranchi
Muri

W E S T   B E N G A L

Chandil
Adiyapur
Jamshedpur
Chaibasa
Ghatsila

BANGLADESH
BANGLADESH

Note:
District-wise market potential has been estimated on the basis of the following indicators: Area, Number of households, Population, Population growth rate, Sex ratio, Density of population, Male literacy, Female literacy, Urbanisation, Workforce, Number of traders, Bank offices/100,000 population, Per capita bank credit to agriculture, Per capita bank credit to industries, Total cropped area, Production of major crops, Number of motor vehicles, Post offices/100,000 population, Circulation of newspapers, Cinema houses/100,000 population, Number of radio sets, T.V. sets.

O R I S S A

Mouths of the Ganga

BAY OF BENGAL

Scale 1:2 500 000   1 cm = 25 kms

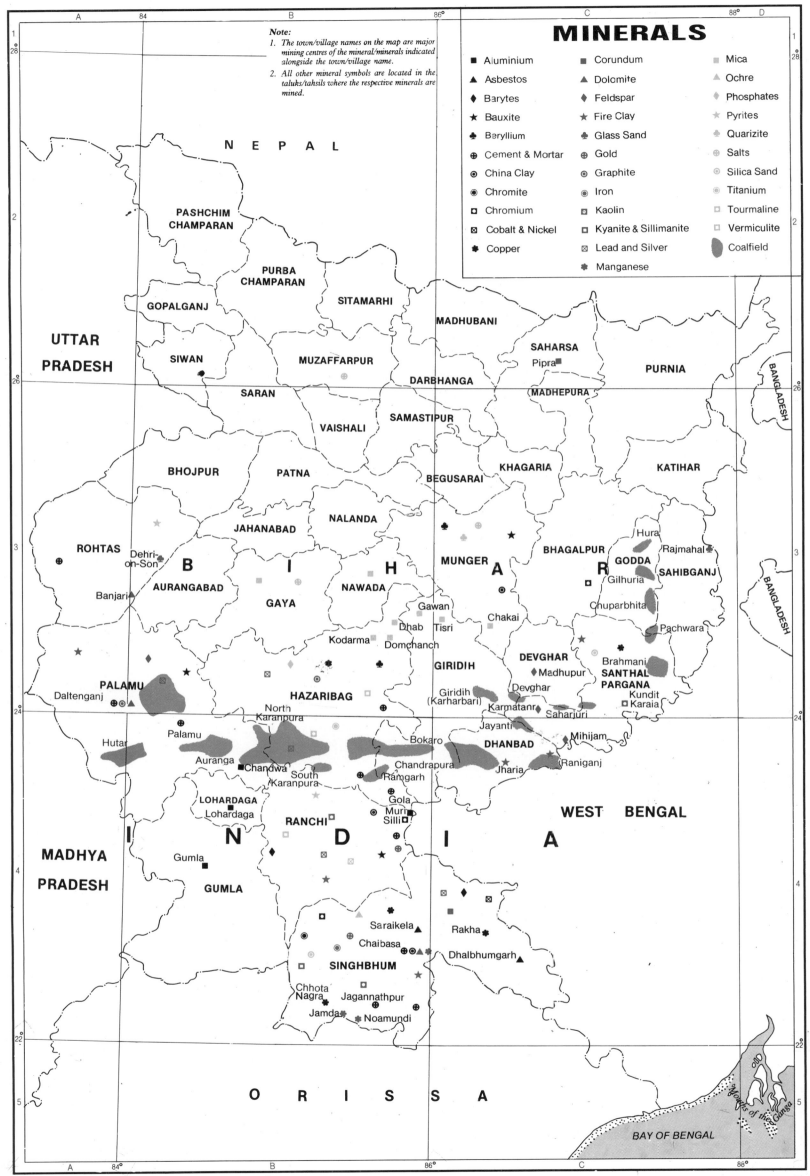

MINERALS

Note:
1. The town/village names on the map are major mining centres of the mineral/minerals indicated alongside the town/village name.
2. All other mineral symbols are located in the taluks/tahsils where the respective minerals are mined.

■ Aluminium	■ Corundum	▪ Mica
▲ Asbestos	▲ Dolomite	▲ Ochre
◆ Barytes	◆ Feldspar	◆ Phosphates
★ Bauxite	★ Fire Clay	★ Pyrites
♣ Beryllium	♣ Glass Sand	♣ Quarizite
⊕ Cement & Mortar	◉ Gold	⊕ Salts
⊙ China Clay	◉ Graphite	◉ Silica Sand
◉ Chromite	◉ Iron	◉ Titanium
□ Chromium	▣ Kaolin	□ Tourmaline
⊠ Cobalt & Nickel	▣ Kyanite & Sillimanite	▢ Vermiculite
✦ Copper	▣ Lead and Silver	⬤ Coalfield
	✱ Manganese	

NEPAL

UTTAR PRADESH

PASHCHIM CHAMPARAN

PURBA CHAMPARAN

GOPALGANJ

SITAMARHI

MADHUBANI

SAHARSA
Pipra

PURNIA

BANGLADESH

SIWAN

MUZAFFARPUR

SARAN

DARBHANGA

MADHEPURA

KATIHAR

VAISHALI

SAMASTIPUR

BHOJPUR

PATNA

KHAGARIA

BEGUSARAI

ROHTAS
Dehri-on-Son
Banjari

BIHAR

JAHANABAD

NALANDA

MUNGER

BHAGALPUR

GODDA
Hura
Rajmahal

SAHIBGANJ

AURANGABAD

GAYA

NAWADA

Gilhuria

Chuparbhita

Pachwara

Gawan

Kodarma
Dhab Tisri
Domchanch

Chakai

DEVGHAR

Brahmani

SANTHAL PARGANA

GIRIDIH

Madhupur

PALAMU
Daltenganj

HAZARIBAG

Giridih (Karharbari)

Devghar

Karmatanr

Kundit Karaia

North Karanpura

Saharjuri

Hutar
Palamu

Jayanti

Bokaro

Mihijam

Auranga
Chandwa
South Karanpura

Chandrapura
Ramgarh

DHANBAD

Jharia

Raniganj

WEST BENGAL

LOHARDAGA
Lohardaga

Gola

RANCHI

Muri
Silli

Gumla

GUMLA

MADHYA PRADESH

INDIA

Saraikela

Rakha

Chaibasa

Dhalbhumgarh

SINGHBHUM

Chhota Nagra
Jagannathpur

Jamda
Noamundi

ORISSA

Mounts of the Ganga

BAY OF BENGAL

Scale 1:2 500 000   1 cm = 25 kms

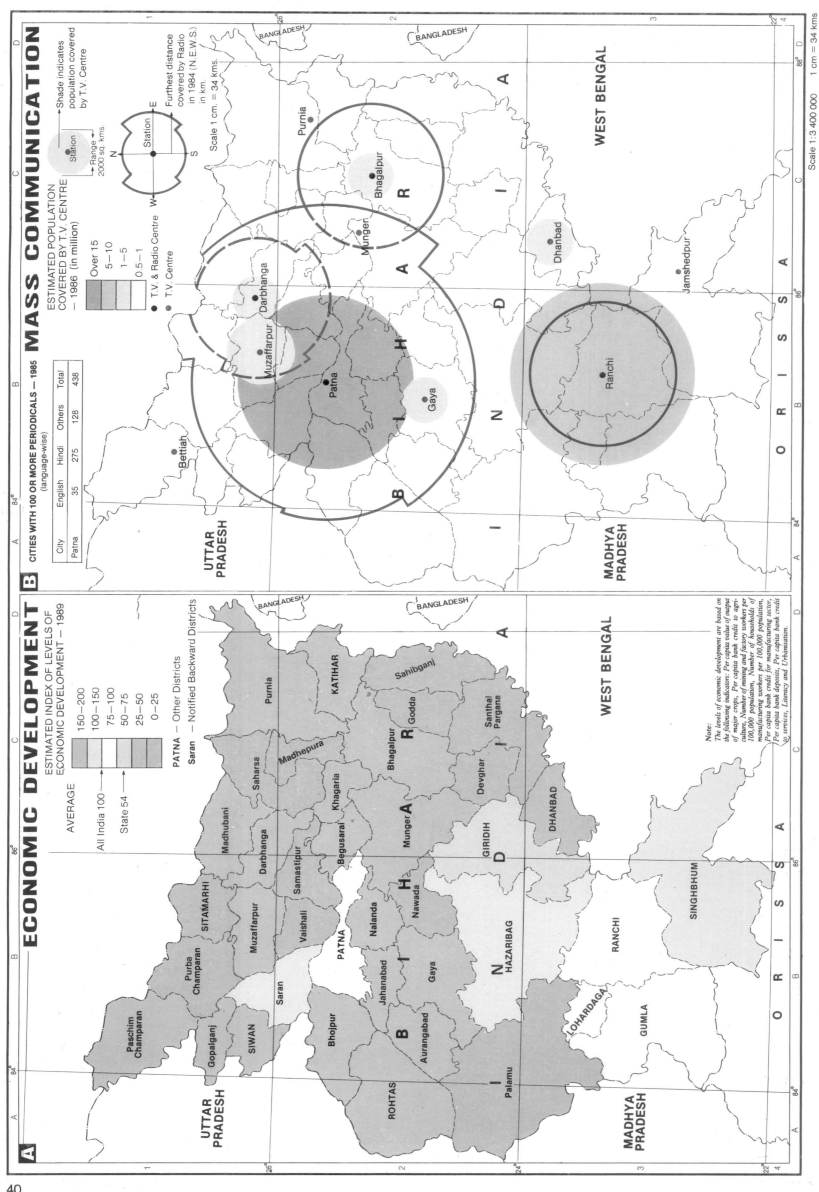

ECONOMIC DEVELOPMENT

A

ESTIMATED INDEX OF LEVELS OF
ECONOMIC DEVELOPMENT — 1989

150 – 200	
100 – 150	
75 – 100	
50 – 75	
25 – 50	
0 – 25	

AVERAGE

All India 100
State 54

PATNA — Other Districts
Saran — Notified Backward Districts

UTTAR
PRADESH

BANGLADESH

BANGLADESH

Paschim
Champaran

Purba
Champaran

Gopalganj

SIWAN

SITAMARHI

Muzaffarpur

Saran

Vaishali

Darbhanga

Samastipur

Madhubani

Saharsa

Madhepura

Purnia

KATIHAR

Sahibganj

Godda

R

Santhal
Pargana

Khagaria

Begusarai

Bhagalpur

Munger **A**

Devghar

DHANBAD

Nalanda

PATNA

Bhojpur

ROHTAS

Jahanabad

Gaya

Nawada

H

GIRIDIH

I

B

Aurangabad

Palamu

N

HAZARIBAG

RANCHI

LOHARDAGA

GUMLA

SINGHBHUM

D

ORISSA

WEST BENGAL

MADHYA
PRADESH

Note:
*The levels of economic development are based on
the following indicators: Per capita value of output
of major crops, Per capita bank credit to agriculture, Number of mining and factory workers per
100,000 population, Number of households of
manufacturing workers per 100,000 population,
Per capita bank credit for manufacturing sector,
Per capita bank deposits, Per capita bank credit
to services, Literacy and Urbanisation.*

MASS COMMUNICATION

B

ESTIMATED POPULATION
COVERED BY T.V. CENTRE
— 1986 (in million)

Over 15	
5 – 10	
1 – 5	
0.5 – 1	

● T.V. & Radio Centre
● T.V. Centre

CITIES WITH 100 OR MORE PERIODICALS — 1985
(language-wise)

City	English	Hindi	Others	Total
Patna	35	275	128	438

Shade indicates
population covered
by T.V. Centre

Furthest distance
covered by Radio
in 1984 (N.E.W.S.)
in km.

Scale 1 cm. = 34 kms.

Station
Range
2000 sq. kms.

Station

N
E
W
S

UTTAR
PRADESH

BANGLADESH

BANGLADESH

Bettiah

Darbhanga

Muzaffarpur

Purnia

Bhagalpur

Munger

B   **I**   **H**   **A**   **R**

Patna

Gaya

Dhanbad

Jamshedpur

Ranchi

I   **N**   **D**   **I**   **A**

ORISSA

WEST BENGAL

MADHYA
PRADESH

Scale 1:3 400 000   1 cm = 34 kms

GOA

Scale 1:510 000   1 cm = 5.1 kms

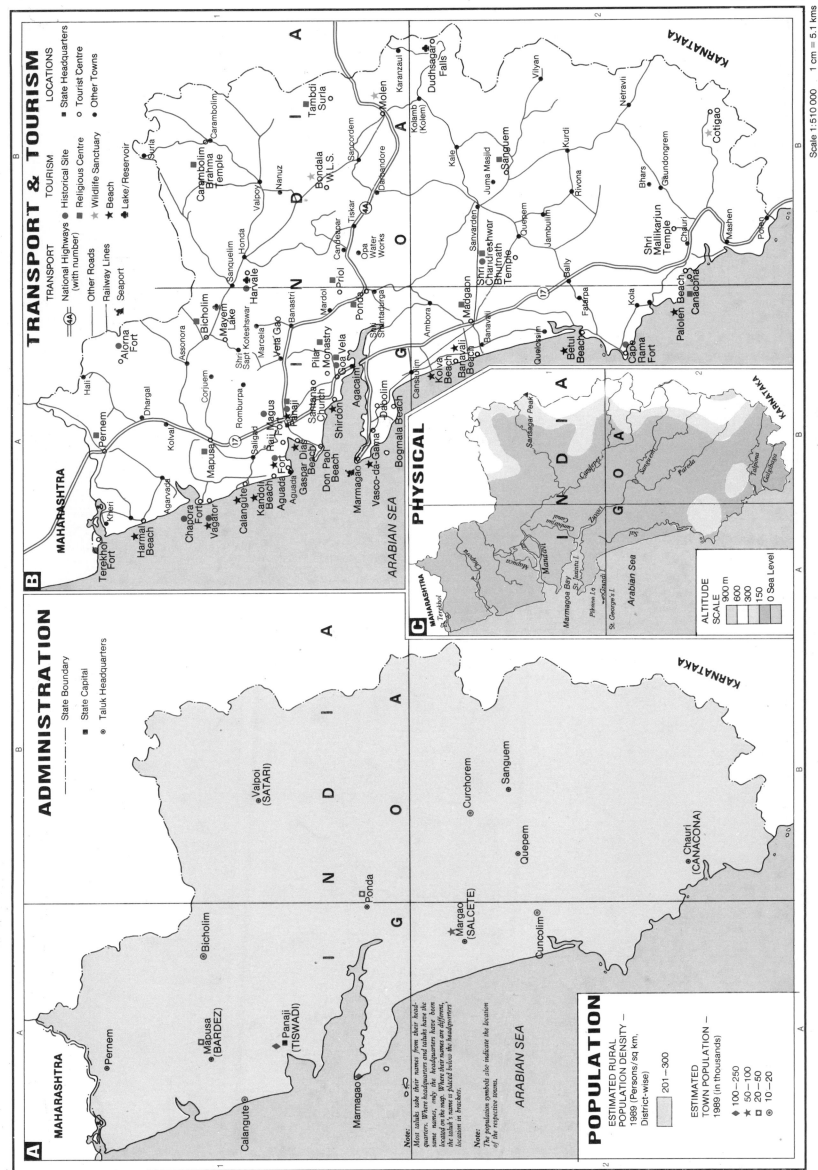

TRANSPORT & TOURISM

LOCATIONS
- ■ State Headquarters
- ○ Tourist Centre
- • Other Towns

TRANSPORT
- 4A= National Highways (with number)
- — Other Roads
- Railway Lines
- Seaport

TOURISM
- • Historical Site
- ■ Religious Centre
- ★ Wildlife Sanctuary
- ★ Beach
- ▲ Lake/Reservoir

MAHARASHTRA

Terekhol Fort
Harmal Beach
Kheri
Agarvada
Chapora Fort
Vagator
Kolval
Pernem
Dhargal
Hali
Coriuem
Assonora
Alorna Fort
Bicholim
Mayem Lake
Harvale
Shri Sapt Koteshwar
Marcela
Sanquelim
Honda
Valpoi
Nanuz
Surla
Carambolim
Carambolim Brahma Temple
Tambdi Surla
Molen
Karanzaul
Kolamb (Kolem)
Dudhsagar Falls
Sapcordem
Darbandore
Tiskar
Candeapar
Opa Water Works
Priol
Mardol
Shri Shantadurga
Ponda
Banastri
Veral Gao
Pilar
Monastery
Goa Vela
Agacaim
Dabolim
Shirdon
Santana Church
Panaji
Reis Magus
Saligad
Mapusa
Kandolim Beach
Aguada Fort
Aguada
Calangute
Gaspar Dias Beach
Don Paol Beach
Marmagao
Vasco-da-Gama
Bogmala Beach
Romburpa
ARABIAN SEA
Bondala W.L.S.
Vilyan
Netravli
Cotigao
Kurdi
Gaundongrem
Bhars
Rivona
Juma Masjid
Sanguem
Kale
Sanvarden
Quepem
Jambulim
Bally
Shri Chandreshwar Bhutnath Temple
Madgaon
Ambora
Banavali
Quelossim
Cansolim
Kolva Beach
Banavali Beach
Betul Beach
Cape Rama Fort
Faturpa
Kola
Paloleñ Beach
Canacona
Poiro
Mashen
Shri Mallikarjun Temple
Chauri
KARNATAKA

PHYSICAL

Santsagar Peak
Candepar
MAHARASHTRA
Terekhol
Chapora
Mandovi
Cumbharjua Canal
Zuari
Marmagoa Bay
St. Jacinto I.
Piloca I.
St. Grandi I.
St. George's I.
Sal
Sanguem
Pareda
Talpona
Galgibaga
ARABIAN SEA
KARNATAKA

ALTITUDE SCALE
- 900 m
- 600
- 300
- 150
- 0 Sea Level

ADMINISTRATION

- — State Boundary
- ■ State Capital
- ◉ Taluk Headquarters

MAHARASHTRA

Pernem
Calangute
Mapusa (BARDEZ)
Panaji (TISWADI)
Bicholim
Valpoi (SATARI)
Ponda
Marmagao
Margao (SALCETE)
Curchorem
Sanguem
Quepem
Cuncolim
Chauri (CANACONA)
ARABIAN SEA
KARNATAKA

Note:
Most taluks take their names from their headquarters. Where headquarters and taluks have the same names, only the headquarters have been located on the map. Where their names are different, the taluk's name is placed below the headquarters' location in brackets.

Note:
The population symbols also indicate the location of the respective towns.

POPULATION

ESTIMATED RURAL POPULATION DENSITY — 1989 (Persons/sq km, District-wise)
- 201–300
- 100–250

ESTIMATED TOWN POPULATION — 1989 (in thousands)
- ◆ 100–100
- ★ 50–100
- ☐ 20–50
- ◉ 10–20

41

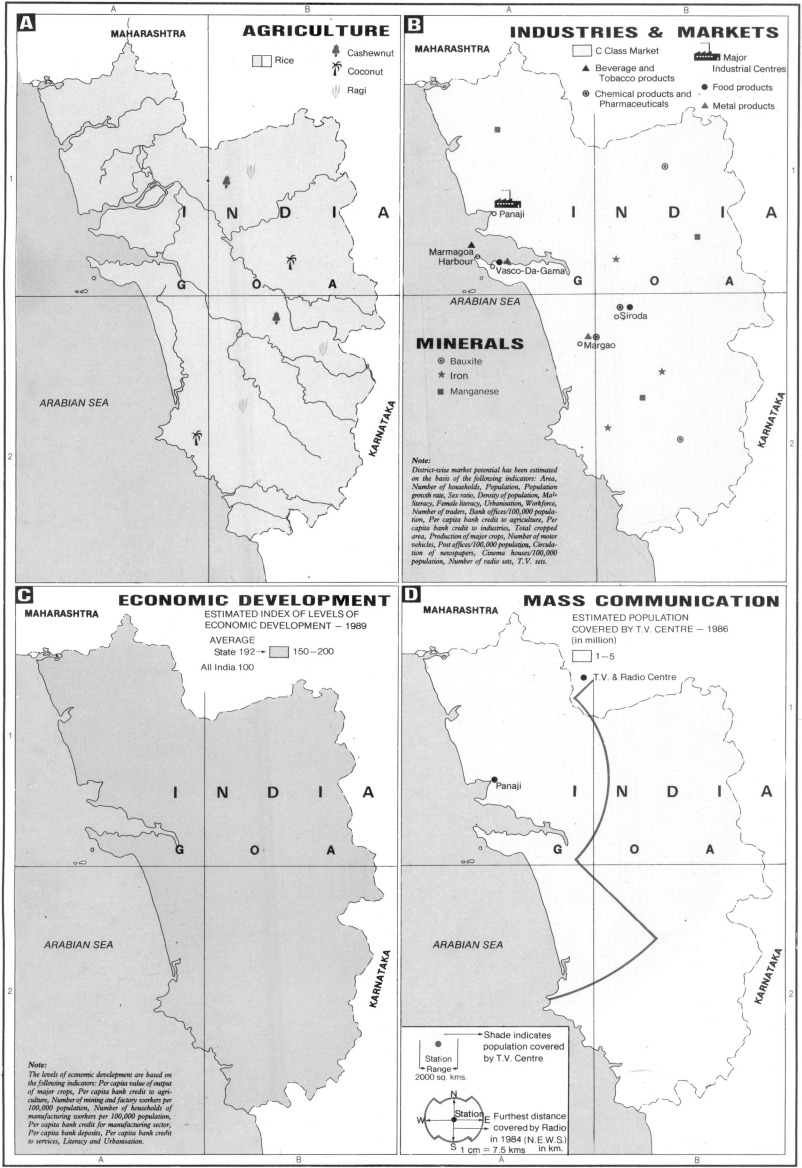

A   AGRICULTURE

MAHARASHTRA

	Rice	🌲 Cashewnut
		🌴 Coconut
		🌾 Ragi

I N D I A

G O A

ARABIAN SEA

KARNATAKA

B   INDUSTRIES & MARKETS

MAHARASHTRA

☐ C Class Market   🏭 Major Industrial Centres

▲ Beverage and Tobacco products   ● Food products

◉ Chemical products and Pharmaceuticals   ▲ Metal products

I N D I A

Panaji

Marmagoa Harbour
Vasco-Da-Gama

G O A

ARABIAN SEA

Siroda

MINERALS

Margao

◉ Bauxite
★ Iron
■ Manganese

KARNATAKA

Note:
District-wise market potential has been estimated on the basis of the following indicators: Area, Number of households, Population, Population growth rate, Sex ratio, Density of population, Male literacy, Female literacy, Urbanisation, Workforce, Number of traders, Bank offices/100,000 population, Per capita bank credit to agriculture, Per capita bank credit to industries, Total cropped area, Production of major crops, Number of motor vehicles, Post offices/100,000 population, Circulation of newspapers, Cinema houses/100,000 population, Number of radio sets, T.V. sets.

C   ECONOMIC DEVELOPMENT

MAHARASHTRA

ESTIMATED INDEX OF LEVELS OF ECONOMIC DEVELOPMENT — 1989

AVERAGE
State 192 → ☐ 150–200
All India 100

I N D I A

G O A

ARABIAN SEA

KARNATAKA

Note:
The levels of economic development are based on the following indicators: Per capita value of output of major crops, Per capita bank credit to agriculture, Number of mining and factory workers per 100,000 population, Number of households of manufacturing workers per 100,000 population, Per capita bank credit for manufacturing sector, Per capita bank deposits, Per capita bank credit to services, Literacy and Urbanisation.

D   MASS COMMUNICATION

MAHARASHTRA

ESTIMATED POPULATION COVERED BY T.V. CENTRE — 1986
(in million)

☐ 1–5

● T.V. & Radio Centre

I N D I A

Panaji

G O A

ARABIAN SEA

KARNATAKA

→ Shade indicates population covered by T.V. Centre

● Station
Range
2000 sq. kms.

N
W — Station — E
S   1 cm = 7.5 kms

E→ Furthest distance covered by Radio in 1984 (N.E.W.S.) in km.

Scale 1:750 000   1 cm = 7.5 kms

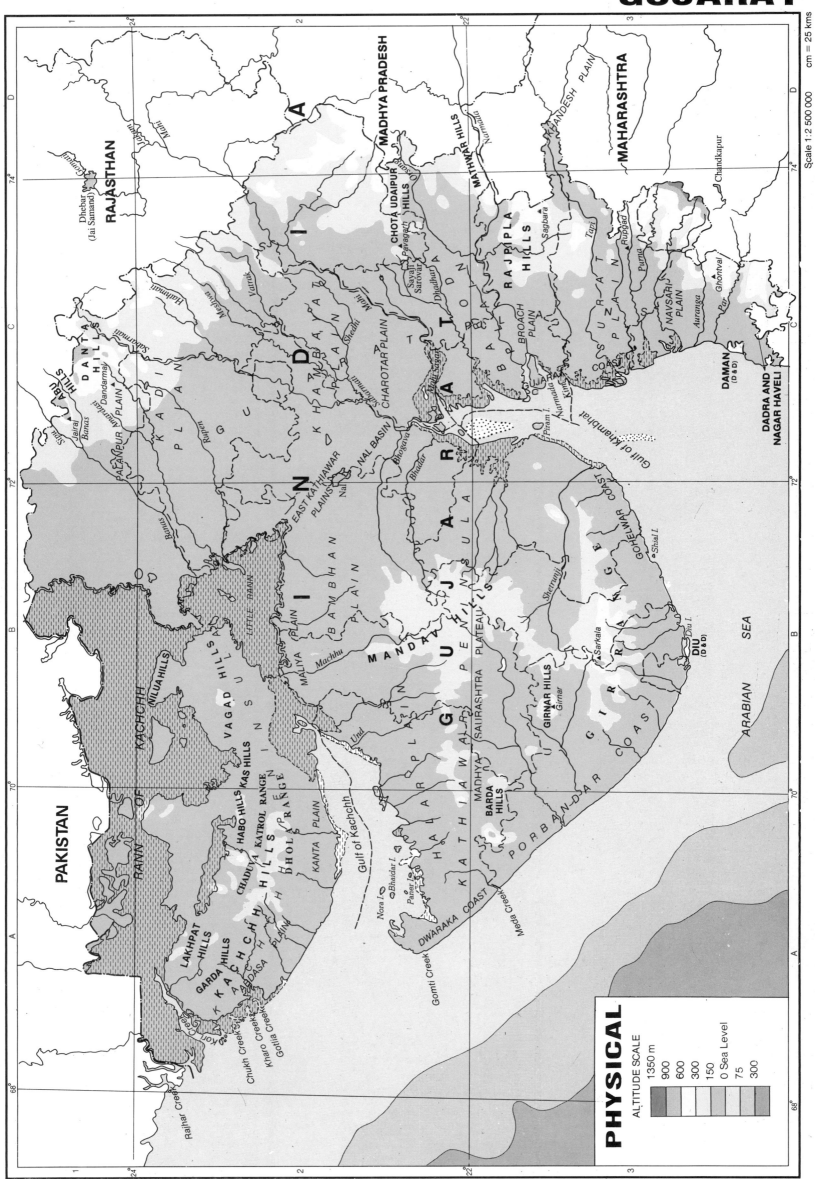

GUJARAT

Scale 1 : 2 500 000

cm = 25 kms

PHYSICAL

ALTITUDE SCALE

| 1350 m |
| 900 |
| 600 |
| 300 |
| 150 |
| 75 |
| 0 Sea Level |
| 300 |

RAJASTHAN

PAKISTAN

MADHYA PRADESH

MAHARASHTRA

DAMAN (D & D)

DADRA AND NAGAR HAVELI

DIU (D & Di)

ARABIAN SEA

RANN OF KACHCHH

GULF OF KACHCHH

Gulf of Khambhat

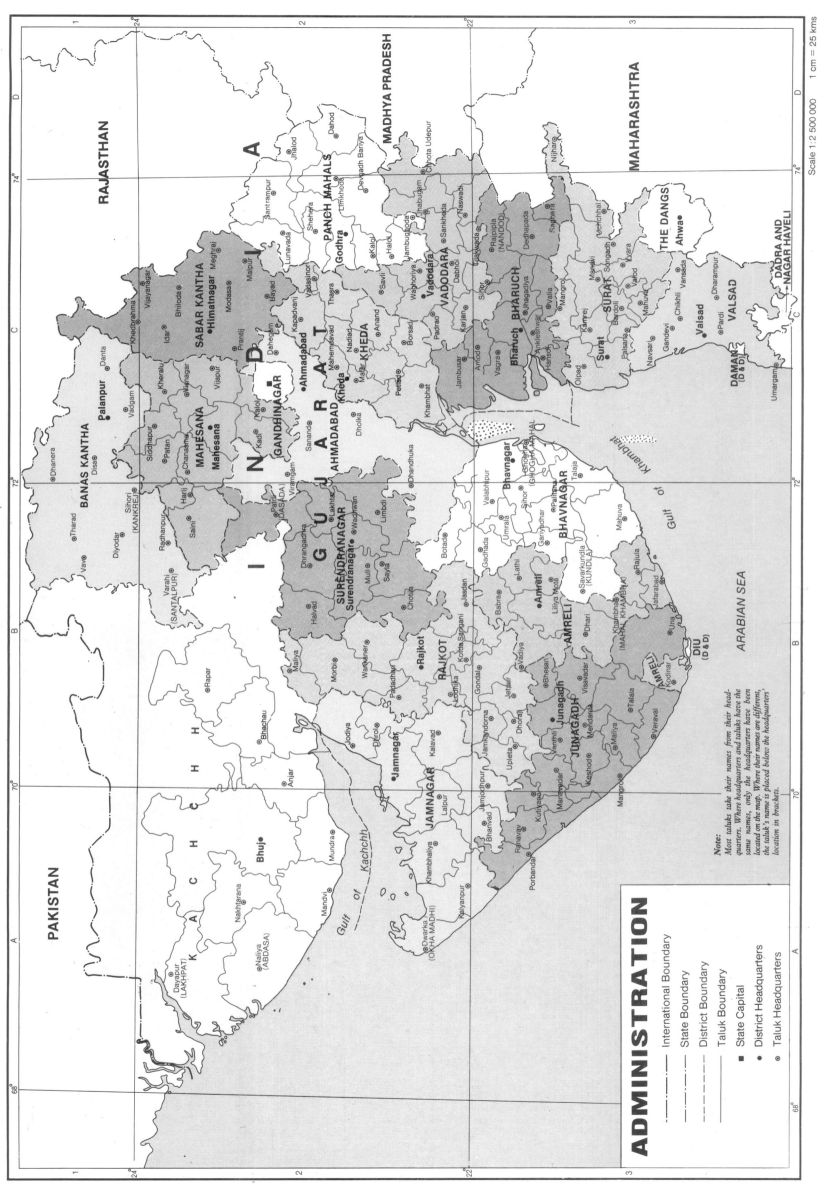

ADMINISTRATION

- ------ International Boundary
- ------ State Boundary
- ----- District Boundary
- ——— Taluk Boundary
- ■ State Capital
- ● District Headquarters
- ◉ Taluk Headquarters

Note:
Most taluks take their names from their head-quarters. Where headquarters and taluks have the same names, only the headquarters have been located on the map. Where their names are different, the taluk's name is placed below the headquarters' location in brackets.

Scale 1:2 500 000   1 cm = 25 kms

PAKISTAN

RAJASTHAN

MADHYA PRADESH

MAHARASHTRA

ARABIAN SEA

Gulf of Kachchh

Gulf of Khambhat

KACHCHH

Bhuj ■

Nakhtarana
Rapar ◉
Bhachau ◉
Anjar ◉
Mundra ◉
Mandvi ◉
Devapur (LAKHPAT) ◉
Naliya (ABDASA) ◉

BANAS KANTHA
Palanpur ■
Dhanera ◉
Disa ◉
Tharad ◉
Vav ◉
Diyodar ◉
Danta ◉
Varahi (SANTALPUR) ◉
Radhanpur ◉
Sihori (KANKREJ) ◉
Deesa

MAHESANA
Mahesana ◉
Kheralu ◉
Siddhapur ◉
Patan ◉
Vadgam ◉
Vijapur ◉
Visnagar ◉
Harij ◉
Saini ◉
Patdi (DASADA) ◉

SABAR KANTHA
Himatnagar ■
Khedbrahma ◉
Vijayanagar ◉
Bhiloda ◉
Idar ◉
Meghraj ◉
Malpur ◉
Modasa ◉
Bayad ◉
Prantij ◉

GANDHINAGAR ■
Kalol ◉
Kadi ◉
Dehgam ◉
Mansa ◉

AHMADABAD
Ahmadabad ●
Sanand ◉
Viramgam ◉
Dholka ◉
Dholka ◉
Dhandhuka ◉

KHEDA
Kheda ●
Kapadvanj ◉
Thasra ◉
Nadiad ◉
Anand ◉
Borsad ◉
Mahemdavad ◉
Matar ◉
Petlad ◉
Mehmedabad
Balasinor ◉
Khambhat ◉

PANCH MAHALS
Godhra ●
Lunavada ◉
Santrampur ◉
Shehra ◉
Dahod ◉
Jhalod ◉
Limkheda ◉
Devgadh Bariya ◉
Halol ◉
Kalol ◉

VADODARA
Vadodara ●
Chhota Udepur ◉
Savli ◉
Waghodia ◉
Jambughoda ◉
Sankheda ◉
Dabhoi ◉
Sinor ◉
Naswadi ◉
Padra ◉
Karjan ◉

BHARUCH
Bharuch ●
Jhagadia ◉
Amod ◉
Vagra ◉
Jambusar ◉
Ankleshwar ◉
Hansot ◉
Valia ◉

SURAT
Surat ●
Mangrol ◉
Mandvi ◉
Songadh ◉
Valod ◉
Bardoli ◉
Palsana ◉
Olpad ◉
Vyara ◉
Kamrej ◉
Chorasi

THE DANGS
Ahwa ●

VALSAD
Valsad ●
Dharampur ◉
Chikhli ◉
Vansada ◉
Navsari ◉
Gandevi ◉
Pardi ◉
Umargam ◉
Nijhar ◉
Uchchhal ◉

DAMAN (D & D)

DADRA AND NAGAR HAVELI

SURENDRANAGAR
Surendranagar ●
Dhrangadhra ◉
Wadhwan ◉
Muli ◉
Limbdi ◉
Sayla ◉
Halvad ◉
Chotila ◉
Lakhtar ◉

RAJKOT
Rajkot ●
Morbi ◉
Maliya ◉
Wankaner ◉
Paddhari ◉
Jasdan ◉
Kotda Sangani ◉
Lodhika ◉
Gondal ◉
Jamkandorna ◉
Upleta ◉
Dhoraji ◉
Jetpur ◉

JAMNAGAR
Jamnagar ●
Jodiya ◉
Dhrol ◉
Kalavad ◉
Lalpur ◉
Jamjodhpur ◉
Bhanvad ◉
Khambhaliya ◉
Kalyanpur ◉
Dwarka (OKHA MADHI) ◉
Okhamandal

JUNAGADH
Junagadh ●
Porbandar ◉
Ranavav ◉
Kutiyana ◉
Manavadar ◉
Vanthali ◉
Mendarda ◉
Keshod ◉
Mangrol ◉
Maliya ◉
Talala ◉
Veraval ◉
Visavadar ◉
Una ◉
Kodinar ◉
Patan-Veraval

AMRELI
Amreli ●
Babra ◉
Lathi ◉
Liliya Mota ◉
Dhari ◉
Savarkundla (KUNDLA) ◉
Khambha ◉
Rajula ◉
Jafarabad ◉
Bhesan ◉

BHAVNAGAR
Bhavnagar ●
Valabhipur ◉
Sihor ◉
Ghogha (GHOGHA MAHAL) ◉
Palitana ◉
Talaja ◉
Umrala ◉
Gadhada ◉
Botad ◉
Gariyadhar ◉
Mahuva ◉

DIU (D & D)

Khabhat (MAHAL KHAMBHA)

INDIA GUJARAT

44

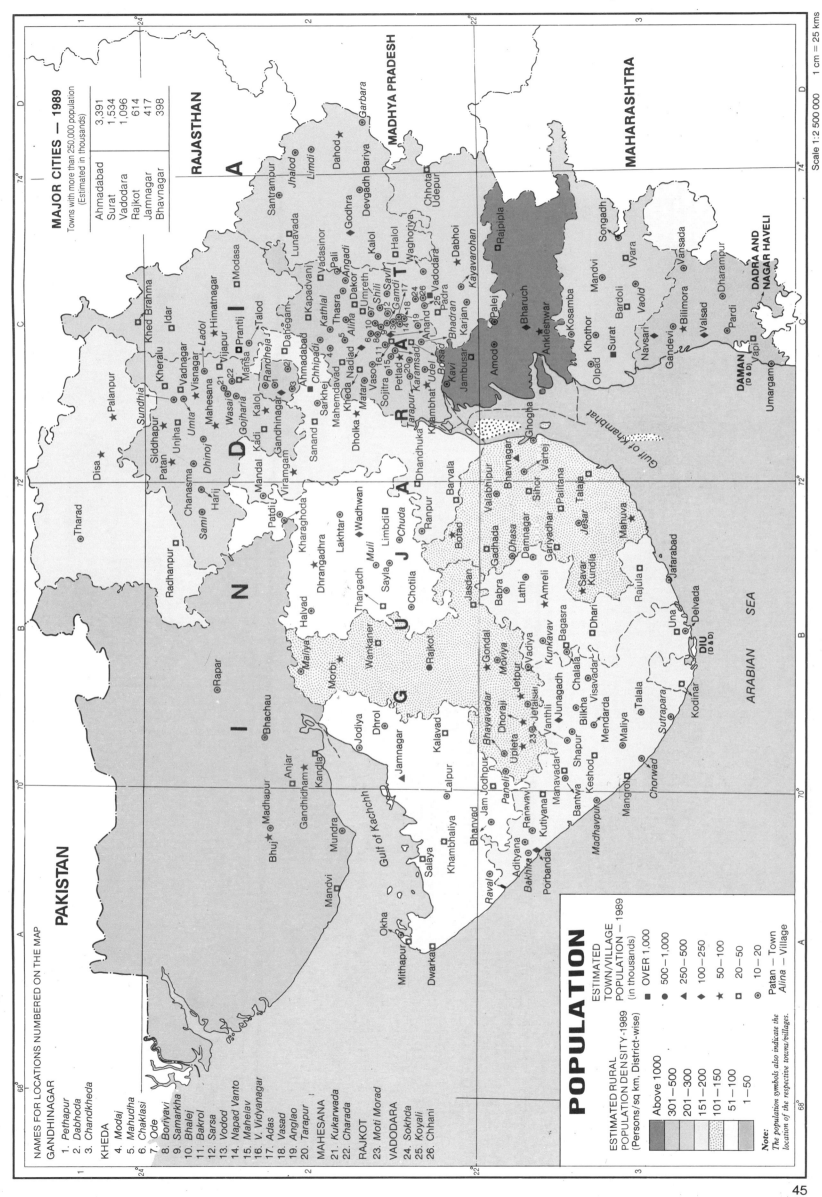

POPULATION

ESTIMATED TOWN/VILLAGE POPULATION — 1989
(in thousands)

- ■ OVER 1,000
- ● 500–1,000
- ▲ 250–500
- ◆ 100–250
- ★ 50–100
- □ 20–50
- ◎ 10–20

Patan — Town
Alina — Village

ESTIMATED RURAL POPULATION DENSITY-1989
(Persons/sq km, District-wise)

- Above 1000
- 301–500
- 201–300
- 151–200
- 101–150
- 51–100
- 1–50

Note:
The population symbols also indicate the location of the respective towns/villages.

MAJOR CITIES — 1989
Towns with more than 250,000 population
(Estimated in thousands)

Ahmadabad	3,391
Surat	1,534
Vadodara	1,096
Rajkot	614
Jamnagar	417
Bhavnagar	398

NAMES FOR LOCATIONS NUMBERED ON THE MAP

GANDHINAGAR
1. *Pethapur*
2. *Dabhoda*
3. *Chandkheda*

KHEDA
4. *Modaj*
5. *Mahudha*
6. *Napad Vanto*
7. *Adas*
8. *Vasad*
9. *Anglao*
10. *Bhalej*
11. *Bakrol*
12. *Sarsa*
13. *Vodod*
14. *Napad Vanto*
15. *Mahelav*
16. *V. Vidyanagar*
17. *Adas*
18. *Vasad*
19. *Anglao*
20. *Tarapur*

MAHESANA
21. *Kukarwada*
22. *Charada*

RAJKOT
23. *Moti Morad*

VADODARA
24. *Sokhda*
25. *Koyali*
26. *Chhani*

PAKISTAN

RAJASTHAN

MADHYA PRADESH

MAHARASHTRA

DADRA AND NAGAR HAVELI

ARABIAN SEA

Gulf of Khambhat

Gulf of Kachchh

Scale 1:2 500 000   1 cm = 25 kms

45

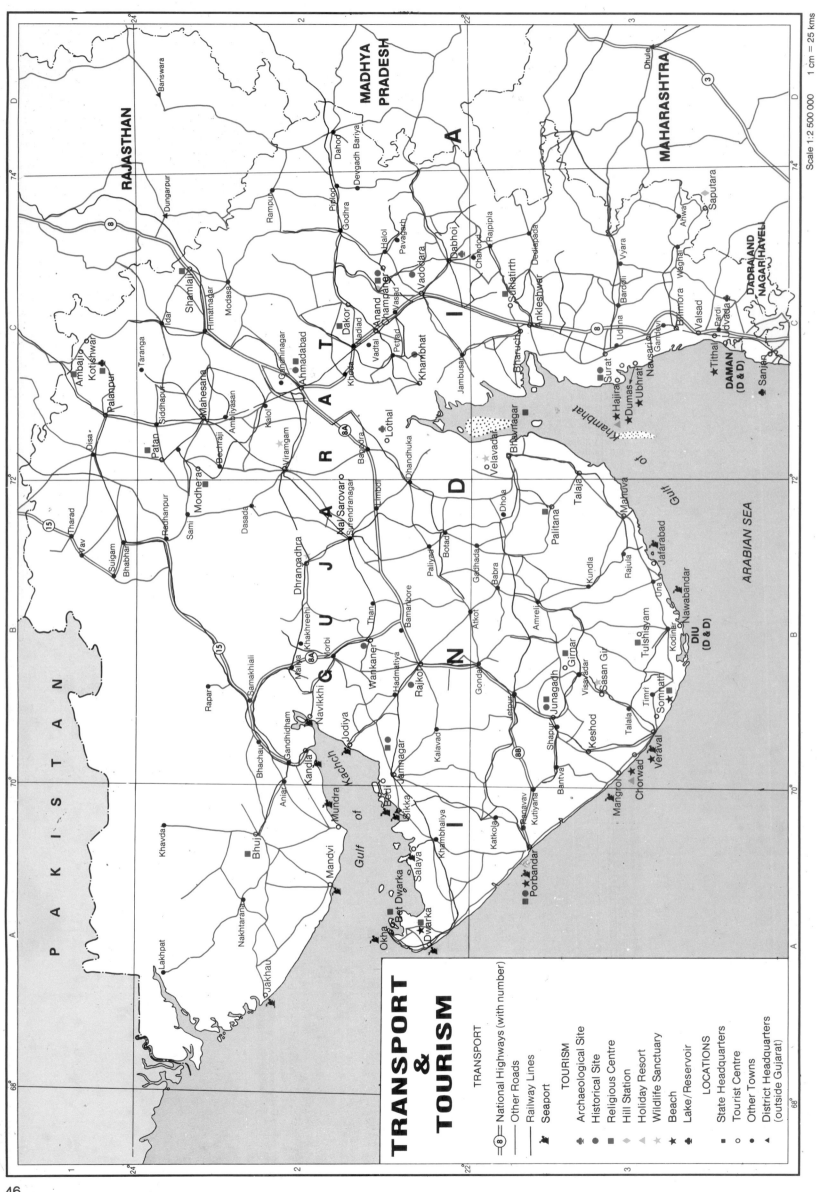

TRANSPORT & TOURISM

TRANSPORT

- ⑧ National Highways (with number)
- Other Roads
- Railway Lines
- ⚓ Seaport

TOURISM

- ♣ Archaeological Site
- ● Historical Site
- ■ Religious Centre
- ◆ Hill Station
- ▲ Holiday Resort
- ★ Wildlife Sanctuary
- ♣ Beach
- ◆ Lake/Reservoir

LOCATIONS

- ■ State Headquarters
- ○ Tourist Centre
- ● Other Towns
- ▲ District Headquarters (outside Gujarat)

Scale 1:2 500 000   1 cm = 25 kms

RAJASTHAN

MADHYA PRADESH

MAHARASHTRA

DADRA AND NAGAR HAVELI

DAMAN (D & D)

DIU (D & D)

PAKISTAN

ARABIAN SEA

Gulf of Kachchh

Gulf of Khambhat

G U J A R A T

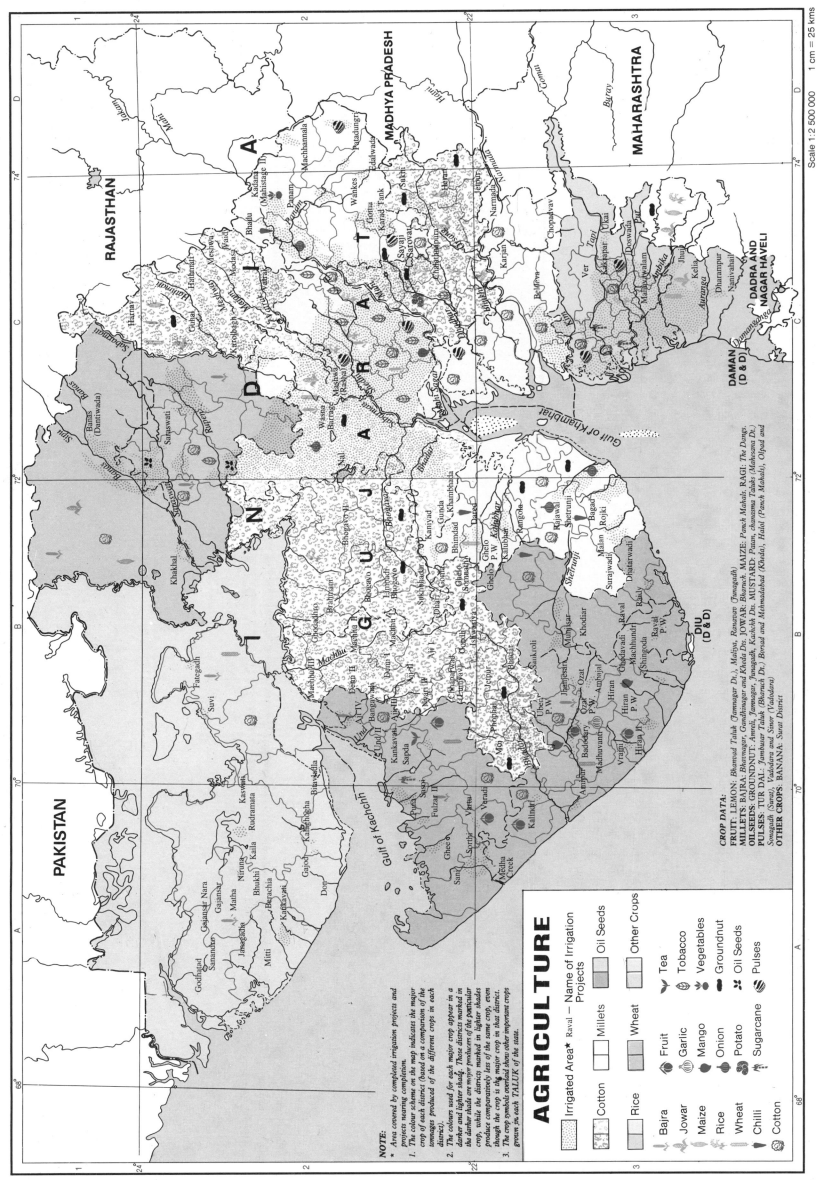

AGRICULTURE

NOTE:
* Area covered by completed irrigation projects and projects nearing completion.
1. The colour scheme on the map indicates the major crop of each district (based on a comparison of the tonnages produced of the different crops in each district).
2. The colours used for each major crop appear in a darker and lighter shade. Those districts marked in the darker shade are major producers of the particular crop, while the districts marked in lighter shades produce comparatively less of the same crop, even though the crop is the major crop in that district.
3. The crop symbols overlaid show other important crops grown in each TALUK of the state.

Irrigated Area* **Raval**★ — Name of Irrigation Projects

Cotton	Millets	Oil Seeds
Rice	Wheat	Other Crops

Bajra · Jowar · Maize · Rice · Wheat · Chilli · Cotton

Fruit · Garlic · Mango · Onion · Potato · Sugarcane
Tea · Tobacco · Vegetables · Groundnut · Oil Seeds · Pulses

CROP DATA:
FRUIT: LEMON: *Bhanvad Taluk (Jamnagar Dt.), Maliya, Ranavav, Ranavav (Junagadh)*
MILLETS: BAJRA: *Bhavnagar, Gandhinagar and Kheda Dts. Panch Mahals.* RAGI: *The Dangs.*
MAIZE: *Panch Mahals.* JOWAR: *Bharuch.*
OILSEEDS: GROUNDNUT: *Amreli, Jamnagar, Junagadh, Kachchh Dts.* MUSTARD: *Patan, chanasma Taluks (Mahesana Dt.)*
PULSES: TUR DAL: *Jambusar Taluk (Bharuch Dt.) Borsad and Mehmadabad (Kheda), Halol (Panch Mahals), Olpad and Sonagadh (Surat), Vadodara and Sinor (Vadodara)*
OTHER CROPS: BANANA: *Surat District*

Scale 1:2 500 000   1 cm = 25 kms

PAKISTAN

RAJASTHAN

MADHYA PRADESH

MAHARASHTRA

DADRA AND NAGAR HAVELI

DAMAN (D & D)

DIU (D & D)

Gulf of Kachchh

Gulf of Khambhat

47

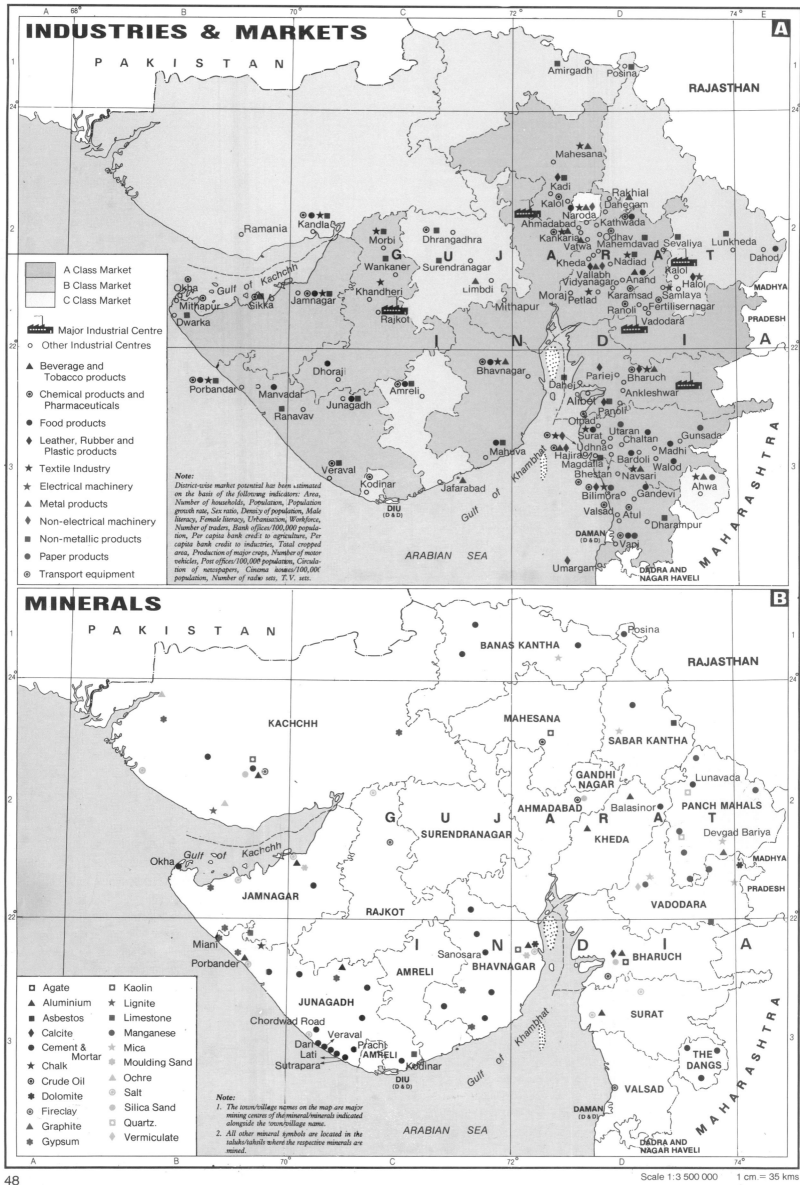

INDUSTRIES & MARKETS

PAKISTAN

RAJASTHAN

G U J A R A T

I N D I A

MADHYA PRADESH

MAHARASHTRA

ARABIAN SEA

Amirgadh • Posina ■
Mahesana ★★
Kadi ◆■
Kalol
Naroda ★
Ahmadabad ■
Kankaria
Vatwa
Kheda
Vidyanagar
Anand
Moraj ◆ Petlad
Karamsad
Ranoli
Vadodara

Ramania •
Kandla ◉●★■
Morbi ★
Dhrangadhra ■
Wankaner ★
Surendranagar
Limbdi ▲

Okha •
Mithapur
Dwarka
Sikka ●■
Jamnagar ●■
Khandheri ★
Rajkot

Rakhial ■
Dahegam •
Kathwada ◉
Odhav
Mahemdavad
Sevaliya ■
Lunkheda ■
Nadiad
Kalol ◆
Halol
Samlaya
Fertilisernagar •
Dahod •

Porbandar ◆◉★
Manvadar
Junagadh •
Ranavav

Dhoraji •
Amreli ◉★

Bhavnagar ◉★▲
Dahej ■
Pariej •
Bharuch ◆▲
Ankleshwar ■
Alibet •
Panoli
Otpad
Surat
Chaltan
Gunsada •
Udhna
Madhi •
Hajira ◉
Magdalla
Bardoli
Walod
Bhestan
Navsari ▲
Bilimora ◆
Gandevi •
Valsad ●
Atul
Ahwa ★▲
Dharampur
DAMAN (D & D)
Vapi •
Umargam ◆

Mahuva •
Veraval •
Kodinar
Jafarabad ▲
DIU (D & D)
Gulf of Khambhat
Gulf of
Gulf of Kachchh

Legend

- A Class Market (shaded)
- B Class Market (shaded)
- C Class Market (shaded)

🏭 Major Industrial Centre
○ Other Industrial Centres
▲ Beverage and Tobacco products
◉ Chemical products and Pharmaceuticals
● Food products
◆ Leather, Rubber and Plastic products
★ Textile Industry
★ Electrical machinery
▲ Metal products
◆ Non-electrical machinery
■ Non-metallic products
● Paper products
◉ Transport equipment

Note:
District-wise market potential has been estimated on the basis of the following indicators: Area, Number of households, Population, Population growth rate, Sex ratio, Density of population, Male literacy, Female literacy, Urbanisation, Workforce, Number of traders, Bank offices/100,000 population, Per capita bank credit to agriculture, Per capita bank credit to industries, Total cropped area, Production of major crops, Number of motor vehicles, Post offices/100,000 population, Circulation of newspapers, Cinema houses/100,000 population, Number of radio sets, T.V. sets.

MINERALS

PAKISTAN

RAJASTHAN

G U J A R A T

I N D I A

MADHYA PRADESH

MAHARASHTRA

ARABIAN SEA

Posina •
BANAS KANTHA ●

KACHCHH

MAHESANA ■
SABAR KANTHA
GANDHI NAGAR
AHMADABAD
Balasinor ▲
KHEDA
Lunavada •
PANCH MAHALS
Devgad Bariya •

SURENDRANAGAR

RAJKOT

JAMNAGAR
Okha ◆
Gulf of Kachchh

Miani
Porbander
Sanosara
Sutrapara
Chordwad Road
Veraval
Darri Lati Prachi
AMRELI
Kodinar
DIU (D & D)

AMRELI
JUNAGADH

BHAVNAGAR

VADODARA
BHARUCH
SURAT
THE DANGS
VALSAD
DAMAN (D & D)
DADRA AND NAGAR HAVELI

Gulf of Khambhat

Legend

- □ Agate
- ▲ Aluminium
- ■ Asbestos
- ◆ Calcite
- ● Cement & Mortar
- ★ Chalk
- ◉ Crude Oil
- ✳ Dolomite
- ◉ Fireclay
- ▲ Graphite
- ✳ Gypsum
- □ Kaolin
- ★ Lignite
- ■ Limestone
- ● Manganese
- ★ Mica
- ✳ Moulding Sand
- ▲ Ochre
- ◉ Salt
- ● Silica Sand
- □ Quartz
- ◆ Vermiculate

Note:
1. The town/village names on the map are major mining centres of the mineral/minerals indicated alongside the town/village name.
2. All other mineral symbols are located in the taluks/tahsils where the respective minerals are mined.

Scale 1:3 500 000   1 cm.= 35 kms

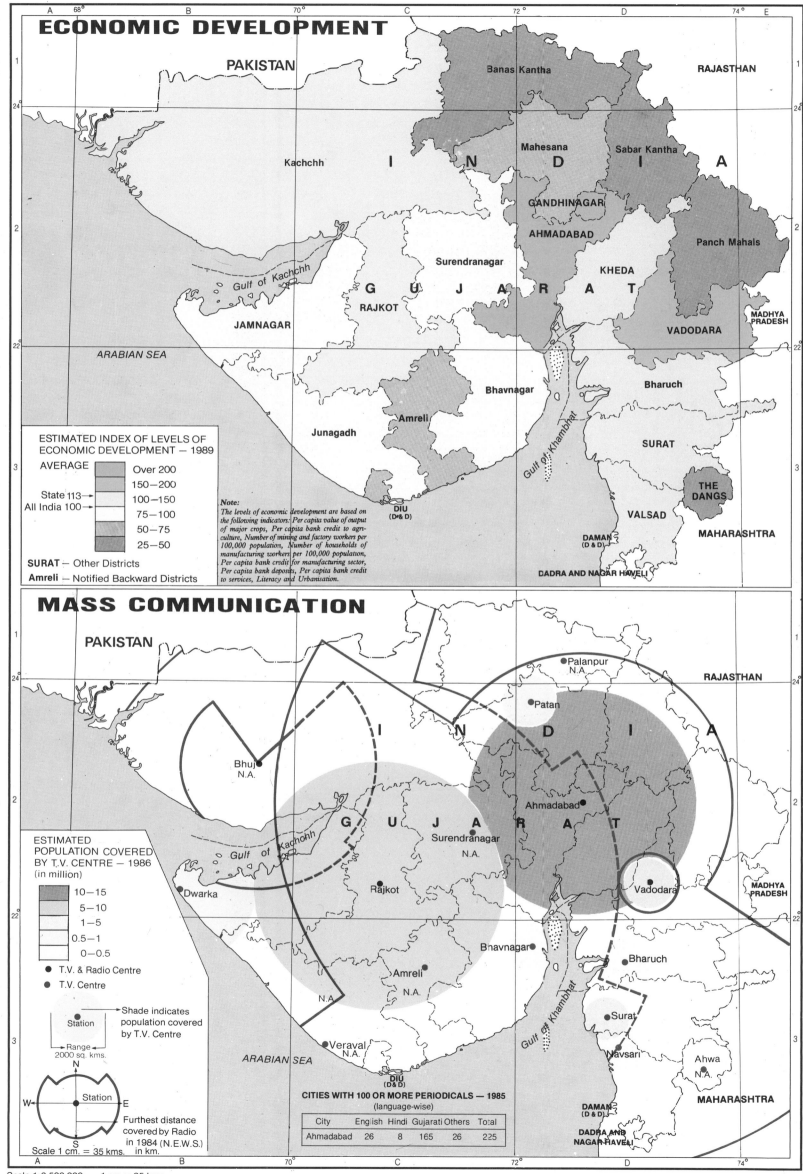

ECONOMIC DEVELOPMENT

PAKISTAN

RAJASTHAN

Banas Kantha

I N D I A

Kachchh

Mahesana

Sabar Kantha

GANDHINAGAR

AHMADABAD

Panch Mahals

Surendranagar

KHEDA

G U J A R A T

RAJKOT

VADODARA

MADHYA
PRADESH

JAMNAGAR

ARABIAN SEA

Bhavnagar

Bharuch

Amreli

Junagadh

SURAT

THE
DANGS

VALSAD

DIU
(D&D)

DAMAN
(D & D)

MAHARASHTRA

DADRA AND NAGAR HAVELI

**ESTIMATED INDEX OF LEVELS OF
ECONOMIC DEVELOPMENT — 1989**

AVERAGE

Over 200

150—200

State 113 →
All India 100 →

100—150

75—100

50—75

25—50

SURAT — Other Districts

Amreli — Notified Backward Districts

Note:
*The levels of economic development are based on
the following indicators: Per capita value of output
of major crops, Per capita bank credit to agriculture, Number of mining and factory workers per
100,000 population, Number of households of
manufacturing workers per 100,000 population,
Per capita bank credit for manufacturing sector,
Per capita bank deposits, Per capita bank credit
to services, Literacy and Urbanisation.*

MASS COMMUNICATION

PAKISTAN

Palanpur
N.A.

RAJASTHAN

Patan

I N D I A

Bhuj
N.A.

Ahmadabad

G U J A R A T

Surendranagar
N.A.

Vadodara

MADHYA
PRADESH

Dwarka

Rajkot

**ESTIMATED
POPULATION COVERED
BY T.V. CENTRE — 1986**
(in million)

10—15

5—10

1—5

0.5—1

0—0.5

● T.V. & Radio Centre

● T.V. Centre

Bhavnagar

Bharuch

Amreli
N.A.

N.A.

Surat

Navsari

Ahwa
N.A.

Veraval
N.A.

ARABIAN SEA

DIU
(D&D)

DAMAN
(D & D)

MAHARASHTRA

DADRA AND
NAGAR HAVELI

Station

→ Shade indicates
population covered
by T.V. Centre

→ Range
2000 sq. kms.

N

W   E

Station

S

Furthest distance
covered by Radio
in 1984 (N.E.W.S.)
in km.

Scale 1 cm. = 35 kms.

CITIES WITH 100 OR MORE PERIODICALS — 1985
(language-wise)

City	English	Hindi	Gujarati	Others	Total
Ahmadabad	26	8	165	26	225

HARYANA

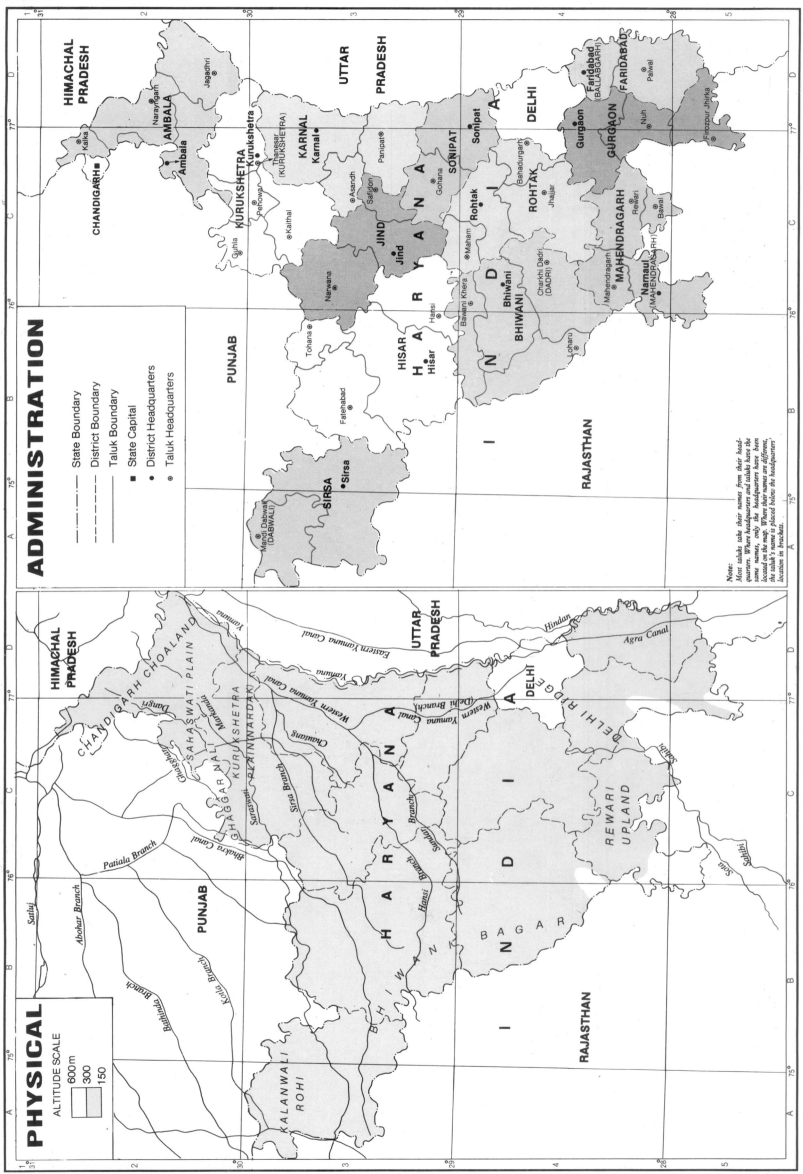

ADMINISTRATION

Scale 1:2 000 000   1 cm = 20 kms

HIMACHAL PRADESH

UTTAR PRADESH

DELHI

Narayngarh
Jagadhri
Kalka
AMBALA
Ambala
CHANDIGARH■

KURUKSHETRA
Kurukshetra
Taneesar (KURUKSHETRA)
Pehowa
Guhla
Kaithal

KARNAL
Karnal ●
Asandh
Satjon
Panipat

Sonipat
SONIPAT
Gohana
Samalkha

JIND
Jind
Narwana

Rohtak
ROHTAK
Maham
Jhajjar
Bahadurgarh

Gurgaon
GURGAON
Faridabad (BALLABGARH)
FARIDABAD
Nuh
Palwal
Firozpur Jhirka

HISAR
Hisar
Hansi
Tohana
Fatehabad

BHIWANI
Bhiwani
Bawani Khera
Charkhi Dadri (DADRI)
Loharu

MAHENDRAGARH
Mahendragarh
Rewari
Bawal
Narnaul (MAHENDRAGARH)

SIRSA
Sirsa
Mandi Dabwali (DABWALI)

PUNJAB

RAJASTHAN

Legend:
— ‧ — ‧ — State Boundary
— — — — District Boundary
———— Taluk Boundary
■ State Capital
● District Headquarters
⊙ Taluk Headquarters

Note:
Most taluks take their names from their head-quarters. Where headquarters and taluks have the same names, only the headquarters have been located on the map. Where their names are different, the taluk's name is placed below the headquarters' location in brackets.

PHYSICAL

ALTITUDE SCALE
600m
300
150

HIMACHAL PRADESH

UTTAR PRADESH

DELHI

CHANDIGARH CHOALAND
Dangri
Ghaggar
SARASWATI PLAIN
GHAGGAR NAL
KURUKSHETRA PLAIN (NARDAK)
Markanda
Saraswati
Chautang

Yamuna
Eastern Yamuna Canal
Western Yamuna Canal
Western Yamuna Canal (Delhi Branch)
Hindan
Agra Canal

Satluj
Abohar Branch
Patiala Branch
Kotla Branch
Bhakra Canal
Bathinda Branch

Sirsa Branch
Hansi Branch
Sutpani
Butana Branch

HARYANA

DELHI RIDGE

REWARI UPLAND

KALANWALI ROHI
BHIWANI
BAGAR

Jaitpur
Sahibi
Soni

PUNJAB

RAJASTHAN

INDIA

POPULATION

ESTIMATED RURAL POPULATION DENSITY-1989
(Persons/sq km, District-wise)

- 501—1000
- 301—500
- 201—300
- 151—200
- 101—150

ESTIMATED TOWN/VILLAGE POPULATION — 1989
(in thousands)

- ▲ 250—500
- ◆ 100—250
- ★ 50—100
- ▫ 20—50
- ⊙ 10—20

Hansi — Town
Siwan — Village

Note:
The population symbols also indicate the location of the respective towns/villages.

MAJOR CITIES — 1989
Towns with more than 250,000 population
(Estimated in thousands)

Faridabad	487

HIMACHAL   PRADESH

PUNJAB

CHANDIGARH ←

Kalka
Panchkula
Narayangarh
Sadhaura
Ambala ◆
Barara
Shahabad
Yamunanagar ◆
Cheeka
Siwan   Pehowa   Thanesar ★   Ladwa
Keorak
Kaul
Nilokheri
Tirawari
Kaithal ★
Nisang   Kunjpura
Tohana   Pundri
Kalayat   Pai   Karnal ◆
Balu
Hijarawan   Ratia   Karora   Gondar
Khurd
Hijarawan Kalan   Narwana   Rajaund   Jundla   Uncha Siwana
Mandi Kalan   Asandh   Salwan   Gharaunda
Fatehabad   Bhuna   Alewa   Moana   Bala
Gorakhpur   Pabra   Safidon   Panipat ◆
Bhattu   Barwala
Kalan   Sadalpur   Samalkha
H A R Y A N A
Adampur   Jind ★
Bir Hisar   Gohana   Ganaur
N   ◆ Hisar   D   Julana   I   A
Balsamand   Morthal Khas
Hansi ★   Dhana   Sonipat ◆
Barsi   Nindana   Bohar   Sisana
Bawani Khera   Dhanana   Maham   Rohtak   Kharkhauda
Chang   Bahelba   Mokhrakhas
Siwani   Baliyal   Kelanga   Kalanaur
Bapora   Tigrana   Kharak   Dighal   Mandothi
Bhiwani ◆   Kalan   Beri   Chhara
Bond Kalan   Bahadurgarh ★   DELHI
Badli
Charkhi Dadri   Jhajjar

UTTAR
PRADESH

RAJASTHAN

Gurgaon ◆
Jharsa
Haileymandi   Bahora Kalan   Faridabad ▲
Pataudi   Tigaon
Tigra
Mahendragarh   Sohna
Rewari ★
Narnaul ★
Palwal ★
Hodal
Firozpur Jhirka

Mandi
Dabwali
Chotala
Kalanwali
Rori
Rania
Jiwannagar   Sirsa ◆
Ellenabad

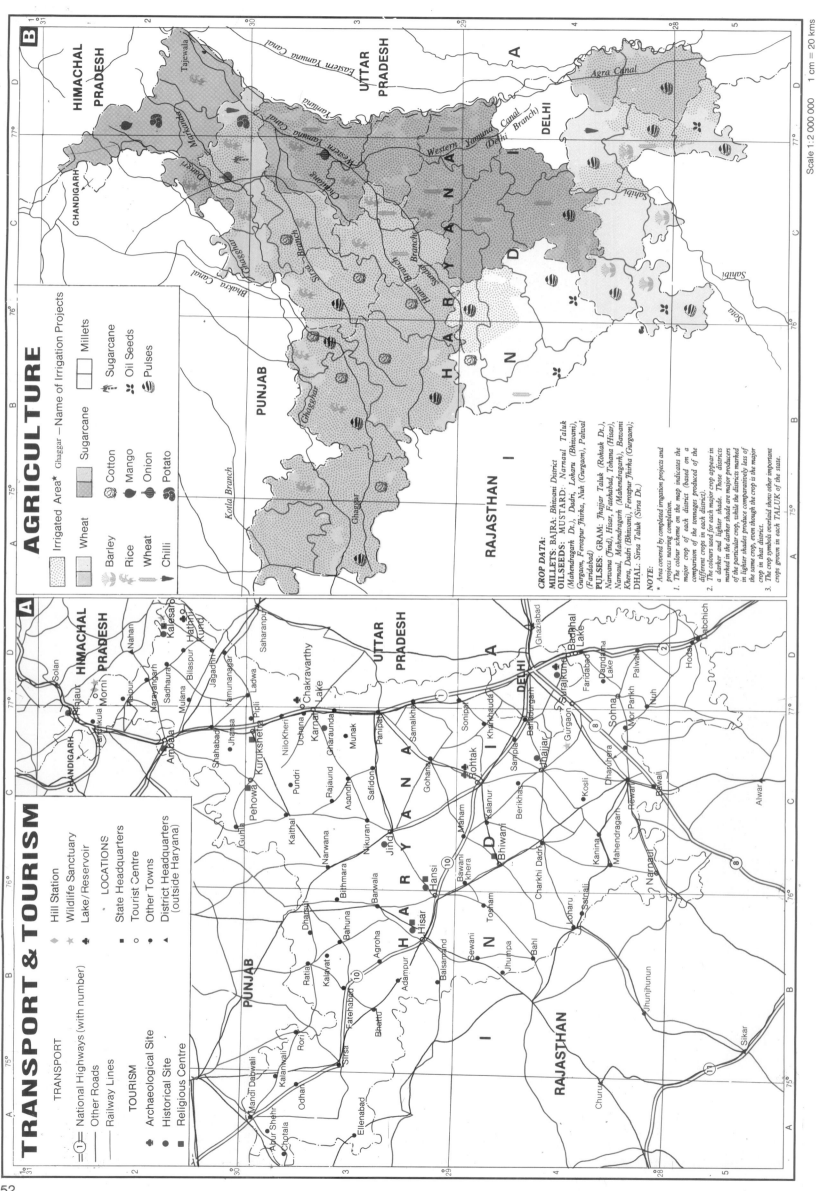

TRANSPORT & TOURISM

A

TRANSPORT
- =1= National Highways (with number)
- —— Other Roads
- —— Railway Lines

TOURISM
- ◆ Archaeological Site
- ● Historical Site
- ■ Religious Centre
- ◆ Hill Station
- ★ Wildlife Sanctuary
- ♣ Lake/Reservoir

LOCATIONS
- ■ State Headquarters
- ○ Tourist Centre
- ● Other Towns
- ▲ District Headquarters (outside Haryana)

HIMACHAL PRADESH

CHANDIGARH

PUNJAB

HARYANA

RAJASTHAN

UTTAR PRADESH

DELHI

Solan · Nahan · Kalesar · Hattni Kund · Panchkula · Morni · Pinjaur · Raipur · Paonta · Marayangarh · Sadhaura · Jagadhri · Bilaspur · Yamunanagar · Saharanpur · Ambala · Shahabad · Jhinha · Ladwa · Pipli · Kurukshetra · Chakravarthy Lake · Karnal · NiloKheri · Uchana · Gharaunda · Munak · Panipat · Samalkha · Sonipat · Ghaziabad

Guhla · Pehowa · Pundri · Rajaund · Asandh · Safidon · Gohana · Rohtak · Khankhauda · Bahadurgarh · DELHI · Badkhal Lake · Surajkund · Dandahera · Faridabad · Palwal · Ballabgarh

Kaithal · Narwana · Nikuran · Jind · Maham · Kalanur · Berikhas · Jhajjar · Gurgaon · Sohna · Mor Parikh · Nuh · Hodal · Kabchich

Bithmara · Barwala · Hansi · Bawani Khera · Bhiwani · Charkhi Dadri · Kosli · Dhakhpera · Rewari · Alwar

Dhasri · Bahuna · Agroha · Hisar · Tosham · Loharu · Kanina · Mahendragarh · Narnaul · Sirsal

Ratia · Kalayat · Sewani · Jhumpa · Bahl

Fatehabad · Adampur · Balsapand · Bhattu · Bahlad · Churu

Abur Sheni · Choptala · Rori · Odhan · Kalanwali · Ellenabad · Sirsa · Mandi Dabwali

Jhunjhunun · Sikar

B

AGRICULTURE

HIMACHAL PRADESH

CHANDIGARH

PUNJAB

HARYANA

RAJASTHAN

UTTAR PRADESH

DELHI

Tajewala · Markanda · Dangri · Chautang · Western Yamuna Canal · Eastern Yamuna Canal · Yamuna · Ghaggar Canal · Bhakra Canal · Sirsa Branch · Jhuai Branch · Ghaggar · Kotla Branch · Delhi Branch · Agra Canal · Sahibi · Sota

Irrigated Area * Ghaggar — Name of Irrigation Projects

Legend	
Irrigated Area*	Millets
Wheat	Sugarcane
Sugarcane	Oil Seeds
	Pulses
Barley	Cotton
Rice	Mango
Wheat	Onion
Chilli	Potato

CROP DATA:
MILLETS: BAJRA: Bhiwani District
OILSEEDS: MUSTARD: Narnaul Taluk (Mahendragarh Dt.), Dadri, Loharu (Bhiwani), Gurgaon, Ferozpur Jhirka, Nuh (Gurgaon), Palwal (Faridabad).
PULSES: GRAM: Jhajjar Taluk (Rohtak Dt.), Narwana (Jind), Hisar, Fatehabad, Tohana (Hisar), Narnaul, Mahendragarh (Mahendragarh), Bawani Khera, Dadri (Bhiwani), Ferozpur Jhirka (Gurgaon);
DHAL: Sirsa Taluk (Sirsa Dt.)

NOTE:
* Area covered by completed irrigation projects and projects nearing completion.
1. The colour scheme on the map indicates the major crop of each district (based on a comparison of the tonnages produced of the different crops in each district).
2. The colours used for each major crop appear in a darker and lighter shade. Those districts marked in the darker shade are major producers of the particular crop, while the districts marked in lighter shades produce comparatively less of the same crop, even though the crop is the major crop in that district.
3. The crop symbols overlaid show other important crops grown in each TALUK of the state.

Scale 1:2 000 000   1 cm = 20 kms

52

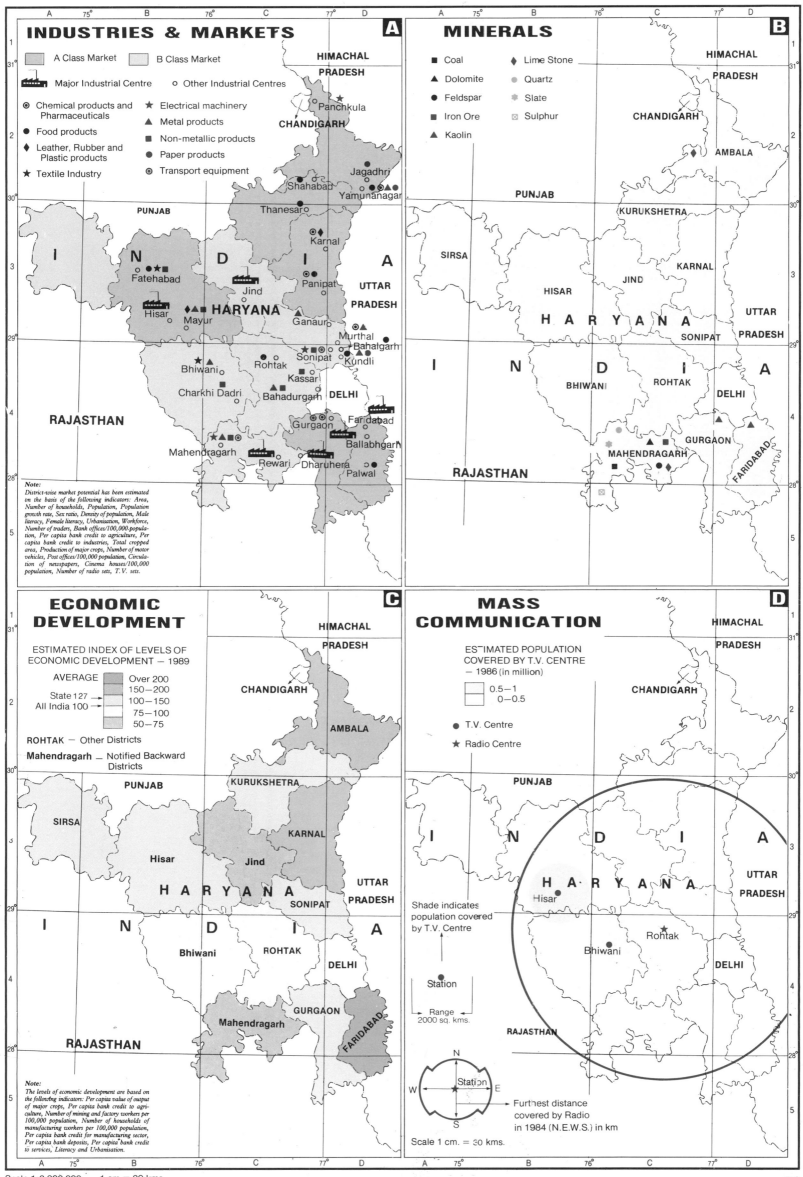

INDUSTRIES & MARKETS

A

A Class Market   B Class Market

Major Industrial Centre   ○ Other Industrial Centres

⊙ Chemical products and Pharmaceuticals   ★ Electrical machinery
● Food products   ▲ Metal products
◆ Leather, Rubber and Plastic products   ■ Non-metallic products
★ Textile Industry   ● Paper products
    ⊙ Transport equipment

HIMACHAL PRADESH
Panchkula
CHANDIGARH
Jagadhri
Shahabad   Yamunanagar
Thanesar
PUNJAB
Karnal
I N D I A
Fatehabad
Jind   Panipat   UTTAR PRADESH
Hisar   HARYANA   Ganaur
Mayur   Murthal
    Bahalgarh
Bhiwani   Rohtak   Sonipat   Kundli
Charkhi Dadri   Kassar
    Bahadurgarh   DELHI
RAJASTHAN   Gurgaon   Faridabad
Mahendragarh   Ballabhgarh
Rewari   Dharuhera   Palwal

Note:
District-wise market potential has been estimated on the basis of the following indicators: Area, Number of households, Population, Population growth rate, Sex ratio, Density of population, Male literacy, Female literacy, Urbanisation, Workforce, Number of traders, Bank offices/100,000 population, Per capita bank credit to agriculture, Per capita bank credit to industries, Total cropped area, Production of major crops, Number of motor vehicles, Post offices/100,000 population, Circulation of newspapers, Cinema houses/100,000 population, Number of radio sets, T.V. sets.

MINERALS

B

■ Coal   ◆ Lime Stone
▲ Dolomite   ● Quartz
● Feldspar   ✳ Slate
■ Iron Ore   ⊠ Sulphur
▲ Kaolin

HIMACHAL PRADESH
CHANDIGARH
    ◆ AMBALA
PUNJAB
    KURUKSHETRA
SIRSA   KARNAL
    JIND
HISAR   UTTAR PRADESH
H A R Y A N A   SONIPAT
I N   BHIWANI   ROHTAK   D   DELHI   I A
    GURGAON
    MAHENDRAGARH   FARIDABAD
RAJASTHAN

ECONOMIC DEVELOPMENT

C

ESTIMATED INDEX OF LEVELS OF ECONOMIC DEVELOPMENT — 1989

AVERAGE   Over 200
    150—200
State 127 →   100—150
All India 100 →   75—100
    50—75

ROHTAK — Other Districts
Mahendragarh — Notified Backward Districts

HIMACHAL PRADESH
CHANDIGARH
AMBALA
PUNJAB
    KURUKSHETRA
SIRSA   KARNAL
Hisar   Jind
H A R Y A N A
    SONIPAT   UTTAR PRADESH
I N D I A
Bhiwani   ROHTAK   DELHI
    GURGAON
Mahendragarh   FARIDABAD
RAJASTHAN

Note:
The levels of economic development are based on the following indicators: Per capita value of output of major crops, Per capita bank credit to agriculture, Number of mining and factory workers per 100,000 population, Number of households of manufacturing workers per 100,000 population, Per capita bank credit for manufacturing sector, Per capita bank deposits, Per capita bank credit to services, Literacy and Urbanisation.

MASS COMMUNICATION

D

ESTIMATED POPULATION COVERED BY T.V. CENTRE — 1986 (in million)
    0.5—1
    0—0.5

● T.V. Centre
★ Radio Centre

HIMACHAL PRADESH
CHANDIGARH
PUNJAB
I N D I A
H A R Y A N A   UTTAR PRADESH
Hisar
Shade indicates population covered by T.V. Centre
    Bhiwani   Rohtak
    DELHI
Station
Range 2000 sq. kms.

RAJASTHAN

Furthest distance covered by Radio in 1984 (N.E.W.S.) in km

Scale 1 cm. = 30 kms.

Scale 1:3 000 000   1 cm = 30 kms   53

HIMACHAL PRADESH

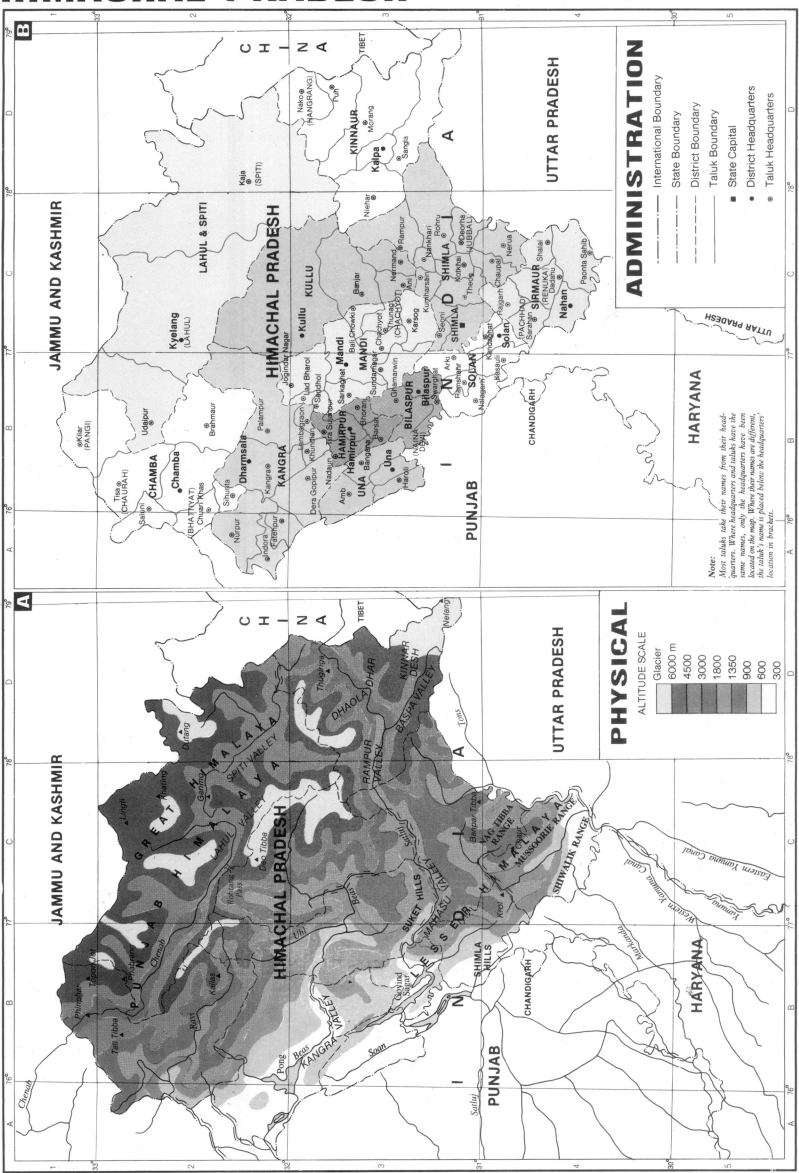

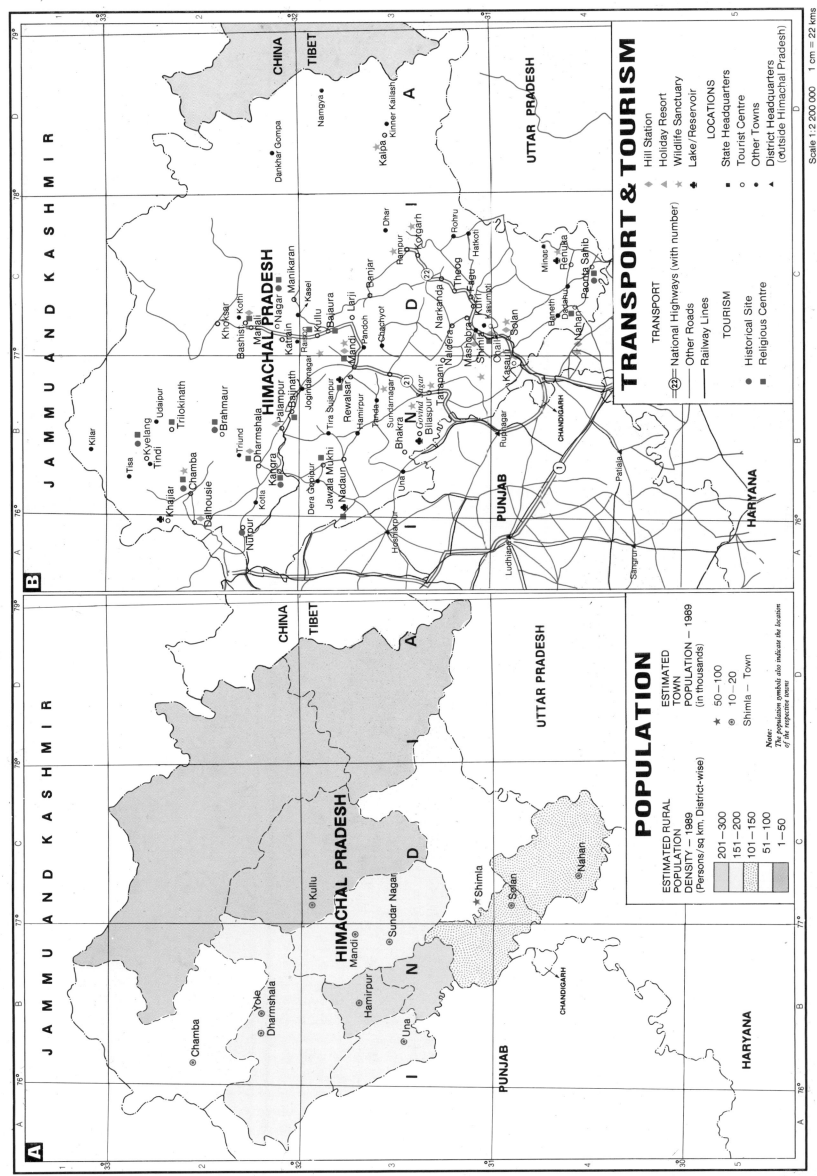

TRANSPORT & TOURISM

TRANSPORT

◆ Hill Station
▲ Holiday Resort
★ Wildlife Sanctuary
✚ Lake/Reservoir

LOCATIONS

■ State Headquarters
○ Tourist Centre
• Other Towns
◄ District Headquarters (outside Himachal Pradesh)

TRANSPORT

(22) National Highways (with number)
━━━ Other Roads
─── Railway Lines

TOURISM

● Historical Site
■ Religious Centre

Scale 1:2 200 000   1 cm = 22 kms

POPULATION

ESTIMATED RURAL POPULATION DENSITY – 1989
(Persons/sq km, District-wise)

201–300
151–200
101–150
51–100
1–50

ESTIMATED TOWN POPULATION – 1989
(in thousands)

★ 50–100
⊚ 10–20

Shimla — Town

Note:
The population symbols also indicate the location of the respective towns

CHINA
TIBET
JAMMU AND KASHMIR
HIMACHAL PRADESH
UTTAR PRADESH
PUNJAB
HARYANA
CHANDIGARH

55

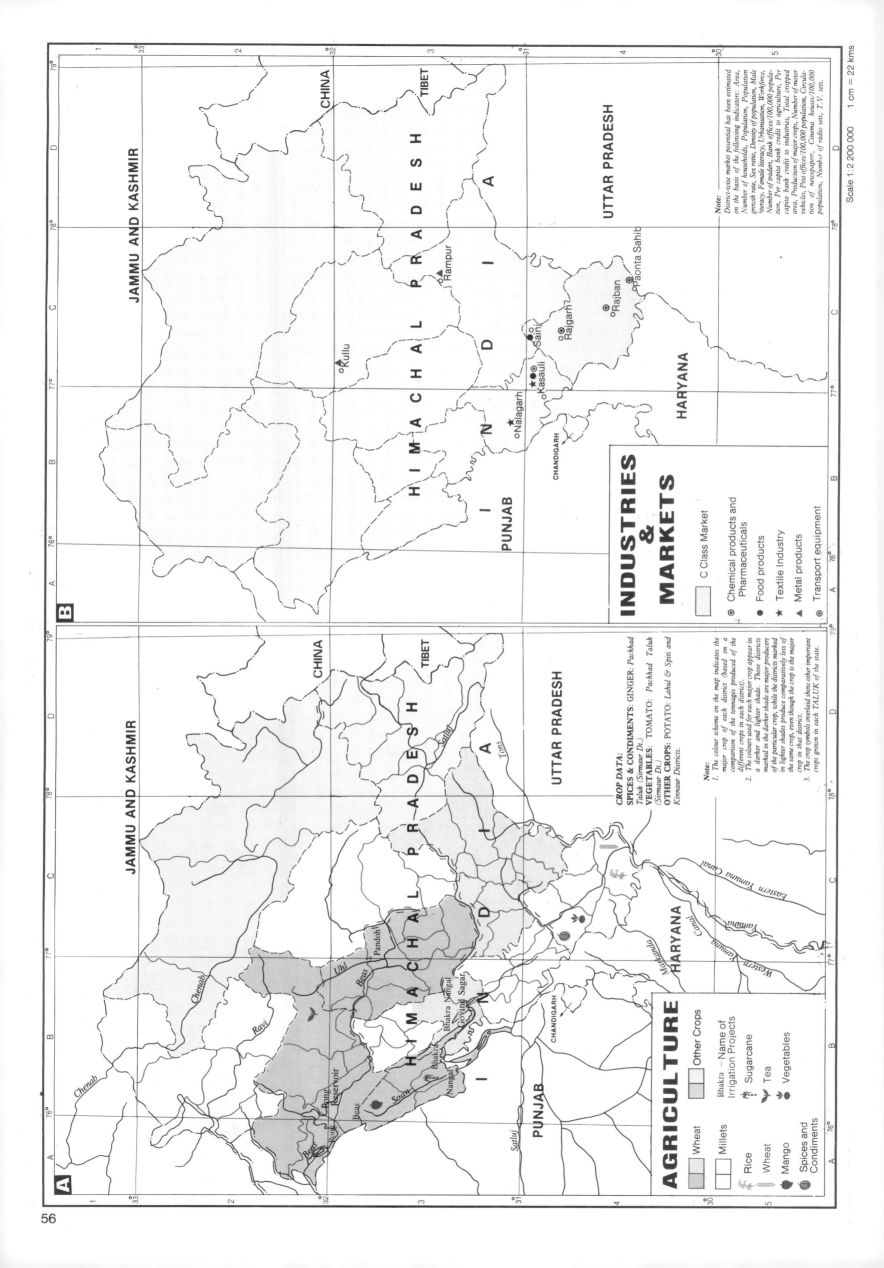

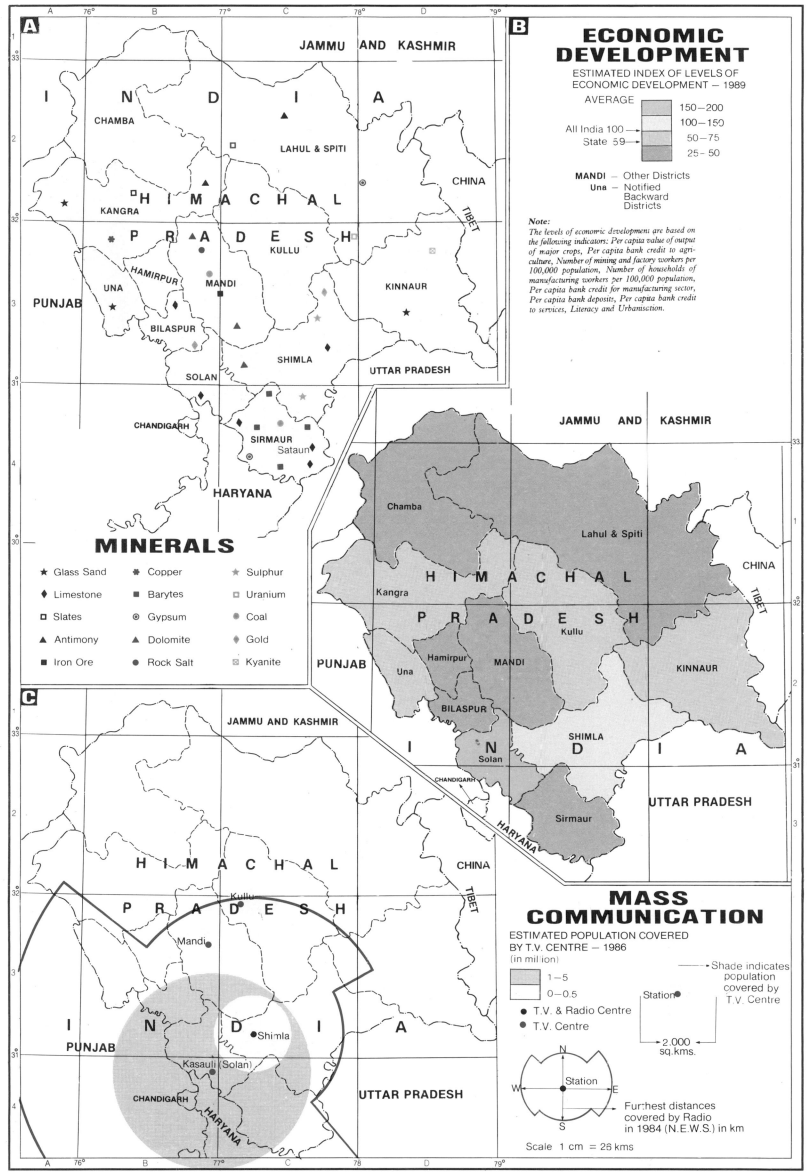

A

JAMMU AND KASHMIR

I N D I A

CHAMBA

LAHUL & SPITI

H I M A C H A L

KANGRA

P R A D E S H

KULLU

CHINA

TIBET

HAMIRPUR

UNA

MANDI

PUNJAB

KINNAUR

BILASPUR

SOLAN

SHIMLA

UTTAR PRADESH

CHANDIGARH

SIRMAUR
Sataun

HARYANA

MINERALS

★ Glass Sand	✣ Copper	★ Sulphur
◆ Limestone	■ Barytes	☐ Uranium
☐ Slates	◉ Gypsum	● Coal
▲ Antimony	▲ Dolomite	◆ Gold
■ Iron Ore	● Rock Salt	⊠ Kyanite

B

ECONOMIC DEVELOPMENT

ESTIMATED INDEX OF LEVELS OF
ECONOMIC DEVELOPMENT — 1989

AVERAGE

	150—200
	100—150
All India 100 →	50—75
State 59 →	25—50

MANDI — Other Districts
Una — Notified
Backward
Districts

Note:
The levels of economic development are based on the following indicators: Per capita value of output of major crops, Per capita bank credit to agriculture, Number of mining and factory workers per 100,000 population, Number of households of manufacturing workers per 100,000 population, Per capita bank credit for manufacturing sector, Per capita bank deposits, Per capita bank credit to services, Literacy and Urbanisation.

JAMMU AND KASHMIR

Chamba

Lahul & Spiti

CHINA

H I M A C H A L

Kangra

P R A D E S H

Kullu

TIBET

PUNJAB

Una

Hamirpur

Mandi

Kinnaur

BILASPUR

I N D I A

Solan

SHIMLA

CHANDIGARH

UTTAR PRADESH

HARYANA

Sirmaur

C

JAMMU AND KASHMIR

H I M A C H A L

P R A D E S H

Kullu

CHINA

Mandi

TIBET

I N D I A

PUNJAB

Shimla

Kasauli (Solan)

CHANDIGARH

HARYANA

UTTAR PRADESH

MASS COMMUNICATION

ESTIMATED POPULATION COVERED
BY T.V. CENTRE — 1986
(in million)

	1—5
	0—0.5

● T.V. & Radio Centre
● T.V. Centre

→ Shade indicates population covered by T.V. Centre

Station ●

2.000 sq.kms.

Furthest distances covered by Radio in 1984 (N.E.W.S.) in km

N
W — Station — E
S

Scale 1 cm = 26 kms

JAMMU & KASHMIR

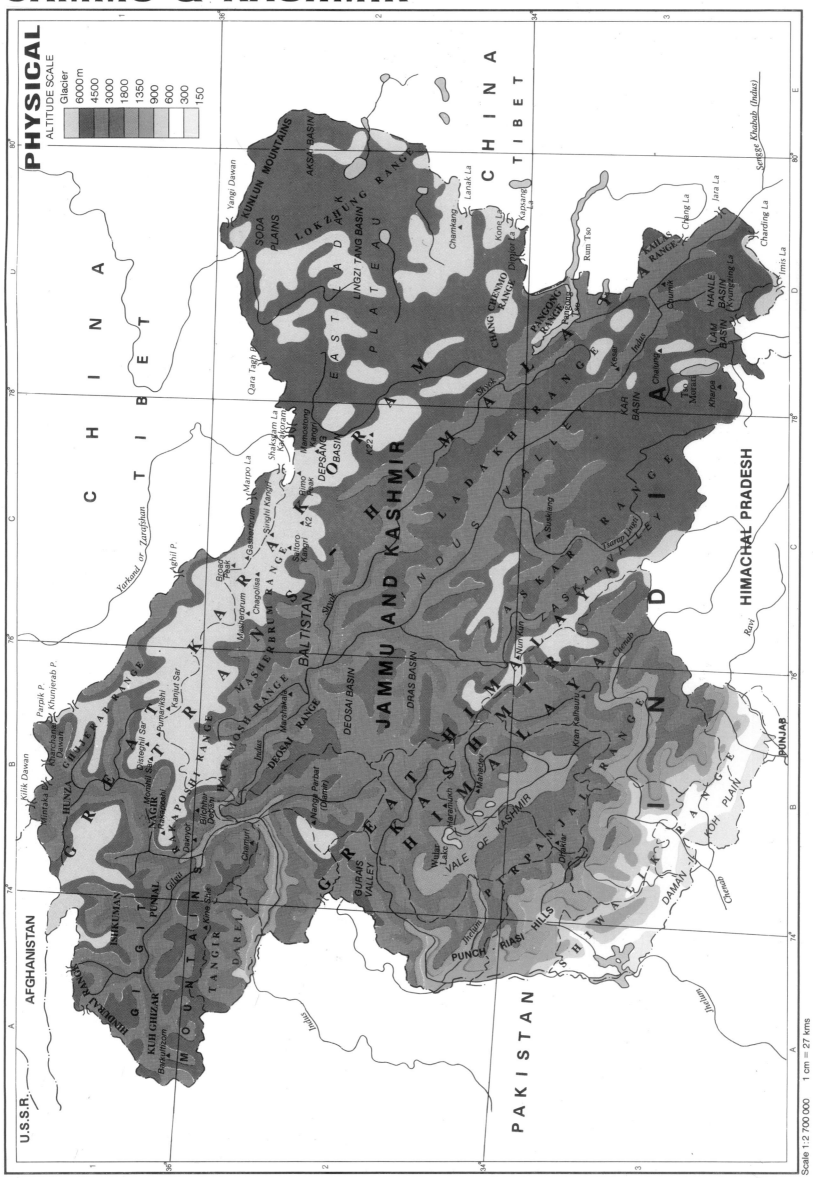

PHYSICAL

ALTITUDE SCALE

Glacier	
	6000m
	4500
	3000
	1800
	1350
	900
	600
	300
	150

Scale 1:2 700 000   1 cm = 27 kms

U.S.S.R.

AFGHANISTAN

C H I N A

T I B E T

C H I N A

T I B E T

PAKISTAN

PUNJAB

HIMACHAL PRADESH

JAMMU AND KASHMIR

KUNLUN MOUNTAINS

SODA PLAINS

LOKZHUNG

AKSAI BASIN

LINGZI TANG BASIN

EAST LADAKH PLATEAU

CHANG CHENMO RANGE

PANGONG RANGE

KAILAS RANGE

HANLE BASIN

LAM BASIN

KARAKORAM RANGE

LADAKH RANGE

ZASKAR RANGE

ZASKAR VALLEY

INDUS VALLEY

DRAS BASIN

DEOSAI BASIN

DEOSAI RANGE

BALTISTAN

NUBRA RANGE

MASHERBRUM RANGE

HARAMOSH RANGE

RAKAPOSHI RANGE

GHUJERAB RANGE

GREAT KARAKORAM MOUNTAINS

HINDU RAJ RANGE

KUH GHIZAR

ISHKUMAN

HUNZA

NAGIR

PUNIAL

GILGIT

TANGIR

DAREL

GREAT HIMALAYAN RANGE

KASHMIR HIMALAYAN RANGE

PIR PANJAL RANGE

SIWALIK RANGE

DAMAN-I-KOH PLAIN

GURAIS VALLEY

VALE OF KASHMIR

PUNCH RIASI HILLS

RANGE

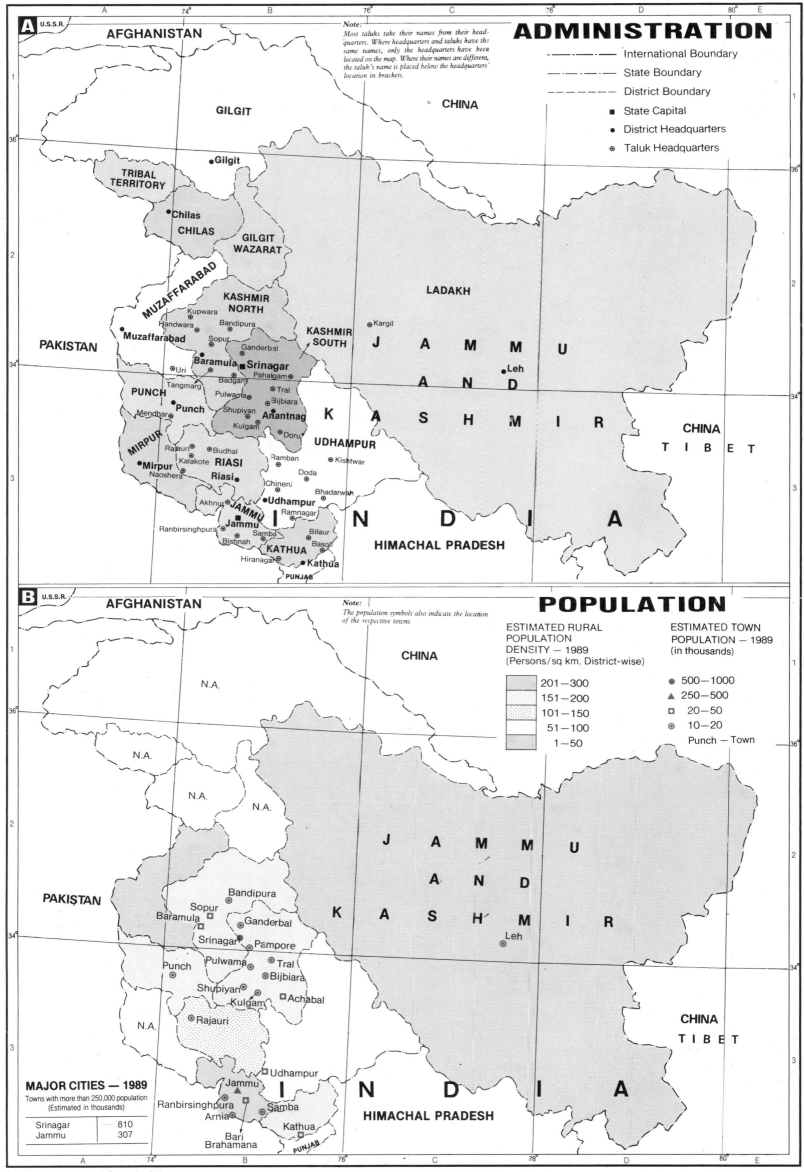

ADMINISTRATION

A U.S.S.R.

AFGHANISTAN

CHINA

Note:
Most taluks take their names from their head-quarters. Where headquarters and taluks have the same names, only the headquarters have been located on the map. Where their names are different, the taluk's name is placed below the headquarters' location in brackets.

———·——·——	International Boundary
———·—·——	State Boundary
———‑‑‑‑———	District Boundary
■	State Capital
●	District Headquarters
◉	Taluk Headquarters

GILGIT

●Gilgit

TRIBAL TERRITORY

●Chilas
CHILAS

GILGIT WAZARAT

LADAKH

MUZAFFARABAD

KASHMIR NORTH

●Kupwara
Handwara ◉Bandipura
●Muzaffarabad Sopur◉ KASHMIR SOUTH
◉Ganderbal
●Baramula ■Srinagar
◉Kargil

PAKISTAN

◉Uri Badgam◉ Pahalgam◉
Tangmarg◉ ◉Tral ●Leh
PUNCH Pulwama◉ ◉Bijbiara
●Punch Shupiyan◉ ●Anantnag
Mendhar◉ Kulgam◉ ◉Doru
UDHAMPUR

J A M M U

A N D

K A S H M I R

CHINA
T I B E T

MIRPUR
◉Rajauri ◉Budhal
◉Kalakote RIASI Ramban◉
●Mirpur ◉Kishtwar
Naoshera◉ Riasi● Doda◉
◉Chineni ●Bhadarwah

Akhnur◉ ◉JAMMU ●Udhampur
Ranbirsinghpura ◉Ramnagar
■Jammu ◉Bilaur
Samba◉ Basoli◉
Bishnah◉ KATHUA
Hiranagar◉ ●Kathua
PUNJAB

I N D I A

HIMACHAL PRADESH

POPULATION

B U.S.S.R.

AFGHANISTAN

CHINA

Note:
The population symbols also indicate the location of the respective towns.

ESTIMATED RURAL POPULATION DENSITY — 1989
(Persons/sq km, District-wise)

	201—300
	151—200
	101—150
	51—100
	1—50

ESTIMATED TOWN POPULATION — 1989
(in thousands)

●	500—1000
▲	250—500
▢	20—50
◉	10—20

Punch — Town

N.A.

N.A.

N.A. N.A.

J A M M U

A N D

K A S H M I R

PAKISTAN

◉Bandipura
◉Sopur
Baramula▢ ◉Ganderbal
Srinagar● ◉Pampore
Pulwama◉ ◉Tral Leh◉
Punch◉ ◉Bijbiara
Shupiyan◉
Kulgam◉ ◉Achabal
N.A. ◉Rajauri

▢Udhampur

▲Jammu
Ranbirsinghpura▢ ◉Samba
Arnia
▲Bari Brahamana Kathua◉
PUNJAB

I N D I A

CHINA
T I B E T

HIMACHAL PRADESH

MAJOR CITIES — 1989
Towns with more than 250,000 population
(Estimated in thousands)

Srinagar	810
Jammu	307

Scale 1:3 750 000   1 Cm = 37.5 Km.

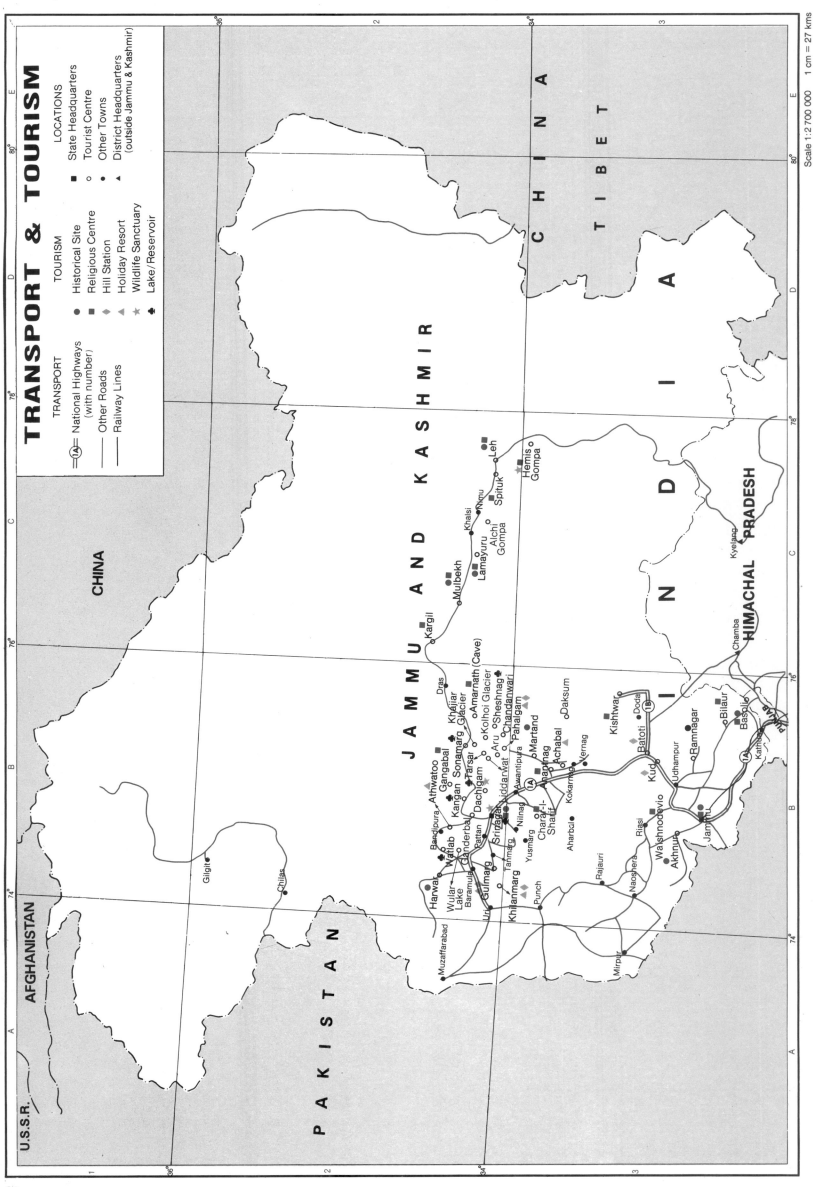

TRANSPORT & TOURISM

U.S.S.R.

AFGHANISTAN

CHINA

PAKISTAN

JAMMU AND KASHMIR

CHINA

TIBET

I N D I A

HIMACHAL PRADESH

Gilgit

Chilas

Muzaffarabad

Harwan
Bandipura
Wular
Lake
Waftab
Baramula
Ganderbal
Uri
Gulmarg
Tanmarg
Srinagar
Pattan
Khilanmarg
Punch
Yusmarg
Charar-i-
Shafif
Nilnag
Athwatoo
Gangabal
Kangan
Sonamarg
Dachigam
Tarsar
Siddarwat
Aru
Kolhoi Glacier
Khajiar
Glacier
Amarnath (Cave)
Sheshnag
Chandanwari
Pahalgam
Anapnag
Achabal
Martand
Kokarnag
Aharbal
Vernag
Daksum

Dras

Kargil

Mulbekh

Khalsi
Lamayuru
Alchi
Gompa

Nimu
Spituk
Leh
Hemis
Gompa

Batoti
Kud
Kishtwar
Doda
Udhampur
Ramnagar

Chamba
Kyelang

Bilaur
Basohli
Kathua
Jammu
Akhnur
Waishnodevio
Riasi
Naoshera
Rajauri
Mirpur

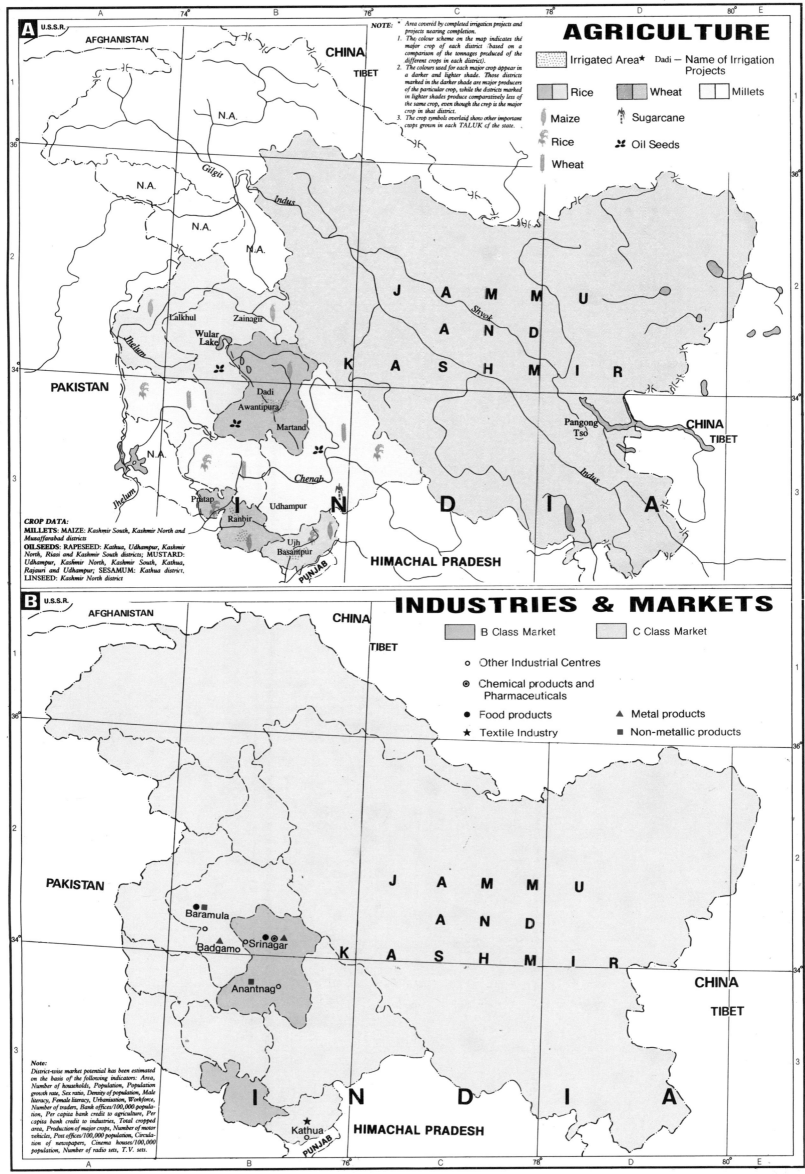

A

U.S.S.R.

AFGHANISTAN

CHINA

TIBET

NOTE: * Area covered by completed irrigation projects and projects nearing completion.
1. The colour scheme on the map indicates the major crop of each district (based on a comparison of the tonnages produced of the different crops in each district).
2. The colours used for each major crop appear in a darker and lighter shade. Those districts marked in the darker shade are major producers of the particular crop, while the districts marked in lighter shades produce comparatively less of the same crop, even though the crop is the major crop in that district.
3. The crop symbols overlaid show other important crops grown in each TALUK of the state.

AGRICULTURE

Irrigated Area★   Dadi — Name of Irrigation Projects

Rice   Wheat   Millets

Maize   Sugarcane

Rice   Oil Seeds

Wheat

N.A.

Gilgit

Indus

N.A.

N.A.

N.A.

J A M M U

Shyok

A N D

K A S H M I R

Jhelum

Lalkhul   Zainagir

Wular Lake

PAKISTAN

Dadi

Awantipura

Martand

Pangong Tso

CHINA

TIBET

N.A.

Chenab

Indus

I   N   D   I   A

Pratap

Ranbir

Udhampur

Ujh Basantpur

HIMACHAL PRADESH

Jhelum

CROP DATA:
MILLETS: MAIZE: *Kashmir South, Kashmir North and Musaffarabad districts*
OILSEEDS: RAPESEED: *Kathua, Udhampur, Kashmir North, Riasi and Kashmir South districts;* **MUSTARD:** *Udhampur, Kashmir North, Kashmir South, Kathua, Rajauri and Udhampur;* **SESAMUM:** *Kathua district,* **LINSEED:** *Kashmir North district*

PUNJAB

B

U.S.S.R.

AFGHANISTAN

CHINA

TIBET

INDUSTRIES & MARKETS

B Class Market   C Class Market

○ Other Industrial Centres

◉ Chemical products and Pharmaceuticals

● Food products   ▲ Metal products

★ Textile Industry   ■ Non-metallic products

PAKISTAN

J A M M U

A N D

K A S H M I R

● ■
Baramula
○
Badgamo ▲
○Srinagar ● ◉ ▲

■
Anantnag ○

CHINA

TIBET

I   N   D   I   A

Note:
District-wise market potential has been estimated on the basis of the following indicators: Area, Number of households, Population, Population growth rate, Sex ratio, Density of population, Male literacy, Female literacy, Urbanisation, Workforce, Number of traders, Bank offices/100,000 population, Per capita bank credit to agriculture, Per capita bank credit to industries, Total cropped area, Production of major crops, Number of motor vehicles, Post offices/100,000 population, Circulation of newspapers, Cinema houses/100,000 population, Number of radio sets, T.V. sets.

★
Kathua

HIMACHAL PRADESH

PUNJAB

Scale 1:3 750 000   1 cm = 37.5 kms

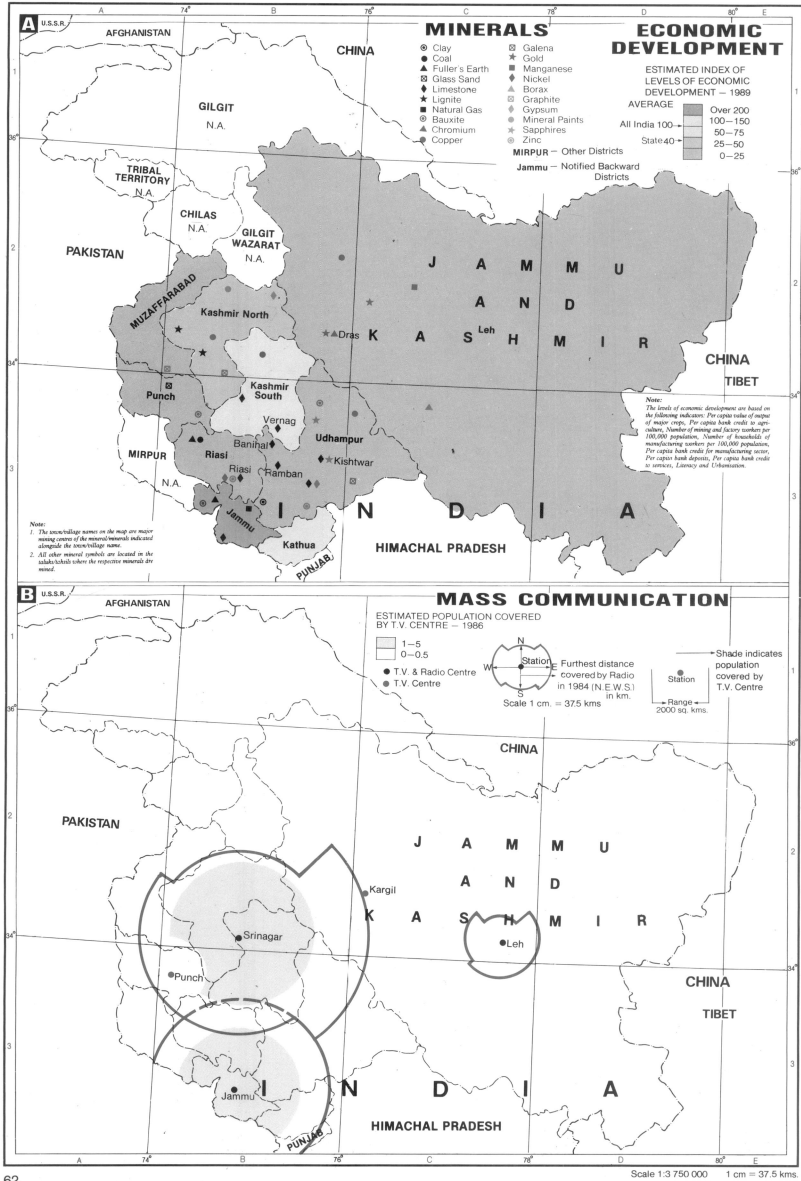

MINERALS

A U.S.S.R.

AFGHANISTAN

CHINA

Minerals legend:
- ⊙ Clay
- ● Coal
- ▲ Fuller's Earth
- ⊠ Glass Sand
- ◆ Limestone
- ◆ Lignite
- ■ Natural Gas
- ⊙ Bauxite
- ▲ Chromium
- ● Copper
- ⊠ Galena
- ★ Gold
- ■ Manganese
- ◆ Nickel
- ▲ Borax
- ⊠ Graphite
- ◆ Gypsum
- ● Mineral Paints
- ★ Sapphires
- ⊙ Zinc

MIRPUR — Other Districts

Jammu — Notified Backward Districts

ECONOMIC DEVELOPMENT

ESTIMATED INDEX OF LEVELS OF ECONOMIC DEVELOPMENT — 1989

AVERAGE

All India 100 →

State 40 →

	Over 200
	100–150
	50–75
	25–50
	0–25

GILGIT N.A.

TRIBAL TERRITORY N.A.

CHILAS N.A.

GILGIT WAZARAT N.A.

PAKISTAN

MUZAFFARABAD

Kashmir North

J A M M U

A N D

K A S H M I R

Leh

Dras

CHINA

TIBET

Note:
The levels of economic development are based on the following indicators: Per capita value of output of major crops, Per capita bank credit to agriculture, Number of mining and factory workers per 100,000 population, Number of households of manufacturing workers per 100,000 population, Per capita bank credit for manufacturing sector, Per capita bank deposits, Per capita bank credit to services, Literacy and Urbanisation.

Punch

Kashmir South

Vernag

Udhampur

MIRPUR

Riasi

Banihal

Riasi   Ramban

Kishtwar

I N D I A

Jammu

Kathua

PUNJAB

HIMACHAL PRADESH

Note:
1. The town/village names on the map are major mining centres of the mineral/minerals indicated alongside the town/village name.
2. All other mineral symbols are located in the taluks/tehsils where the respective minerals are mined.

MASS COMMUNICATION

B U.S.S.R.

AFGHANISTAN

ESTIMATED POPULATION COVERED BY T.V. CENTRE — 1986

	1–5
	0–0.5

● T.V. & Radio Centre
● T.V. Centre

Furthest distance covered by Radio in 1984 (N.E.W.S.) in km.
Scale 1 cm. = 37.5 kms.

Station

Range 2000 sq. kms.

Shade indicates population covered by T.V. Centre

PAKISTAN

CHINA

J A M M U

A N D

K A S H M I R

Kargil

Srinagar

Leh

Punch

CHINA

TIBET

I N D I A

Jammu

HIMACHAL PRADESH

PUNJAB

Scale 1:3 750 000   1 cm = 37.5 kms.

KARNATAKA

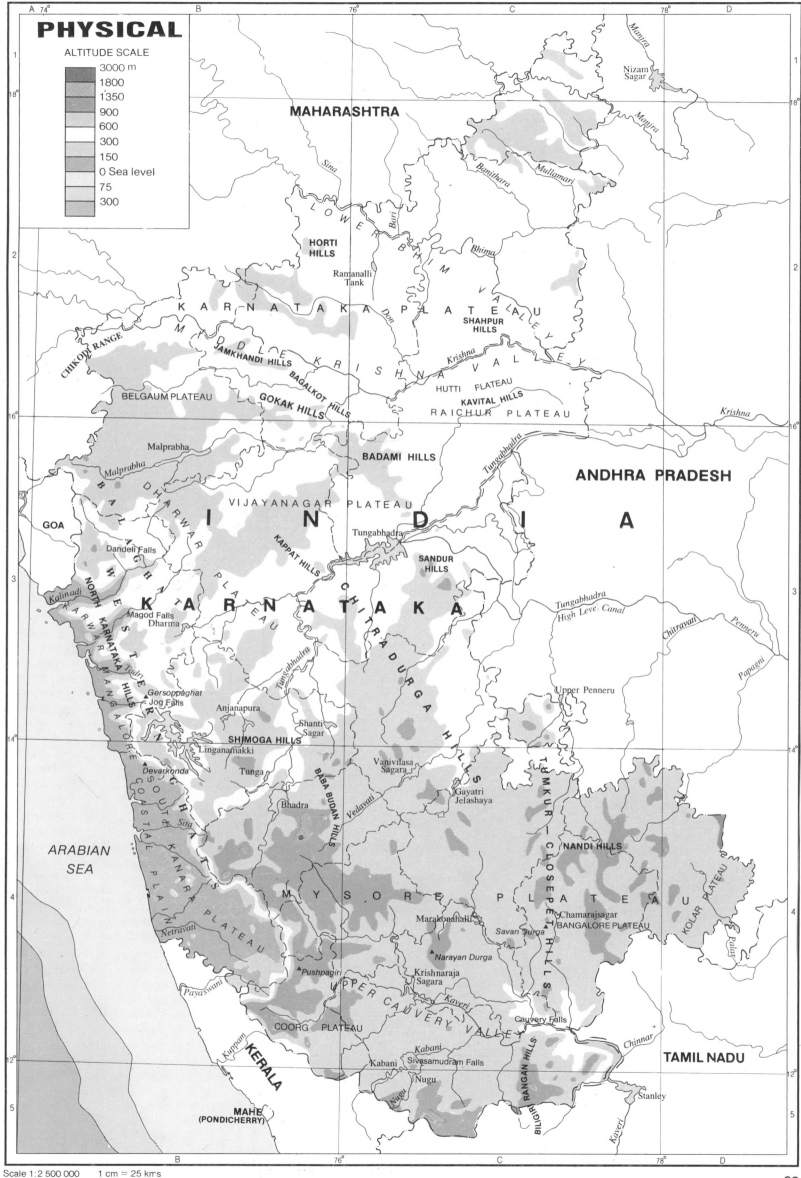

PHYSICAL

ALTITUDE SCALE

	3000 m
	1800
	1350
	900
	600
	300
	150
	0 Sea level
	75
	300

MAHARASHTRA

HORTI HILLS

Ramanalli Tank

Sina

Buri

Don

Bhima

Banithara

Mullamari

Nizam Sagar

Manjra

Monjira

K A R N A T A K A   P L A T E A U

SHAHPUR HILLS

CHIKODI RANGE

JAMKHANDI HILLS

BAGALKOT HILLS

BELGAUM PLATEAU

GOKAK HILLS

Krishna

HUTTI FLATEAU

KAVITAL HILLS

RAICHUR PLATEAU

Krishna

Malprabha

Malprabha

BADAMI HILLS

Tungabhadra

ANDHRA PRADESH

VIJAYANAGAR PLATEAU

GOA

Dandeli Falls

KAPPAT HILLS

Tungabhadra

SANDUR HILLS

Kalinadi

Magod Falls
Dharma

NORTH KANARA HILLS

Tungabhadra

CHITRADURGA HILLS

Tungabhadra High Leve Canal

Chitravati

Penneru

Papagni

Upper Penneru

Gersoppaghat
Jog Falls

Anjanapura

Shanti Sagar

SHIMOGA HILLS

Linganamakki

Tunga

Tunga

Vanivilasa Sagara

Gayatri Jelashaya

TUMKUR-CLOSEPET HILLS

Devarkonda

Sita

Bhadra

BABA BUDAN HILLS

Vedavati

ARABIAN SEA

M Y S O R E   P L A T E A U

NANDI HILLS

Chamarajsagar
BANGALORE PLATEAU

KOLAR PLATEAU

Payaswani

Marakonahalli

Savan Durga

Narayan Durga

Krishnaraja Sagara

Pushpagiri

Kaveri

Cauvery Falls

Chinnar

COORG PLATEAU

U P P E R   C A U V E R Y   V A L L E Y

TAMIL NADU

Kuppam

KERALA

Kabani

Kabani

Nugu

Nugu

Sivasamudram Falls

BILIGIRI RANGAN HILLS

Stanley

Kaveri

MAHE (PONDICHERRY)

Netravati

Scale 1:2 500 000   1 cm = 25 kms

63

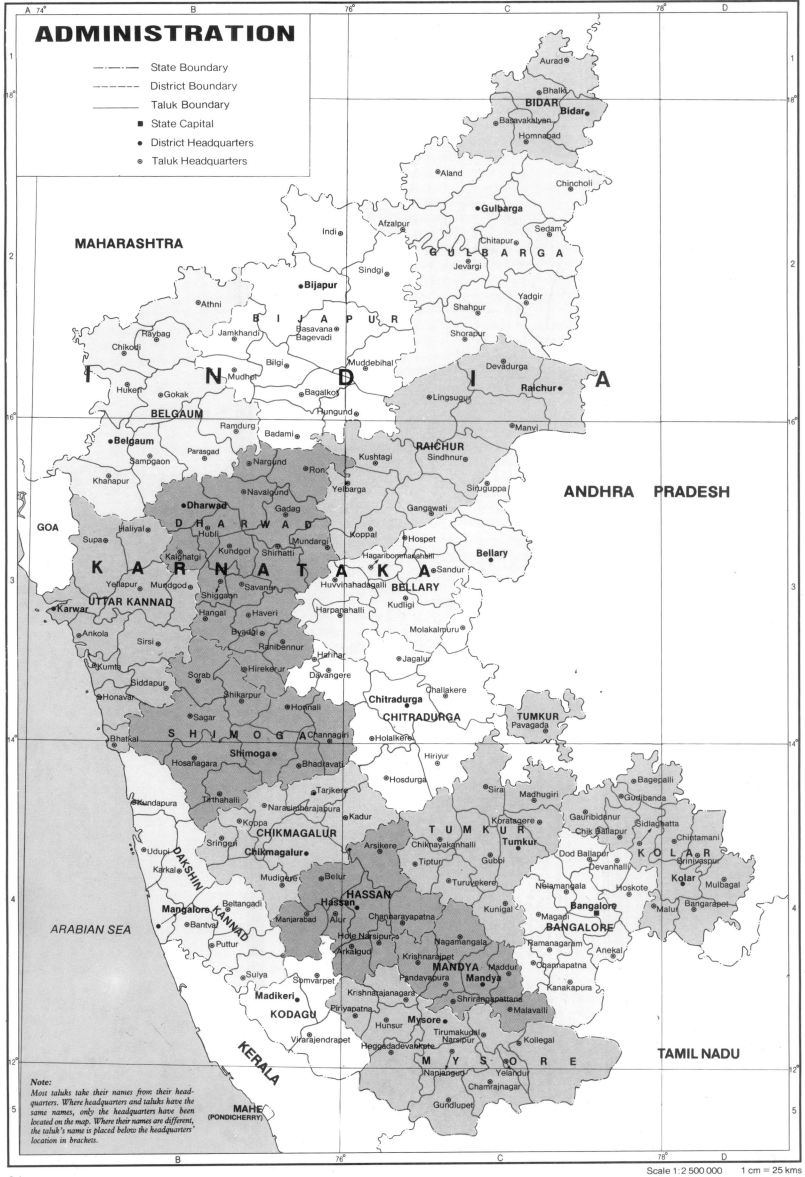

ADMINISTRATION

State Boundary
District Boundary
Taluk Boundary
■ State Capital
● District Headquarters
◉ Taluk Headquarters

MAHARASHTRA

MAHARASHTRA

Aurad

BIDAR
Bhalki
Bidar
Basavakalyan
Homnabad

● Gulbarga
Aland
Chincholi
Afzalpur
Indi
Chitapur
Sedam
G U L B A R G A
Jevargi
Sindgi
Shahpur
Yadgir
● Bijapur
Athni
B I J A P U R
Shorapur
Raybag
Jamkhandi
Basavana Bagevadi
Chikodi
Muddebihal
Bilgi
Devadurga
Hukeri Gokak
Mudhol
Bagalkot
I N D
Lingsugur
Raichur
A
Hungund
BELGAUM
Ramdurg
Badami
Manvi
● Belgaum
Parasgad
Kushtagi
Sampgaon
Nargund
Ron
Yelburga
RAICHUR
Sindhnur
Khanapur
Navalgund
Gadag
Koppal
Siruguppa
GOA
● Dharwad
Gangawati
D H A R W A D
Hubli
Mundargi
Hospet
ANDHRA   PRADESH
Supa
Haliyal
Kundgol
Shirhatti
Hagaribommanahalli
Bellary
Kalghatgi
Sandur
K A R N A T A K A
Yellapur
Mundgod
Savanur
Huvvinahadagalli
Sandur
Karwar
UTTAR KANNAD
Shiggaon
BELLARY
Ankola
Hangal
Haveri
Harpanahalli
Kudligi
Sirsi
Byadgi
Molakalmuru
Kumta
Ranibennur
Harihar
Jagalur
Siddapur
Sorab
Hirekerur
Davangere
Challakere
Honavar
Shikarpur
Honnali
Chitradurga
SHIMOGA
Channagiri
CHITRADURGA
TUMKUR
Bhatkal
Sagar
Holalkere
Pavagada
Hosanagara
Shimoga
Hiriyur
Bhadravati
Hosdurga
Kundapura
Tirthahalli
Tarikere
Bagepalli
Narasimharajapura
Sira
Madhugiri
Gudibanda
Kadur
Koppa
CHIKMAGALUR
Arsikere
T U M K U R
Gauribidanur
Chik Ballapur
Sidlaghatta
Sringeri
Chiknayakanhalli
Koratagere
Udupi
Mudigere
Chikmagalur
Belur
Tiptur
Tumkur
Dod Ballapur
Chintamani
Karkal
Gubbi
K O L A R
Beltangadi
Turuvekere
Nelamangala
Devanhalli
Mangalore
DAKSHIN
Hassan
Hoskote
Kolar
Mulbagal
Bantval
HASSAN
Channarayapatna
Kunigal
Bangalore
Bangarapet
Manjarabad
Alur
Magadi
ARABIAN SEA
KANNAD
Nagamangala
BANGALORE
Malur
Puttur
Hole Narsipur
Ramanagaram
Anekal
Sulya
Arkalgud
Krishnarajpet
Mandya
Channapatna
Madikeri
Somvarpet
Pandavapura
Maddur
Kanakapura
Krishnarajanagara
Mandya
KODAGU
Piriyapatna
MANDYA
Shrirangapattana
Virarajendrapet
Hunsur
Mysore
Malavalli
KERALA
Heggadadevankote
Tirumakudal Narsipur
Kollegal
TAMIL NADU
M Y S O R E
Nanjangud
Yelandur
MAHE
(PONDICHERRY)
Gundlupet
Chamrajnagar

Note:
Most taluks take their names from their head-
quarters. Where headquarters and taluks have the
same names, only the headquarters have been
located on the map. Where their names are different,
the taluk's name is placed below the headquarters'
location in brackets.

Scale 1: 2 500 000   1 cm = 25 kms

POPULATION

ESTIMATED RURAL POPULATION DENSITY — 1989
(Persons/sq km, District-wise)

	301—500
	201—300
	151—200
	101—150
	51—100

ESTIMATED TOWN/VILLAGE POPULATION —1989
(in thousands)

■ OVER 1,000
● 500—1,000
▲ 250—500
◆ 100—250
★ 50—100
☐ 20—50
○ 10—20

Sira — Town
Maski — Village

Note:
The population symbols also indicate the location of the respective towns/villages.

MAHARASHTRA

ANDHRA PRADESH

KARNATAKA

INDIA

GOA

ARABIAN SEA

KERALA

MAHE (PONDICHERRY)

TAMIL NADU

MAJOR CITIES — 1989
Towns with more than 250,000 population
(Estimated in thousands)

Bangalore	4,689
Hubli-Dharwad	690
Mysore	606
Mangalore	408
Belgaum	397
Gulburga	306
Bellary	298
Davangere	294

Scale 1:2 500 000   1 cm = 25 kms

65

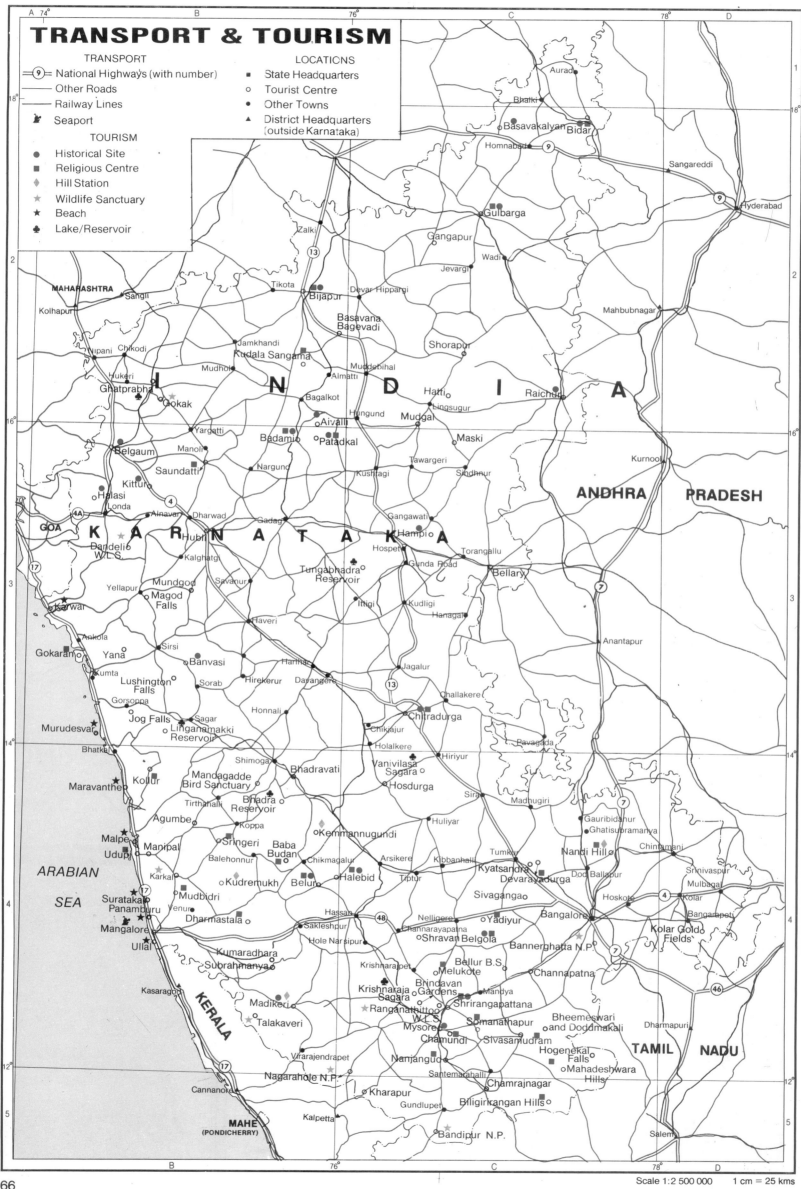

TRANSPORT & TOURISM

TRANSPORT
- ⊨9⊨ National Highways (with number)
- —— Other Roads
- —— Railway Lines
- ⚓ Seaport

TOURISM
- ● Historical Site
- ■ Religious Centre
- ◆ Hill Station
- ★ Wildlife Sanctuary
- ★ Beach
- ♣ Lake/Reservoir

LOCATIONS
- ■ State Headquarters
- ○ Tourist Centre
- ● Other Towns
- ▲ District Headquarters (outside Karnataka)

MAHARASHTRA

GOA

KARNATAKA

INDIA

ANDHRA PRADESH

ARABIAN SEA

KERALA

TAMIL NADU

MAHE (PONDICHERRY)

Sangli
Kolhapur
Nipani Chikodi
Hukeri
Ghatprabha
Gokak
Belgaum
Manoli
Kittur
Halasi
Londa
Alnavar Dharwad
Dandeli W.L.S.
Hubli
Kalghatgi
Yellapur
Karwar
Ankola
Gokaran
Yana
Kumta
Sirsi
Lushington Falls
Gorsoppa
Jog Falls
Sagar
Murudesvar
Linganamakki Reservoir
Bhatkal
Kollur
Maravanthe
Mandagadde Bird Sanctuary
Tirthahalli
Agumbe
Malpe
Manipal
Udupi
Karkal
Suratakal
Panamburu
Mangalore
Ullal
Kumaradhara
Subrahmanya
Kasaragod
Madikeri
Talakaveri
Cannanore
Kalpetta
Virajendrapet
Nagarahole N.P.
Kharapur
Gundlupet
Bandipur N.P.
Biligirirangan Hills

Zalki
Tikota
Bijapur
Devar Hippargi
Basavana Bagevadi
Jamkhandi
Kudala Sangama
Mudhol
Bagalkot
Almatti
Muddebihal
Yargatti
Aivalli
Badami Patadkal
Saundatti
Nargund
Hungund
Kushtagi
Mundgod
Magod Falls
Savanur
Haveri
Banvasi
Sorab
Hirekerur
Davangere
Harihar
Honnali
Shimoga
Bhadravati
Chikjajur
Holalkere
Bhadra Reservoir
Koppa
Sringeri
Baba Budan
Balehonnur
Chikmagalur
Kudremukh
Beluru
Halebid
Mudbidri
Venur
Dharmastala
Hassan
Saklespur
Hole Narsipur
Krishnarajpet
Krishnaraja Sagara
Brindavan Gardens
Ranganathittoo W.L.S.
Mysore
Chamundi
Nanjangud
Santemarahalli
Chamrajnagar

Gulbarga
Gangapur
Wadi
Jevargi
Shorapur
Hatti
Lingsugur
Mudgal
Maski
Tawargeri
Sindhnur
Gangawati
Hampi
Hospet
Gunda Road
Torangallu
Bellary
Ittigi
Kudligi
Hanagal
Jagalur
Challakere
Chitradurga
Pavagada
Vanivilasa Sagara
Hosdurga
Hiriyur
Sira
Madhugiri
Huliyar
Kemmannugundi
Arsikere
Kibbanhalli
Tiptur
Tumkur
Nelligere
Channarayapatna
Shravan Belgola
Yadiyur
Bellur B.S.
Melukote
Mandya
Shrirangapattana
Somanathapur
Sivasamudram
Hogenekal Falls
Mahadeshwara Hills

Aurad
Bhalki
Basavakalyan Bidar
Homnabad
Sangareddi
Hyderabad
Mahbubnagar
Raichur
Kurnool
Anantapur
Gauribidanur
Ghatisubramanya
Nandi Hill
Chintamani
Srinivaspur
Kyatsandra
Devarayadurga
Dod Ballapur
Sivagangao
Hoskote
Bangalore
Kolar
Mulbagal
Bangarapeti
Bannerghatta N.P.
Kolar Gold Fields
Channapatna
Bheemeswari and Doddmakali
Dharmapuri
Salem

Scale 1:2 500 000   1 cm = 25 kms

66

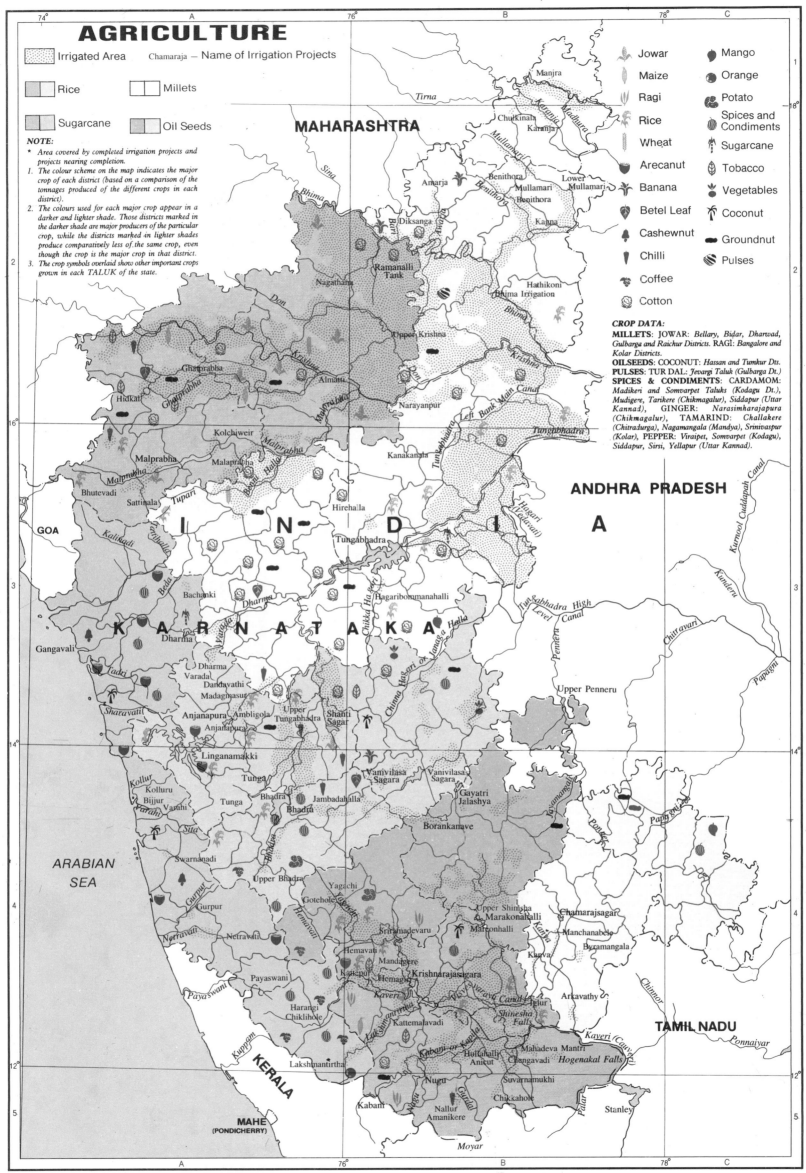

AGRICULTURE

Irrigated Area   Chamaraja — Name of Irrigation Projects

Rice   Millets

Sugarcane   Oil Seeds

NOTE:
* Area covered by completed irrigation projects and projects nearing completion.
1. The colour scheme on the map indicates the major crop of each district (based on a comparison of the tonnages produced of the different crops in each district).
2. The colours used for each major crop appear in a darker and lighter shade. Those districts marked in the darker shade are major producers of the particular crop, while the districts marked in lighter shades produce comparatively less of the same crop, even though the crop is the major crop in that district.
3. The crop symbols overlaid show other important crops grown in each TALUK of the state.

Jowar
Maize
Ragi
Rice
Wheat
Arecanut
Banana
Betel Leaf
Cashewnut
Chilli
Coffee
Cotton

Mango
Orange
Potato
Spices and Condiments
Sugarcane
Tobacco
Vegetables
Coconut
Groundnut
Pulses

CROP DATA:
MILLETS: JOWAR: *Bellary, Bidar, Dharwad, Gulbarga and Raichur Districts.* RAGI: *Bangalore and Kolar Districts.*
OILSEEDS: COCONUT: *Hassan and Tumkur Dts.*
PULSES: TUR DAL: *Jevargi Taluk (Gulbarga Dt.)*
SPICES & CONDIMENTS: CARDAMOM: *Madikeri and Somvarpet Taluks (Kodagu Dt.), Mudigere, Tarikere (Chikmagalur), Siddapur (Uttar Kannad),* GINGER: *Narasimharajapura (Chikmagalur),* TAMARIND: *Challakere (Chitradurga), Nagamangala (Mandya), Srinivaspur (Kolar),* PEPPER: *Viraipet, Somvarpet (Kodagu), Siddapur, Sirsi, Yellapur (Uttar Kannad).*

MAHARASHTRA

Tirna
Sina
Bhima
Manjra
Chulkinala
Karanja
Amarja
Benithora
Benithora
Mullamari
Lower Mullamari
Mullamari
Diksanga
Barri
Kagna
Hathikoni
Bhima Irrigation
Ramanalli Tank
Nagathana
Don
Krishna
Ghataprabha
Hidkal
Ghataprabha
Almatti
Upper Krishna
Bhima
Krishna
Narayanpur
Tungabhadra Left Bank Main Canal
Tungabhadra

ANDHRA PRADESH

Kolchiweir
Malprabha
Malprabha
Malaprabha
Bean Halla
Kanakanala
Bhutevadi
Sattinala
Tupari
Hirehalla
Malprabha
Kalinadi
Hasari (Vedavati)
GOA
Gangavali
Bachanki
Dharma
Hagaribommanahalli
Chikka Hagari
Tungabhadra
I N D I A
Bela
Kurnool Cuddapah Canal
Kunderu
Chitravati
K A R N A T A K A
Dharma
Dharma
Varada
Chinna Hasari or Janaga Halla
Tungabhadra High Level Canal
Papagni
Dandavathi
Madagmasur
Anjanapura
Ambligola
Upper Tungabhadra
Shanti Sagar
Pennar
Upper Penneru
Anjanapura
Tadri
Sharavathi
Linganamakki
Tunga
Vanivilasa Sagara
Vanivilasa Sagara
Gayatri Jalashya
Kollur
Kolluru
Bijjur
Varahi
Tunga
Bhadra
Jambadahalla
Bhadra
Borankamave
Suvarnamukhi
Sita
Swarnanadi
Bhadra
Upper Bhadra
Yagachi
Upper Shimsha & Marakonahalli
Chamarajsagar
Gurpur
Gotehole
Yagachi
Marconhalli
Manchanabele
Byramangala
ARABIAN SEA
Gurpur
Hemavathi
Sriramadevaru
Kanva
Gurpur
Netravati
Netravati
Hemavati
Mandagere
Krishnarajasagara
Kaveri
Hemavati
Hemavati
Payaswani
Kaveri
Arkavathy
Harangi Chiklihole
Lakshmantirtha
Iglur
Shinesha Falls
Payaswani
Kattemalavadi
TAMIL NADU
Kuppam
Lakshmantirtha
Kabani or Kapila
Mahadeva Mantri Anicut
Hullahalli Changavadi
Hogenakal Falls
Kaveri (Cauvery)
KERALA
Nugu
Suvarnamukhi
Ponnaiyar
Kabani
Chikkahole
Stanley
Nallur Amanikere
Pidar
MAHE (PONDICHERRY)
Moyar

Scale 1:2 500 000   1 cm = 25 kms

67

INDUSTRIES & MARKETS

A Class Market   B Class Market   C Class Market

🏭 Major Industrial Centres   ◆ Non-electrical machinery
○ Other Industrial Centres   ■ Non-metallic products
▲ Beverage and Tobacco products   ● Paper products
◉ Chemical products and Pharmaceuticals   ◎ Transport equipment
● Food products
◆ Leather, Rubber and Plastic products
★ Textile Industry
★ Electrical machinery
▲ Metal products

I N D I A

ANDHRA   PRADESH

K A R N A T A K A

GOA

ARABIAN SEA

KERALA

MAHE (PONDICHERRY)

TAMIL   NADU

Bidar▲◎
Hallikhed●

▲★◉★ ○Gulbarga
Malkhaid■
○Chitapur
Wadi○■
Bijapur○■
Ugar Khurd○●
Chikodi○   Shahabad■◆ ○
Sankeshwar○ ●○Hukeri◆   Chicksugar○
○Gokak■★
Hubli○
●▲
○Belgaum

○ Dharwad◆
★●◉▲ Gadag○●   Munirabad◎   ○Vijayanagar
○Hubli○   ▲◆●Hospet
Dandeli○   Hagari○

○Karwar

Gokarn◎○   ◉◆★ Harihar   ★●
○Davangere
Harige○

Shimoga■▲
▲●■Bhadravati○ Hosdurga○ Mathodu○

Chikmagalur○▲   Gubbi○   ●■Tumkur
○Adityapatna   Dod Ballapur
◉Kudremukh   Nagasandra◆   Hoskote
○Belur   Nelamangala ◆●▲ ○Kolar
Hassan○▲   Malleswaram★ Krishnarajapur▲◎
Bangalore   ○ ◆Bangarapet
Penamburu○
▲●● Mangalore○   ●◎○Maddur
Puttur○   Mandya●◎
Belagola●○
Madikeri   Mysore◎○
▲●   Nanjangud○◉◆

Note:
District-wise market potential has been estimated on the basis of the following indicators: Area, Number of households, Population, Population growth rate, Sex ratio, Density of population, Male literacy, Female literacy, Urbanisation, Workforce, Number of traders, Bank offices/100,000 population, Per capita bank credit to agriculture, Per capita bank credit to industries, Total cropped area, Production of major crops, Number of motor vehicles, Post offices/100,000 population, Circulation of newspapers, Cinema houses/100,000 population, Number of radio sets, T.V. sets.

Scale 1:2 500 000   1 cm = 25 kms

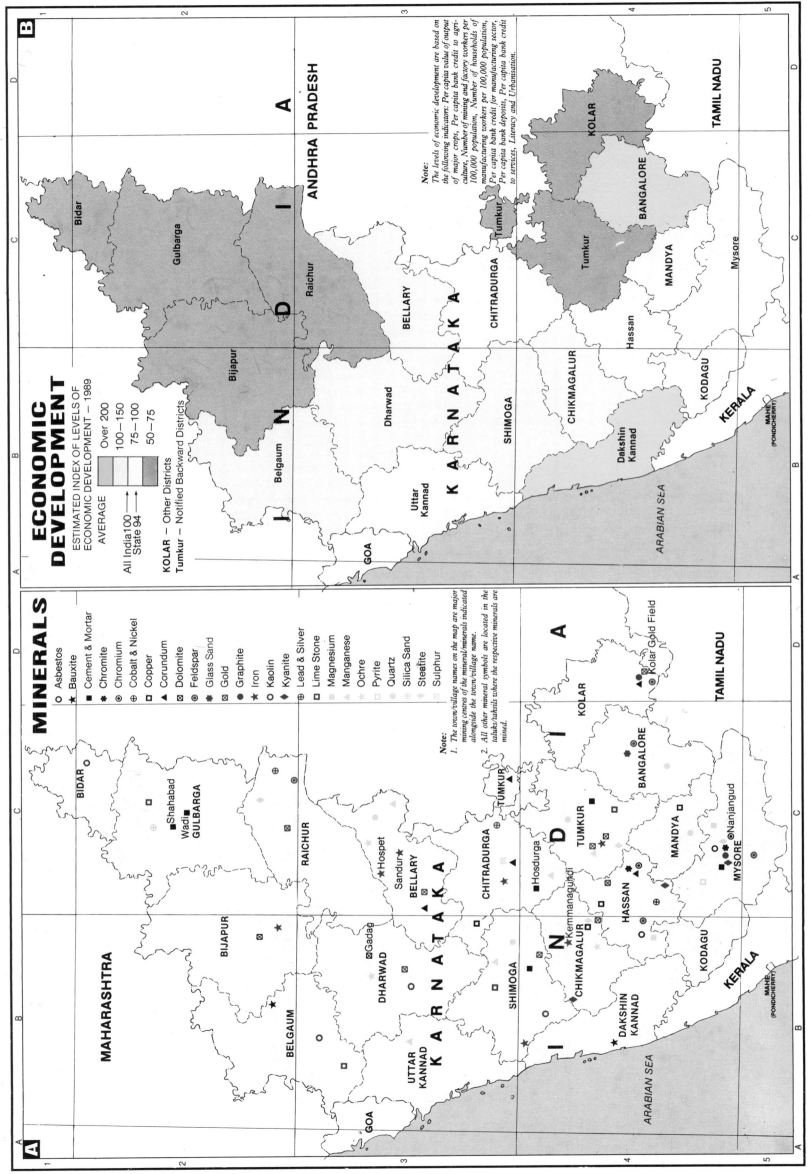

ECONOMIC DEVELOPMENT

ESTIMATED INDEX OF LEVELS OF
ECONOMIC DEVELOPMENT — 1989

AVERAGE

Over 200
100—150
75—100
50—75

All India 100
State 94

KOLAR — Other Districts
Tumkur — Notified Backward Districts

Note:
The levels of economic development are based on
the following indicators: Per capita value of output
of major crops, Per capita bank credit to agriculture, Number of mining and factory workers per
100,000 population, Number of households of
manufacturing workers per 100,000 population,
Per capita bank credit for manufacturing sector,
Per capita bank deposits, Per capita bank credit
to services, Literacy and Urbanisation.

ANDHRA PRADESH

TAMIL NADU

KOLAR

BANGALORE

Tumkur

Tumkur

MANDYA

CHITRADURGA

BELLARY

Hassan

Mysore

SHIMOGA

CHIKMAGALUR

KODAGU

Dharwad

KERALA

MAHE
(PONDICHERRY)

Dakshin
Kannad

ARABIAN SEA

Uttar
Kannad

GOA

Belgaum

Bijapur

Raichur

Gulbarga

Bidar

I N D I A

K A R N A T A K A

MINERALS

○ Asbestos
★ Bauxite
■ Cement & Mortar
✳ Chromite
◉ Chromium
⊕ Cobalt & Nickel
□ Copper
▲ Corundum
⊠ Dolomite
✳ Feldspar
✱ Glass Sand
● Gold
● Graphite
★ Iron
● Kaolin
◆ Kyanite
⊡ Lead & Silver
□ Lime Stone
✦ Magnesium
▶ Manganese
● Ochre
★ Pyrite
▢ Quartz
⊞ Silica Sand
⊕ Steatite
◆ Sulphur

Note:
1. The town/village names on the map are major
mining centres of the mineral/minerals indicated
alongside the town/village name.
2. All other mineral symbols are located in the
taluks/tahsils where the respective minerals are
mined.

MAHARASHTRA

BIDAR ○

■ Shahabad
Wadi ■ GULBARGA

⊕

⊙

RAICHUR

★ Hospet

Sandur ★

BELLARY

BIJAPUR

⊠

★

⊠ Gadag

DHARWAD

⊙

○

BELGAUM

○

□

UTTAR
KANNAD

▶

GOA

ARABIAN SEA

MAHE
(PONDICHERRY)

KERALA

K A R N A T A K A

SHIMOGA

■

⊠

□

CHIKMAGALUR

Kemmanagundi ★

□

⊠

DAKSHIN
KANNAD

★

★

CHITRADURGA

★

▲

■ Hosdurga

⊕

▲ TUMKUR

TUMKUR

■

★

HASSAN

✳

▲

⊙

⊠

□

I N D I A

BANGALORE ◉

● ● ◉ Kolar Gold Field
▲

KOLAR

TAMIL NADU

MANDYA

□

MYSORE

◆

● ✳ ○ Nanjangud

■ ◉

KODAGU

1 cm = 37 kms   Scale 1 : 3 700 000

69

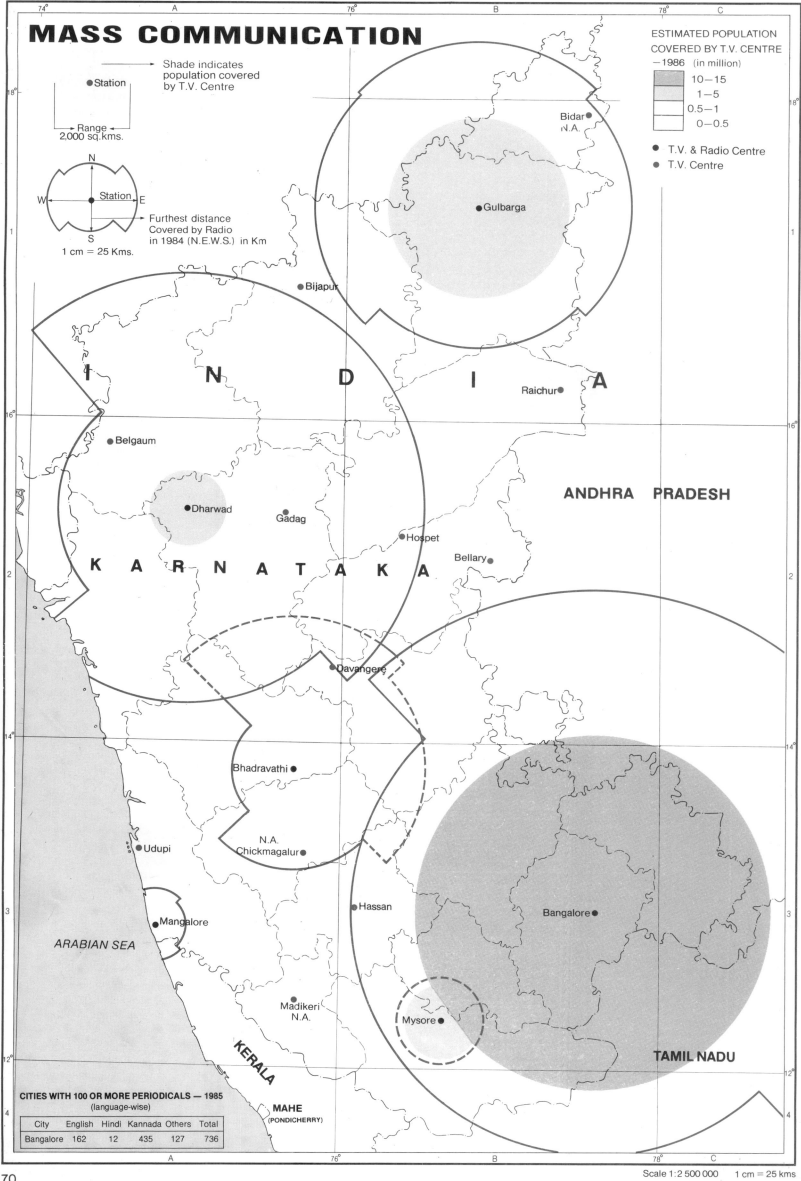

MASS COMMUNICATION

Station

→ Shade indicates population covered by T.V. Centre

← Range
2,000 sq.kms.

Station

→ Furthest distance Covered by Radio in 1984 (N.E.W.S.) in Km

1 cm = 25 Kms.

ESTIMATED POPULATION
COVERED BY T.V. CENTRE
—1986 (in million)

10—15
1—5
0.5—1
0—0.5

● T.V. & Radio Centre
● T.V. Centre

●Bidar
N.A.

●Gulbarga

●Bijapur

I N D I A

Raichur●

●Belgaum

●Dharwad

●Gadag

●Hospet

Bellary●

ANDHRA PRADESH

K A R N A T A K A

●Davangere

Bhadravathi●

●Udupi

N.A.
Chickmagalur●

●Hassan

Bangalore●

●Mangalore

ARABIAN SEA

Madikeri
N.A.

●Mysore

KERALA

TAMIL NADU

MAHE
(PONDICHERRY)

CITIES WITH 100 OR MORE PERIODICALS — 1985
(language-wise)

City	English	Hindi	Kannada	Others	Total
Bangalore	162	12	435	127	736

Scale 1:2 500 000   1 cm = 25 kms.

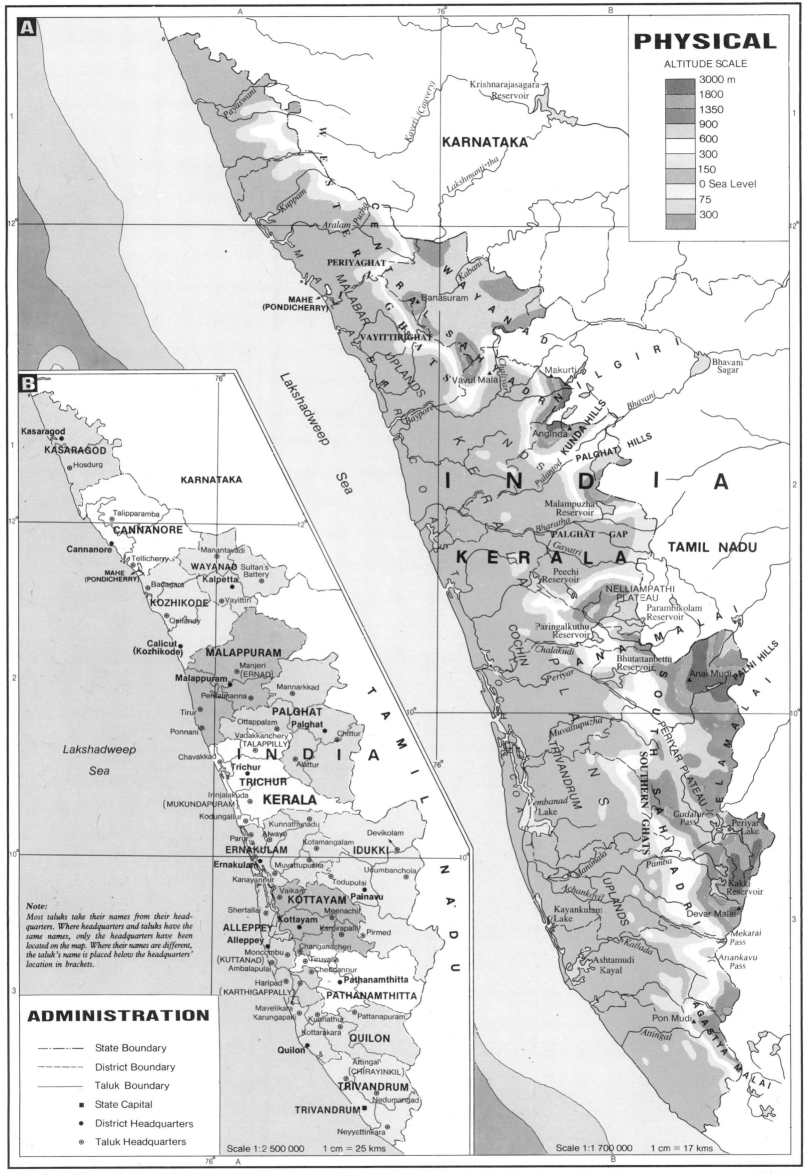

KERALA

PHYSICAL

ALTITUDE SCALE

	3000 m
	1800
	1350
	900
	600
	300
	150
	0 Sea Level
	75
	300

A

KARNATAKA

Krishnarajasagara
Reservoir

Kaveri (Cauvery)

Lakshmanti-tha

Payaswani

Kuppam

Aralam Puzha

PERIYAGHAT

MAHE
(PONDICHERRY)

VAYITTIRIGHAT

Banasuram

WAYANAD

Makurti

Vavul Mala

Kabani

Chaliyar

Bayapore

Anginda

KUNDA HILLS

Bhavani

NILGIRI

Bhavani
Sagar

PALGHAT HILLS

INDIA

Malampuzha
Reservoir

Pulamtod

PALGHAT GAP

Bharatha

Gayatri

KERALA

Peechi
Reservoir

TAMIL NADU

NELLIAMPATHI
PLATEAU

Parambikolam
Reservoir

Paringalkuthu
Reservoir

Chalakudi

COCHIN

Periyar

Bhutattanbettu
Reservoir

Anai Mudi

PALNI HILLS

Muvattupuzha

TRIVANDRUM

PERIYAR PLATEAU

SOUTHERN GHATS

Pamba

Manimala

Vembanad
Lake

UPLANDS

Achankovil

Kailada

Kakki
Reservoir

Kayankulam
Lake

Devar Malai

Mekarai
Pass

Ashtamudi
Kayal

Gudalur
Pass

Periyar
Lake

Ariankavu
Pass

Pon Mudi

AGASTYA MALAI

Attingal

Scale 1:1 700 000   1 cm = 17 kms

B

Lakshadweep
Sea

Kasaragod
KASARAGOD
○ Hosdurg

KARNATAKA

○ Talipparamba

CANNANORE

● Cannanore
○ Tellicherry
○ Manantavadi

MAHE
(PONDICHERRY)

WAYANAD
○ Sultan's
Battery
★ Kalpetta
○ Badagara

KOZHIKODE
○ Vayittiri
○ Quilandy

Calicut
(Kozhikode)

MALAPPURAM
Manjeri
(ERNAD)

● Malappuram
○ Perintalmanna
○ Mannarkkad

○ Tirur

PALGHAT
○ Ottappalam
● Palghat
○ Chittur

○ Ponnani
Vadakkanchery
(TALAPPILLY)

Lakshadweep
Sea

○ Chavakkad

I N D I A

T A M I L   N A D U

★ Trichur
○ Alattur

TRICHUR
(MUKUNDAPURAM)
Irinjalakuda

KERALA

○ Kodungallur

○ Parur
○ Kunnathunadu
○ Alwaye
○ Kotamangalam

Devikolam
IDUKKI

● Ernakulam
ERNAKULAM
○ Kanayannur
○ Muvattupuzha

Udumbanchola

○ Todupulai

★ Painavu

○ Vaikam
KOTTAYAM
○ Meenachil

○ Shertallai

● Kottayam
○ Kanjirapalli

ALLEPPEY
Alleppey
○ Moncombu
(KUTTANAD)
○ Ambalapulai

○ Pirmed

○ Changanacheri
○ Tiruvalla

○ Haripad
(KARTHIGAPPALLY)

● Pathanamthitta
PATHANAMTHITTA

○ Mavelikara
○ Chendannur

○ Karungapali
○ Kunnathur
○ Pattanapuram

○ Kottarakara
QUILON

● Quilon

Attingal
(CHIRAYINKIL)

TRIVANDRUM

■ TRIVANDRUM

○ Nedumangad

○ Neyyattinkara

Note:
Most taluks take their names from their head-
quarters. Where headquarters and taluks have the
same names, only the headquarters have been
located on the map. Where their names are different,
the taluk's name is placed below the headquarters'
location in brackets.

ADMINISTRATION

—·—·—	State Boundary
— — —	District Boundary
———	Taluk Boundary
■	State Capital
●	District Headquarters
○	Taluk Headquarters

Scale 1:2 500 000   1 cm = 25 kms

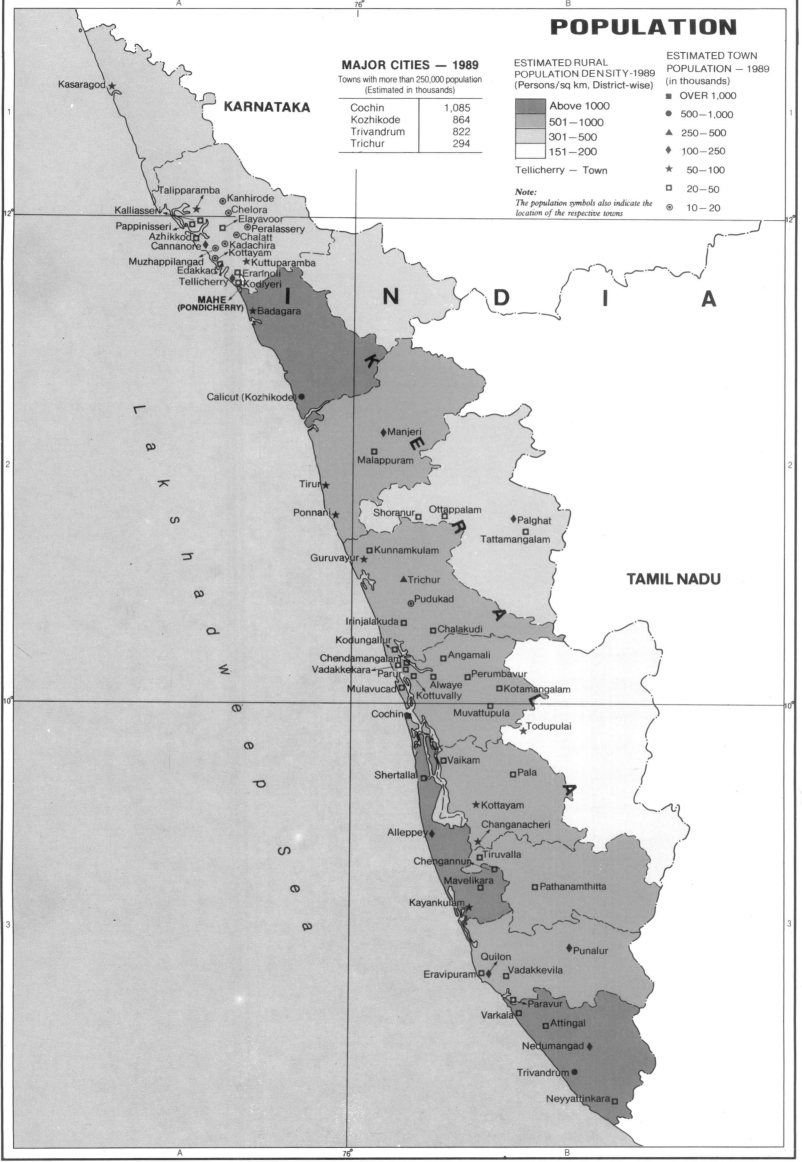

POPULATION

MAJOR CITIES — 1989
Towns with more than 250,000 population
(Estimated in thousands)

Cochin	1,085
Kozhikode	864
Trivandrum	822
Trichur	294

ESTIMATED RURAL POPULATION DENSITY-1989
(Persons/sq km, District-wise)

- Above 1000
- 501–1000
- 301–500
- 151–200

Tellicherry — Town

Note:
The population symbols also indicate the location of the respective towns

ESTIMATED TOWN POPULATION — 1989
(in thousands)

- ■ OVER 1,000
- ● 500–1,000
- ▲ 250–500
- ♦ 100–250
- ★ 50–100
- □ 20–50
- ◎ 10–20

KARNATAKA

Kasaragod

Talipparamba
Kalliasseri
Kanhirode
Chelora
Elayavoor
Pappinisseri
Peralassery
Azhikkod
Chalatt
Cannanore
Kadachira
Kottayam
Muzhappilangad
Kuttuparamba
Edakkad
Erarinoli
Tellicherry
Kodiyeri
MAHE
(PONDICHERRY)
Badagara

I N D I A

Calicut (Kozhikode)

Manjeri
Malappuram

Tirur

Ponnani
Shoranur
Ottappalam
Palghat
Tattamangalam

Guruvayur
Kunnamkulam

TAMIL NADU

Trichur
Pudukad

Irinjalakuda
Chalakudi

Kodungallur
Angamali
Chendamangalam
Vadakkekara
Perumbavur
Paru
Mulavucad
Alwaye
Kotamangalam
Kottuvally
Cochin
Muvattupula

Todupulai

Vaikam
Pala
Shertallai

Kottayam

Changanacheri
Alleppey
Tiruvalla
Chengannur
Mavelikara
Pathanamthitta
Kayankulam

Punalur

Quilon
Eravipuram
Vadakkevila

Paravur
Varkala
Attingal
Nedumangad
Trivandrum
Neyyattinkara

L a k s h a d w e e p   S e a

Scale 1:1 700 000   1 cm = 17 kms

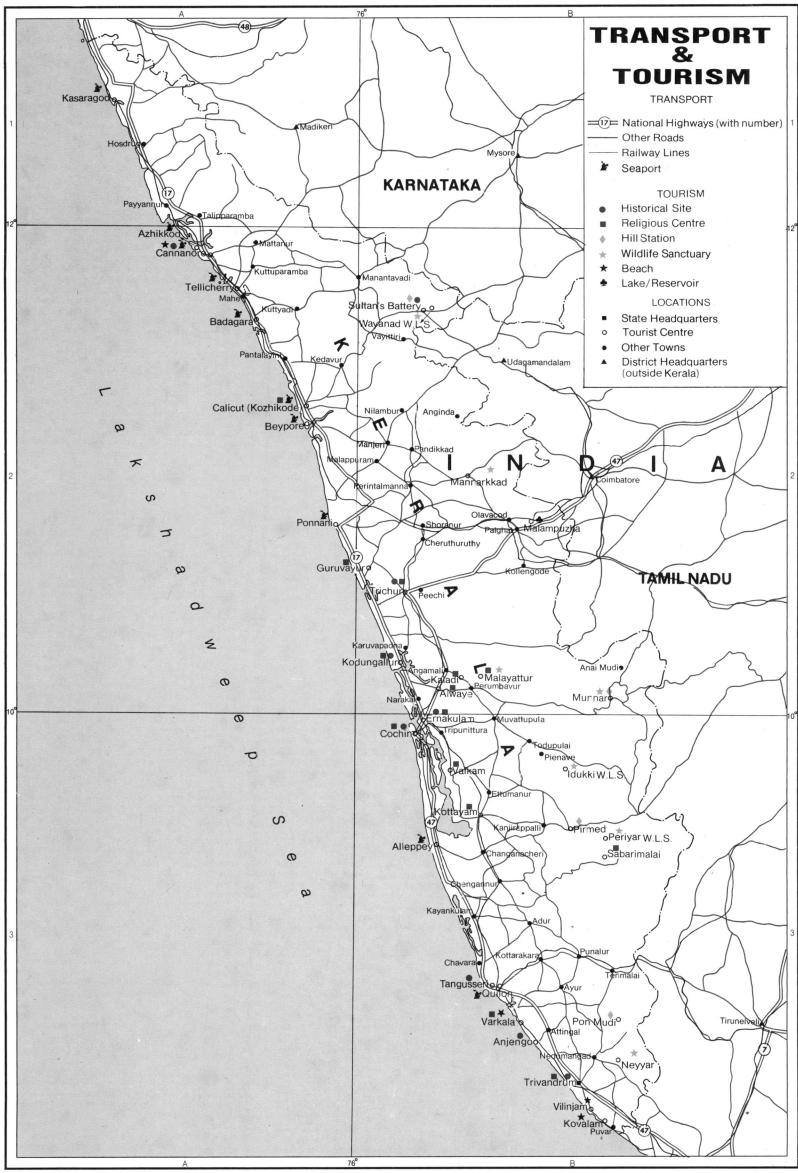

TRANSPORT & TOURISM

TRANSPORT

- ═⟨17⟩═ National Highways (with number)
- ── Other Roads
- ── Railway Lines
- ⚓ Seaport

TOURISM

- ● Historical Site
- ■ Religious Centre
- ◆ Hill Station
- ★ Wildlife Sanctuary
- ★ Beach
- ♣ Lake/Reservoir

LOCATIONS

- ■ State Headquarters
- ○ Tourist Centre
- ● Other Towns
- ▲ District Headquarters (outside Kerala)

KARNATAKA

TAMIL NADU

K E R A L A

INDIA

Lakshadweep Sea

Kasaragod
Hosdrug
Madikeri
Mysore
Payyannur
Talipparamba
Azhikkode
Cannanore
Mattanur
Tellicherry
Kuttuparamba
Mahe
Kuttyadi
Manantavadi
Badagara
Sultan's Battery
Wayanad W.L.S
Vayittiri
Pantalayini
Kedavur
Udagamandalam
Calicut (Kozhikode)
Beypore
Nilambur
Anginda
Manjeri
Pandikkad
Malappuram
Perintalmanna
Mannarkkad
Coimbatore
Ponnani
Olavacod
Shoranur
Palghat
Malampuzha
Cheruthuruthy
Guruvayur
Kollengode
Trichur
Peechi
Karuvapadna
Kodungalluro
Angamali
Anai Mudi
Kaladi
Malayattur
Alwaye
Perumbavur
Munnar
Narakal
Ernakulam
Muvattupula
Cochin
Tripunittura
Todupulai
Pienave
Vaikam
Idukki W.L.S
Ettumanur
Kottayam
Kanjirappalli
Pirmed
Alleppey
Changanacheri
Periyar W.L.S
Sabarimalai
Chengannur
Kayankulam
Adur
Chavara
Kottarakara
Punalur
Tangussero
Quilon
Ayur
Tenmalai
Varkala
Pon Mudi
Tirunelveli
Anjengoo
Attingal
Nedumangad
Neyyar
Trivandrum
Vilinjam
Kovalam
Puvar

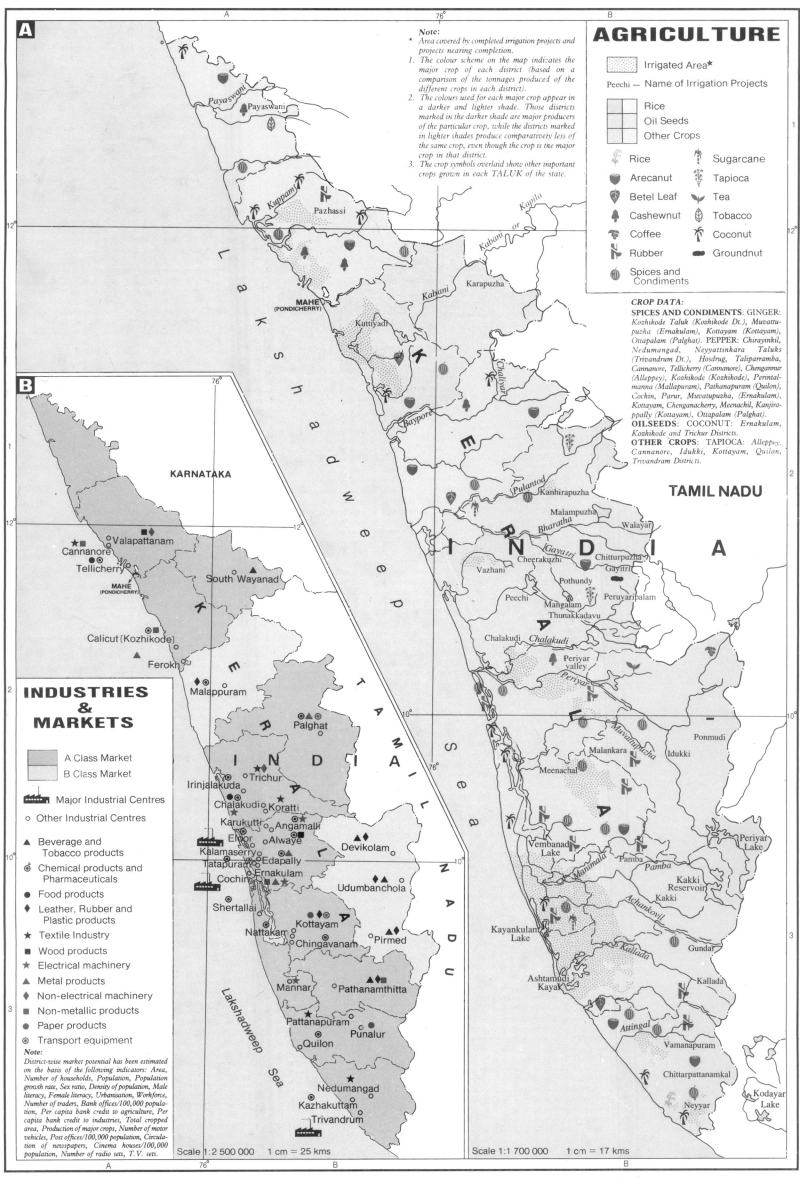

A

Note:
* Area covered by completed irrigation projects and projects nearing completion.
1. The colour scheme on the map indicates the major crop of each district (based on a comparison of the tonnages produced of the different crops in each district).
2. The colours used for each major crop appear in a darker and lighter shade. Those districts marked in the darker shade are major producers of the particular crop, while the districts marked in lighter shades produce comparatively less of the same crop, even though the crop is the major crop in that district.
3. The crop symbols overlaid show other important crops grown in each TALUK of the state.

AGRICULTURE

Irrigated Area★

Peechi — Name of Irrigation Projects

Rice
Oil Seeds
Other Crops

Rice	Sugarcane
Arecanut	Tapioca
Betel Leaf	Tea
Cashewnut	Tobacco
Coffee	Coconut
Rubber	Groundnut
Spices and Condiments	

CROP DATA:
SPICES AND CONDIMENTS: GINGER: *Kozhikode Taluk (Kozhikode Dt.), Muvattupuzha (Ernakulam), Kottayam (Kottayam), Ottapalam (Palghat).* PEPPER: *Chirayinkil, Nedumangad, Neyyattinkara Taluks (Trivandrum Dt.), Hosdrug, Taliparramba, Cannanore, Tellicherry (Cannanore), Chengannur (Alleppey), Kozhikode (Kozhikode), Perintalmanna (Mallapuram), Pathanapuram (Quilon), Cochin, Parur, Muvatupuzha, (Ernakulam), Kottayam, Chenganacherry, Meenachil, Kanjirappally (Kottayam), Ottapalam (Palghat).*
OILSEEDS: COCONUT: *Ernakulam, Kozhikode and Trichur Districts.*
OTHER CROPS: TAPIOCA: *Alleppey, Cannanore, Idukki, Kottayam, Quilon, Trivandram Districts.*

KARNATAKA

TAMIL NADU

B

INDUSTRIES & MARKETS

A Class Market
B Class Market

Major Industrial Centres
o Other Industrial Centres

▲ Beverage and Tobacco products
◉ Chemical products and Pharmaceuticals
● Food products
◆ Leather, Rubber and Plastic products
★ Textile Industry
■ Wood products
✦ Electrical machinery
▲ Metal products
◆ Non-electrical machinery
■ Non-metallic products
● Paper products
◎ Transport equipment

Note:
District-wise market potential has been estimated on the basis of the following indicators: Area, Number of households, Population, Population growth rate, Sex ratio, Density of population, Male literacy, Female literacy, Urbanisation, Workforce, Number of traders, Bank offices/100,000 population, Per capita bank credit to agriculture, Per capita bank credit to industries, Total cropped area, Production of major crops, Number of motor vehicles, Post offices/100,000 population, Circulation of newspapers, Cinema houses/100,000 population, Number of radio sets, T.V. sets.

Scale 1:2 500 000   1 cm = 25 kms

Scale 1:1 700 000   1 cm = 17 kms

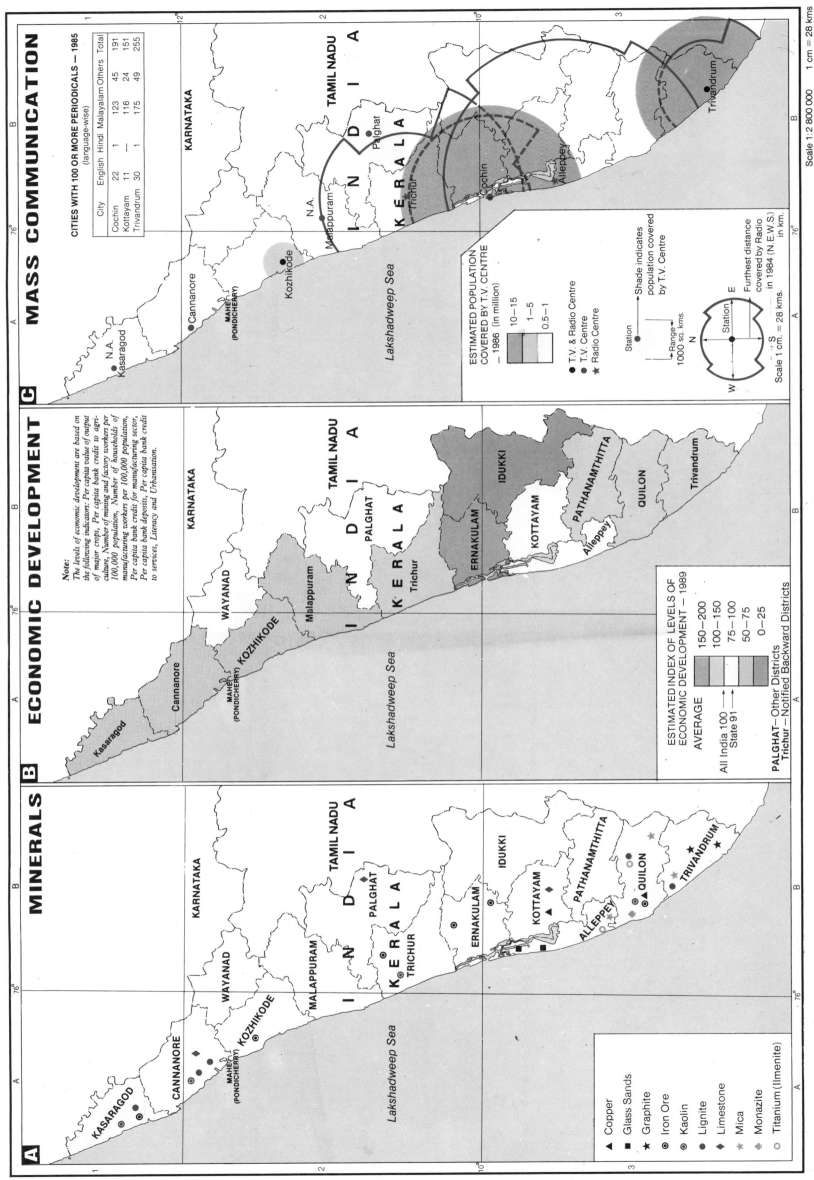

MASS COMMUNICATION

C

CITIES WITH 100 OR MORE PERIODICALS — 1985
(language-wise)

City	English	Hindi	Malayalam	Others	Total
Cochin	22	1	123	45	191
Kottayam	11	—	116	24	151
Trivandrum	30	1	175	49	255

ESTIMATED POPULATION
COVERED BY T.V. CENTRE
— 1986 (in million)

- 10–15
- 1–5
- 0.5–1

- T.V. & Radio Centre
- T.V. Centre
- ★ Radio Centre

Station → Shade indicates
population covered
by T.V. Centre

← Range
1000 sq. kms.

Furthest distance
covered by Radio
in 1984 (N.E.W.S.)
in km.

Scale 1 cm. = 28 kms.

Scale 1:2 800 000   1 cm = 28 kms

ECONOMIC DEVELOPMENT

B

Note:
The levels of economic development are based on
the following indicators: Per capita value of output
of major crops, Per capita bank credit to agriculture, Number of mining and factory workers per
100,000 population, Number of households of
manufacturing workers per 100,000 population,
Per capita bank credit for manufacturing sector,
Per capita bank deposits, Per capita bank credit
to services, Literacy and Urbanisation.

ESTIMATED INDEX OF LEVELS OF
ECONOMIC DEVELOPMENT — 1989

- 150–200
- 100–150
- 75–100
- 50–75
- 0–25

AVERAGE

All India 100
State 91

PALGHAT—Other Districts
Trichur—Notified Backward Districts

MINERALS

A

- ▲ Copper
- ■ Glass Sands
- ★ Graphite
- ◎ Iron Ore
- ◉ Kaolin
- ● Lignite
- ◆ Limestone
- ★ Mica
- ◄ Monazite
- ○ Titanium (Ilmenite)

MADHYA PRADESH

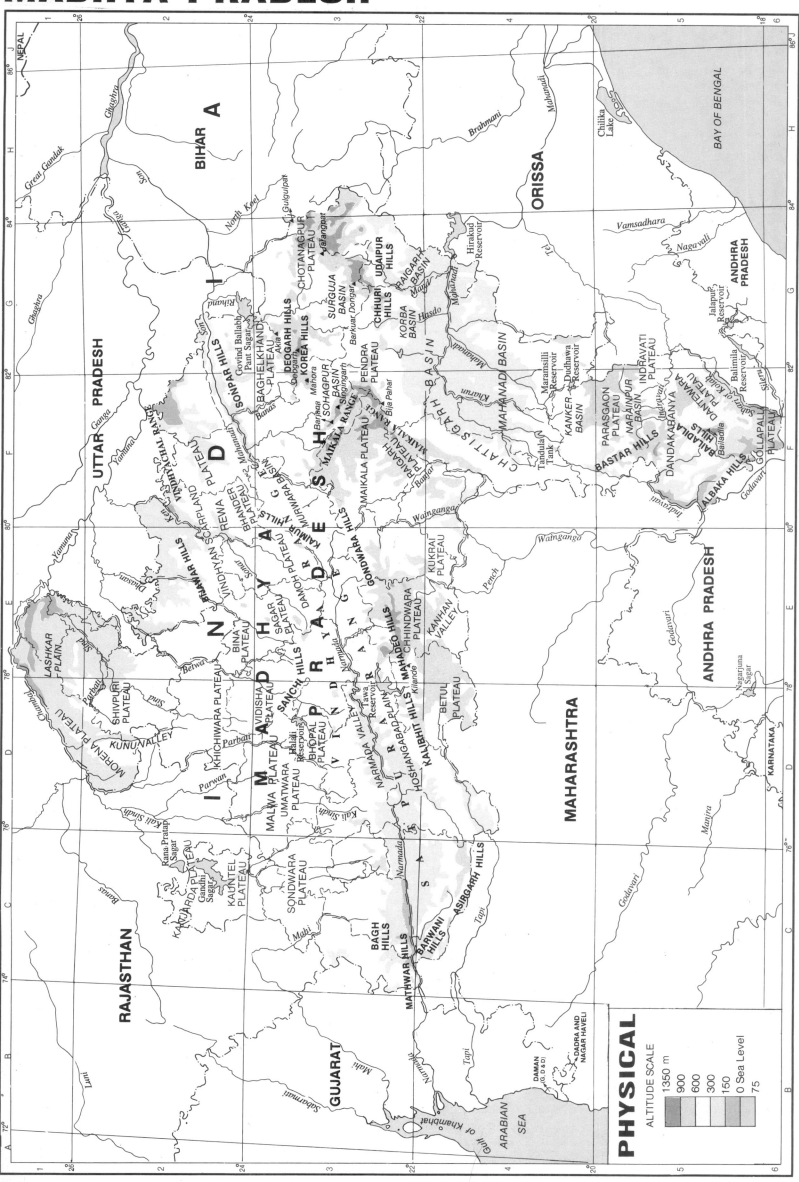

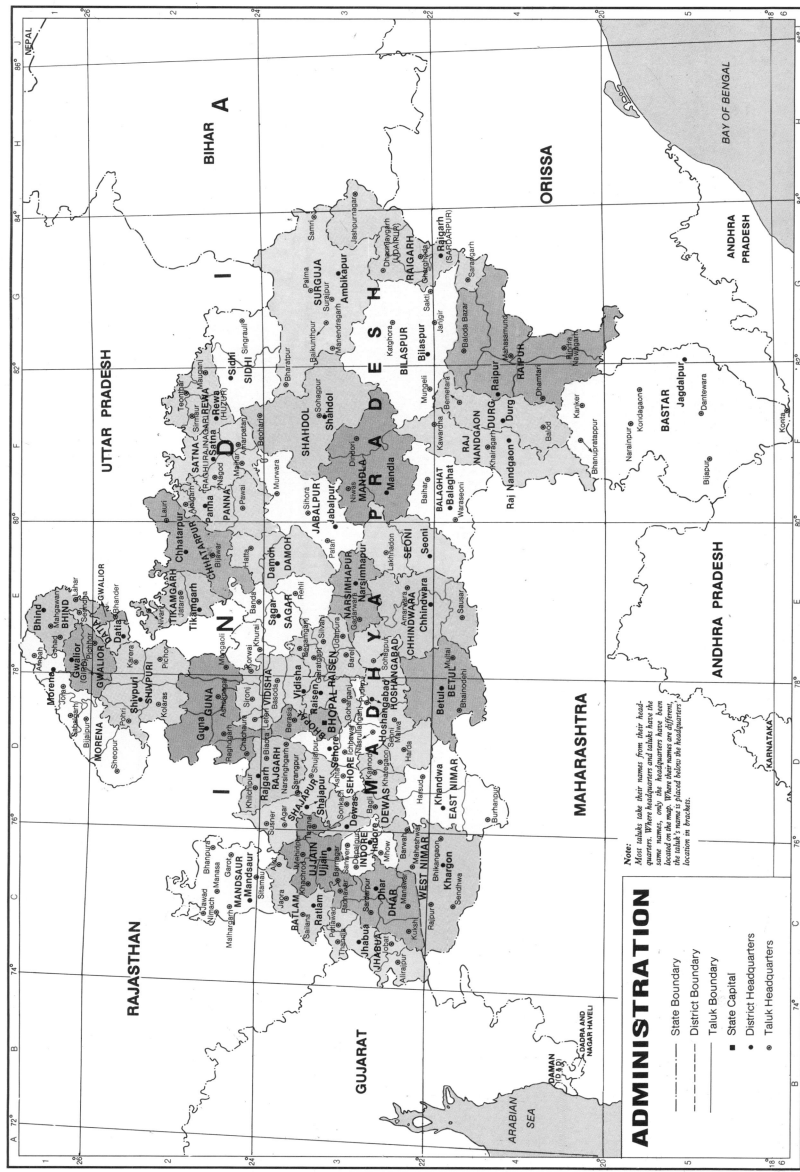

ADMINISTRATION

	State Boundary
	District Boundary
	Taluk Boundary
■	State Capital
●	District Headquarters
◉	Taluk Headquarters

Note:
Most taluks take their names from their head-quarters. Where headquarters and taluks have the same names, only the headquarters have been located on the map. Where their names are different, the taluk's name is placed below the headquarters location in brackets.

Scale 1:4 600 000   1 cm = 46 kms

77

NEPAL

BIHAR

ORISSA

BAY OF BENGAL

UTTAR PRADESH

RAJASTHAN

GUJARAT

MAHARASHTRA

ANDHRA PRADESH

KARNATAKA

ARABIAN SEA

DAMAN (D. & D.)

DADRA AND NAGAR HAVELI

M A D H Y A   P R A D E S H

I N D I A

SURGUJA
Surajpur
Baikunthpur
Palma
Samri
Manendragarh
AMBIKAPUR
Ambikapur
Jashpurnagar

RAIGARH
Dharamjaygarh (UDAIPUR)
Raigarh (SARDARPUR)
Gharghoda
Saraipgarh

Katghora
Sohagpur
Bharatpur

BILASPUR
Bilaspur
Janjgir
Sakti
Mungeli
Baloda Bazar

RAIPUR
Raipur
Mahasamund
Bemetara
Baloda
Kawardha

RAJ NANDGAON
Raj Nandgaon
Khairagarh
DURG
Durg
Dhamtari
Rajim
Nawagarh

BALAGHAT
Balaghat
Waraseoni
Baihar
Lanji

BASTAR
Jagdalpur
Kondagaon
Narainpur
Dantewara
Bhanupratappur
Karker
Bijapur
Konta

SIDHI
Sidhi (HUZUR)
Singrauli
Teonthar
Sirmaur
Mauganj

REWA
Rewa (HUZUR)
RAJNAGAR
TARHRA
Nagod
Satna
SATNA
Amarpatan
Unchehra

PANNA
Panna
Ajaigarh
Pawai
Beohari

SHAHDOL
Shahdol
Sohagpur
Dindori

MANDLA
Mandla
Niwas

JABALPUR
Jabalpur
Sihora
Murwara
Patan

DAMOH
Damoh
Hatta
Rehli

SAGAR
Sagar
Bandri
Rehli
Khurai

CHHATARPUR
Chhatarpur
Bijawar
Lauri

TIKAMGARH
Tikamgarh
Jatara
Nivari

NARSIMHAPUR
Narsimhapur
Gadarwara

SEONI
Seoni
Lakhnadon

CHHINDWARA
Chhindwara
Amarwara
Gadarwara
Sausar

BHIND
Bhind
Gohad
Mehgaon
Lahar
Mahgawan

GWALIOR
Gwalior (GIRD)
Ghander
Sabalgarh

DATIA
Datia
Seondha
Bhander

MORENA
Morena
Ambah
Joura
Sheopur
Bijaipur
Pohri

SHIVPURI
Shivpuri
Karera
Pichhore
Kolaras

GUNA
Guna
Raghogarh
Ashoknagar
Chachaura
Mungaoli

VIDISHA
Vidisha
Basoda
Sironj
Lateri

RAISEN
Raisen
Begamganj
Gairatganj
Silvani
Goharganj
Barel

BHOPAL
Bhopal
Berasia

SEHORE
Sehore
Ashta
Ichhawar
Nasrullaganj
Budhni

HOSHANGABAD
Hoshangabad
Sohagpur
Seoni Malwa
Pipariya
Harda
Timarni

BETUL
Betul
Multai
Bhainsdehi
Bhainodehi

RAJGARH
Rajgarh
Narsinghgarh
Biaora
Khilchipur

SHAJAPUR
Shajapur
Shujalpur
Agar
Susner

DEWAS
Dewas
Sonkatch
Bagli
Kannod
Khategaon

INDORE
Indore
Mhow
Depalpur
Sanwer

UJJAIN
Ujjain
Khachrod
Mahidpur
Tarana
Barnagar
Ghatiya

MANDSAUR
Mandsaur
Neemuch
Manasa
Jawad
Garoth
Bhanpura
Sitamau

RATLAM
Ratlam
Jaora
Alot
Saila

DHAR
Dhar
Sardarpur
Badnawar
Manawar
Kukshi
Dharampuri

JHABUA
Jhabua
Alirajpur
Thandla
Petlawad
Jobat

WEST NIMAR
Kargone
Sendhwa
Maheshwar
Barwaha
Rajpur
Bhikangaon

EAST NIMAR
Khandwa
Harsud
Burhanpur
Barwah

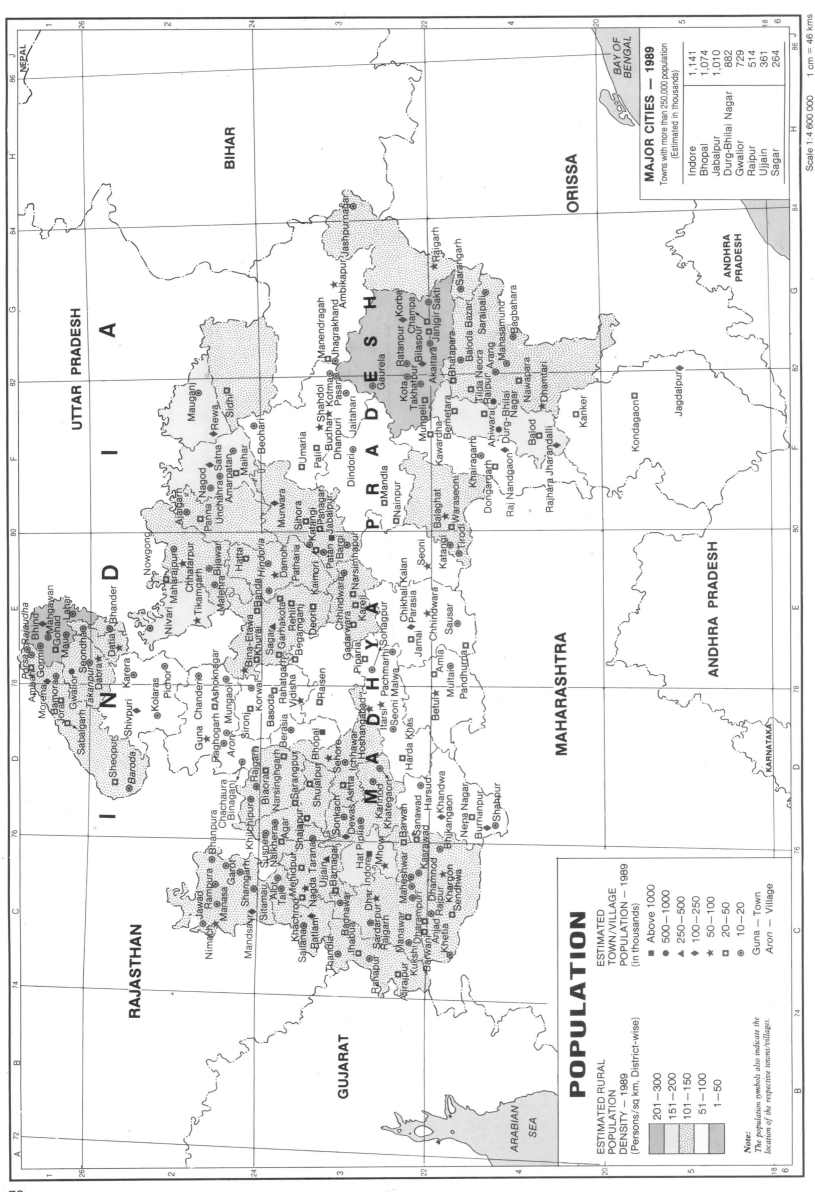

POPULATION

ESTIMATED RURAL
POPULATION
DENSITY – 1989
(Persons/sq km, District-wise)

201–300	
151–200	
101–150	
51–100	
1–50	

ESTIMATED
TOWN/VILLAGE
POPULATION – 1989
(in thousands)

■ Above 1000
● 500–1000
▲ 250–500
◆ 100–250
★ 50–100
◻ 20–50
⊙ 10–20

Guna — Town
Aron — Village

Note:
The population symbols also indicate the
location of the respective towns/villages.

MAJOR CITIES — 1989
Towns with more than 250,000 population
(Estimated in thousands)

Indore	1,141
Bhopal	1,074
Jabalpur	1,010
Durg-Bhilai Nagar	882
Gwalior	729
Raipur	514
Ujjain	361
Sagar	264

Scale 1:4 600 000   1 cm = 46 kms

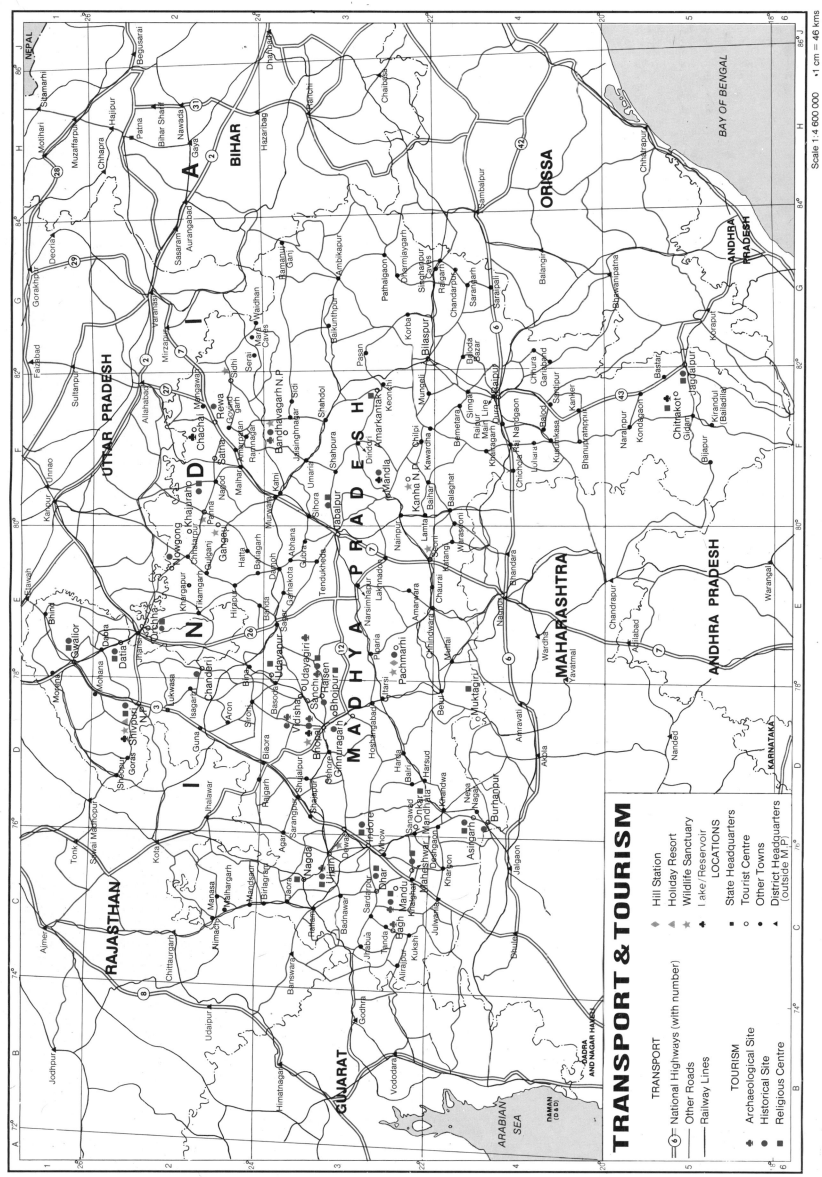

TRANSPORT & TOURISM

TRANSPORT

⬡6⬡ National Highways (with number)
National Highways (with number)
Other Roads
Railway Lines

TOURISM

♣ Archaeological Site
● Historical Site
■ Religious Centre

◆ Hill Station
▲ Holiday Resort
★ Wildlife Sanctuary
♣ Lake/Reservoir

LOCATIONS

■ State Headquarters
○ Tourist Centre
● Other Towns
▲ District Headquarters (outside M.P)

Scale 1:4 600 000   1 cm = 46 kms

79

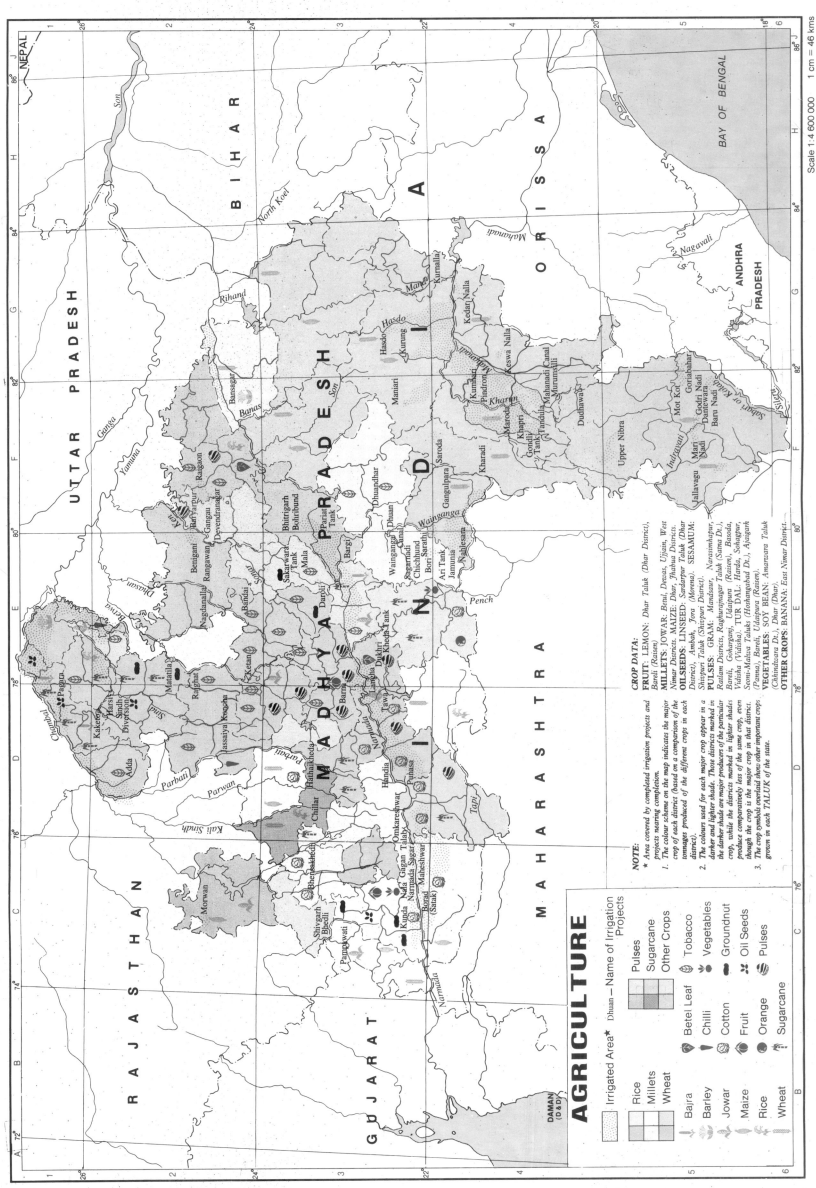

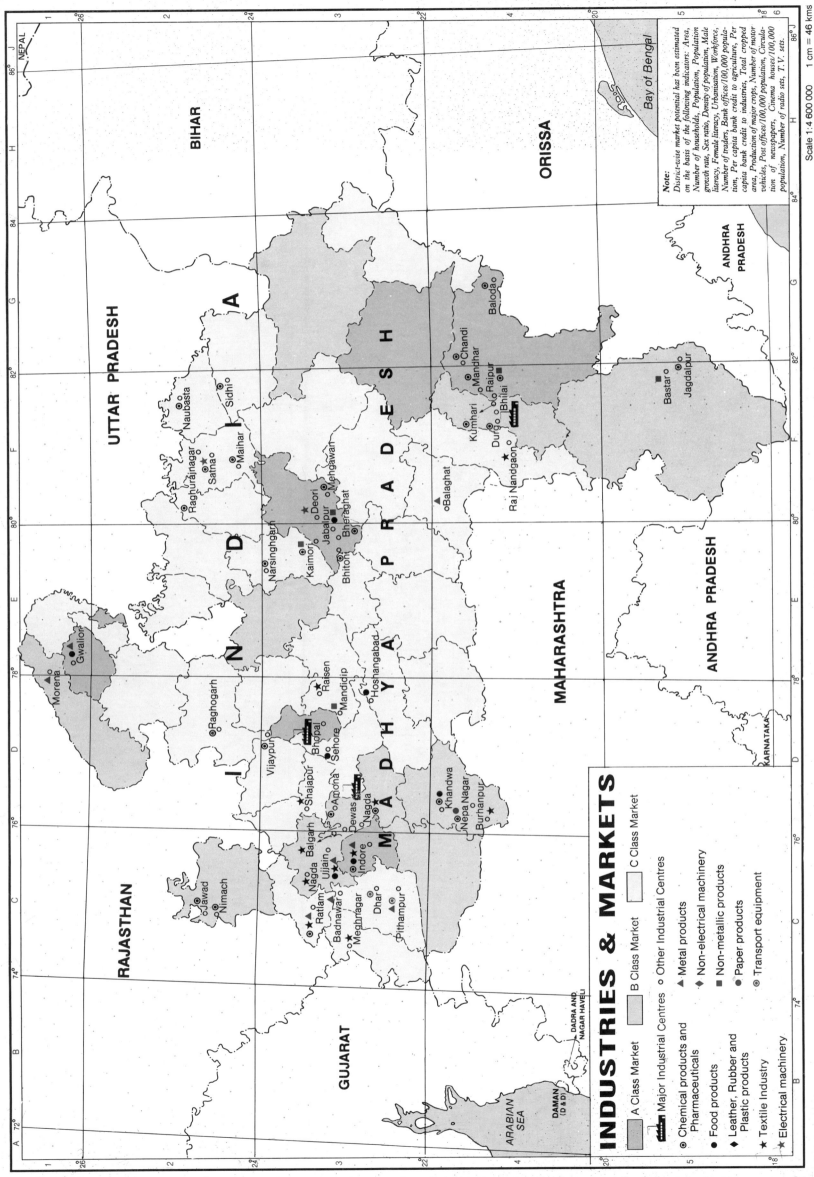

INDUSTRIES & MARKETS

Scale 1:4 600 000   1 cm = 46 kms

Legend:

A Class Market — Major Industrial Centres   ▲ Metal products
B Class Market — ○ Other Industrial Centres   ◆ Non-electrical machinery
C Class Market

⊙ Chemical products and Pharmaceuticals
● Food products
◆ Leather, Rubber and Plastic products
★ Textile Industry
☆ Electrical machinery
■ Non-metallic products
● Paper products
⊙ Transport equipment

States:
NEPAL
BIHAR
ORISSA
UTTAR PRADESH
RAJASTHAN
GUJARAT
MADHYA PRADESH
MAHARASHTRA
ANDHRA PRADESH
KARNATAKA
ANDHRA PRADESH
Bay of Bengal
ARABIAN SEA
DAMAN (D & D)
DADRA AND NAGAR HAVELI

Place names:
Morena, Gwalior, Raghogarh, Vijaypur, Jawad, Nimach, Ratlam, Badnawar, Nagda, Balgarh, Ujjain, Indore, Megbrag, Dhar, Pithampur, Shajapur, Apona, Dewas, Nagda, Sehore, Bhopal, Mandidip, Ralsen, Hoshangabad, Khandwa, Nepa Nagar, Burhanpur, Raghurajnagar, Satna, Maihar, Naubasta, Sidhi, Narsinghgarh, Deori, Mehgawan, Jabalpur, Bheraghat, Bhito, Kaimori, Balaghat, Raj Nandgaon, Kumhari, Durg, Bhilai, Raipur, Mandhar, Chandi, Balod, Bastar, Jagdalpur

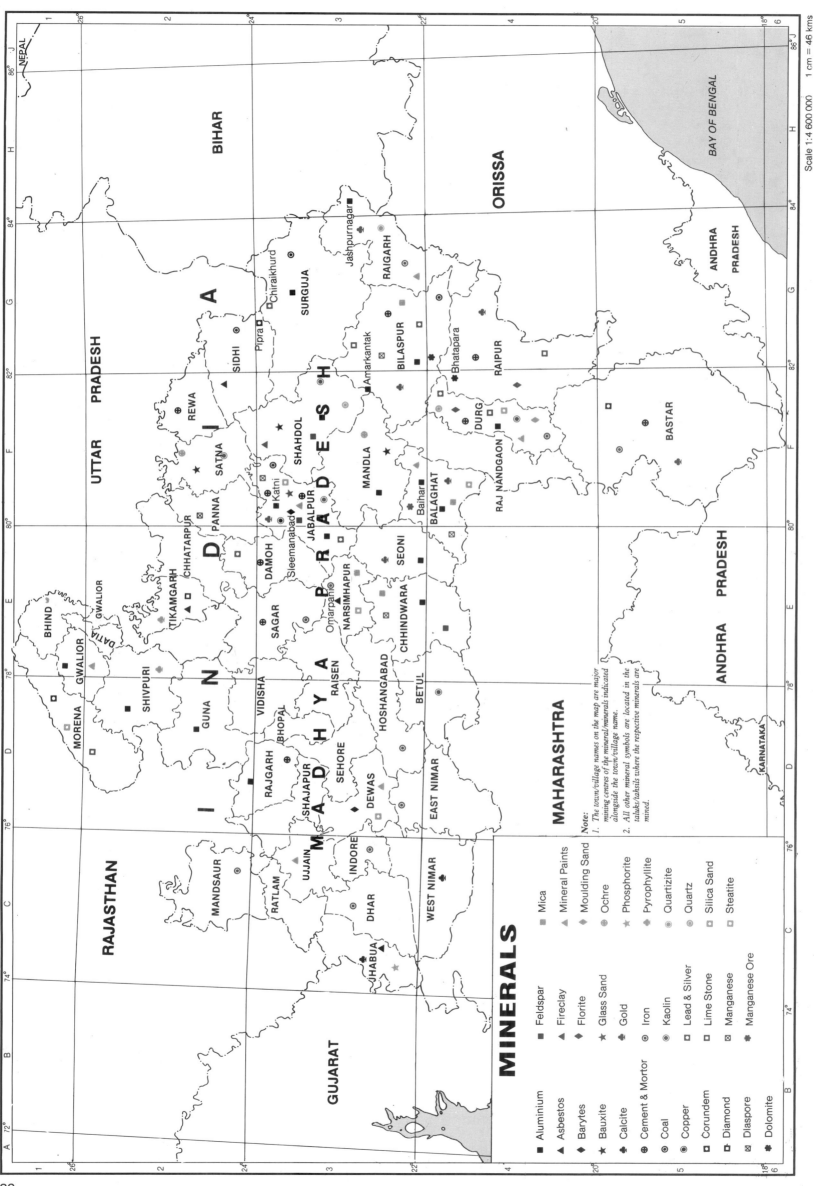

MINERALS

Scale 1 : 4 600 000   1 cm = 46 kms

Note:
1. The town/village names on the map are major mining centres of the mineral/minerals indicated alongside the town/village name.
2. All other mineral symbols are located in the taluks/tahsils where the respective minerals are mined.

Mineral		Mineral	
■	Aluminium	■	Feldspar
▲	Asbestos	▲	Fireclay
◆	Barytes	◆	Florite
★	Bauxite	★	Glass Sand
✦	Calcite	✦	Gold
◎	Cement & Mortar	◎	Iron
◉	Coal	◉	Kaolin
◉	Copper	□	Lead & Silver
□	Corundem	□	Lime Stone
◇	Diamond	⊠	Manganese
⊠	Diaspore	✤	Manganese Ore
✤	Dolomite		

Mineral		Mineral	
■	Mica	◎	Quartzite
▲	Mineral Paints	◎	Quartz
◆	Moulding Sand	□	Silica Sand
◎	Ochre	□	Steatite
✦	Phosphorite		
◆	Pyrophyllite		

82

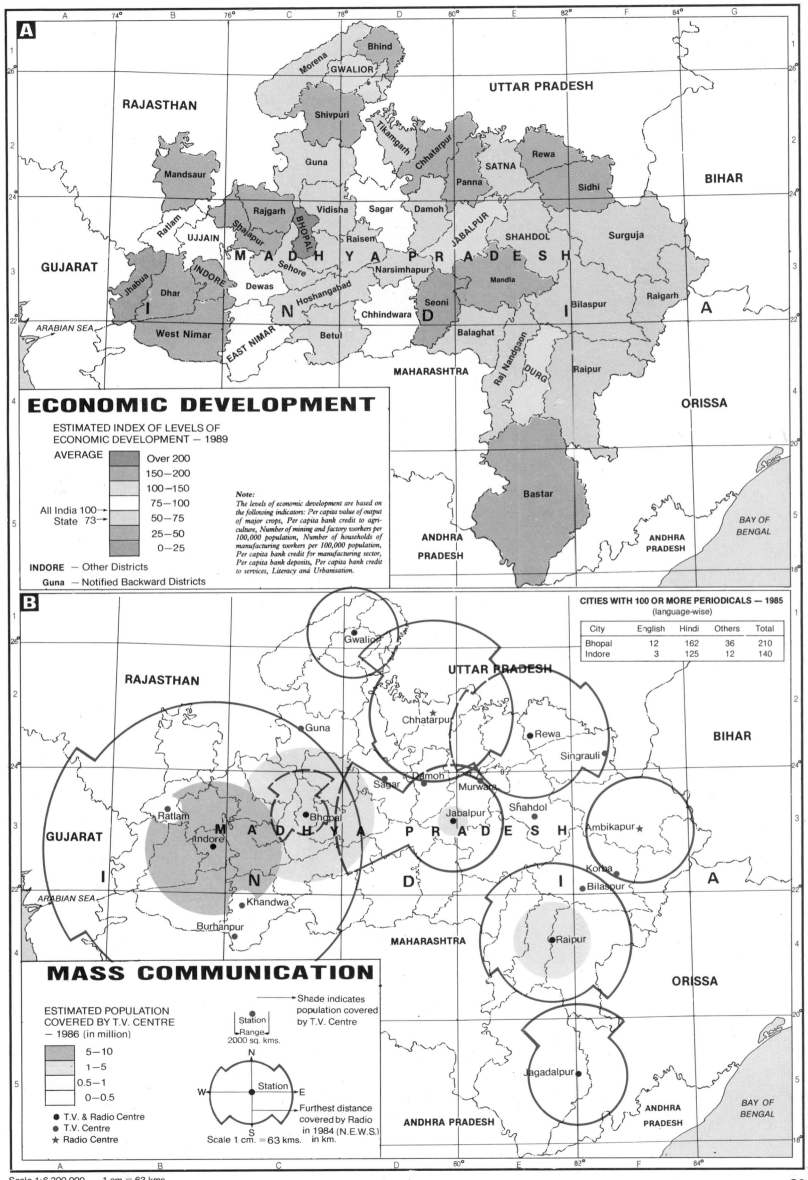

A

ECONOMIC DEVELOPMENT

ESTIMATED INDEX OF LEVELS OF
ECONOMIC DEVELOPMENT — 1989

AVERAGE

	Over 200
	150—200
	100—150
All India 100 →	75—100
State 73 →	50—75
	25—50
	0—25

INDORE — Other Districts

Guna — Notified Backward Districts

Note:
The levels of economic development are based on the following indicators: Per capita value of output of major crops, Per capita bank credit to agriculture, Number of mining and factory workers per 100,000 population, Number of households of manufacturing workers per 100,000 population, Per capita bank credit for manufacturing sector, Per capita bank deposits, Per capita bank credit to services, Literacy and Urbanisation.

B

MASS COMMUNICATION

CITIES WITH 100 OR MORE PERIODICALS — 1985
(language-wise)

City	English	Hindi	Others	Total
Bhopal	12	162	36	210
Indore	3	125	12	140

ESTIMATED POPULATION
COVERED BY T.V. CENTRE
— 1986 (in million)

	5—10
	1—5
	0.5—1
	0—0.5

● T.V. & Radio Centre
● T.V. Centre
★ Radio Centre

Station
Range
2000 sq. kms.

→ Shade indicates
population covered
by T.V. Centre

N
W — Station — E
S

→ Furthest distance
covered by Radio
in 1984 (N.E.W.S.)
in km.

Scale 1 cm. = 63 kms.

Scale 1:6 300 000   1 cm = 63 kms

83

MAHARASHTRA

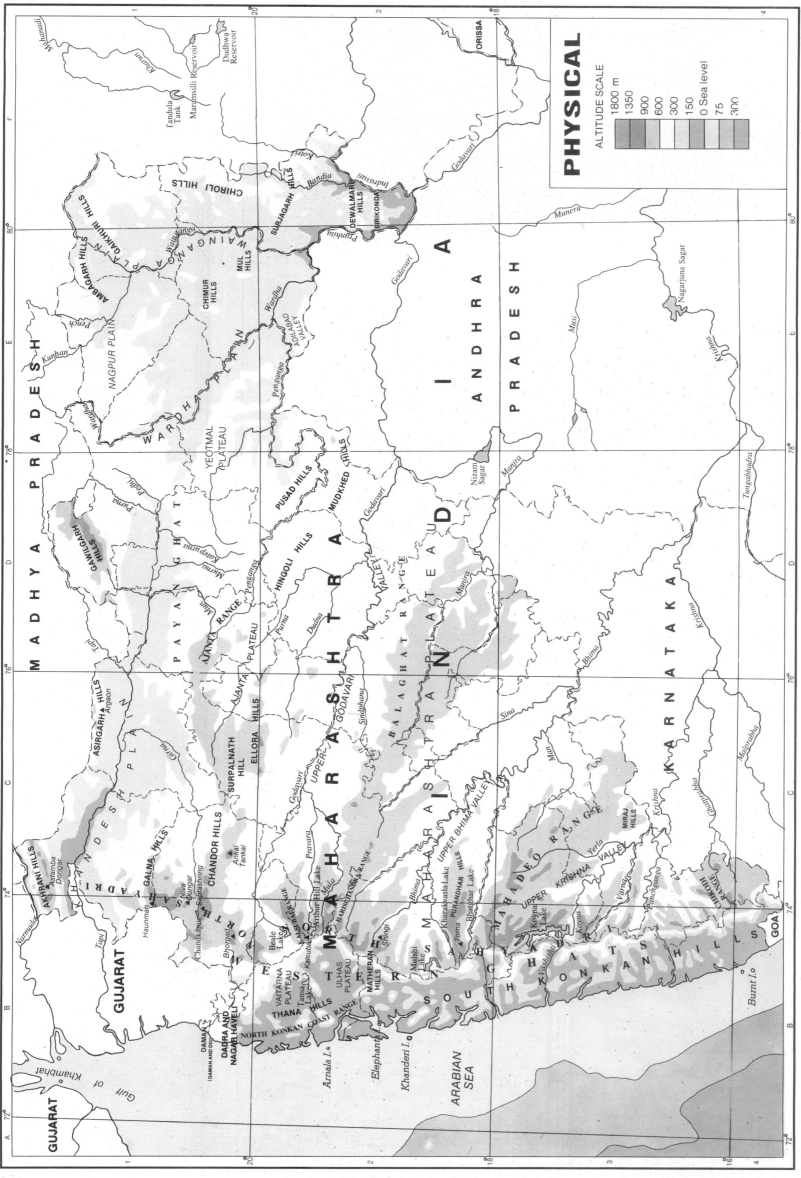

PHYSICAL

ALTITUDE SCALE

| 1800 m | 1350 | 900 | 600 | 300 | 150 | 0 Sea level | 75 | 300 |

GUJARAT

MADHYA PRADESH

ANDHRA PRADESH

KARNATAKA

GOA

ARABIAN SEA

ORISSA

INDIA

Gulf of Khambhat

CHIROLI HILLS

SURJAGARH HILLS

DEWALMARI HILLS

SIRIKONDA

GAIKHURI HILLS

AMBAGARH HILLS

PLATEAU GAIKHURI HILLS

NAGPUR PLAIN

MUL HILLS

CHIMUR HILLS

ADILABAD VALLEY

WARDHA PLAIN

YEOTMAL PLATEAU

GAWILGARH HILLS

PAYANGHAT

AJANTA RANGE

PUSAD HILLS

MUDKHED HILLS

HINGOLI HILLS

BALAGHAT RANGE

M A H A R A S H T R A   P L A T E A U

ASIRGARH HILLS

NDESH PLA

SURPALNATH HILL

ELLORA HILLS

AJANTA PLATEAU

UPPER GODAVARI VALLEY

SINDPHANA

CHANDOR HILLS

GALNA HILLS

Ankai Tankai

AKRANI HILLS

NORTH SAHYADRI

Beale Lake

Arthur Hill Lake

HARISHCHANDRA RANGE

Kharakvasla Lake

PURANDHAR HILLS

UPPER BHIMA VALLEY

MAHADEO RANGE

MIRAJ HILLS

KRISHNA VALLEY

UPPER KRISHNA VALLEY

Koyna Lake

MATHERAN HILLS

ULHAS PLATEAU

VAITARNA PLATEAU

THANA

NORTH KONKAN COAST RANGE

SOUTH SAHYADRI HARISHTS

KONKAN HILLS

CHIMOR RANGE

DAMAN AND DIU

DADRA AND NAGAR HAVELI

Arnala I.

Elephanta I.

Khanderi I.

Burnt I.

GUJARAT

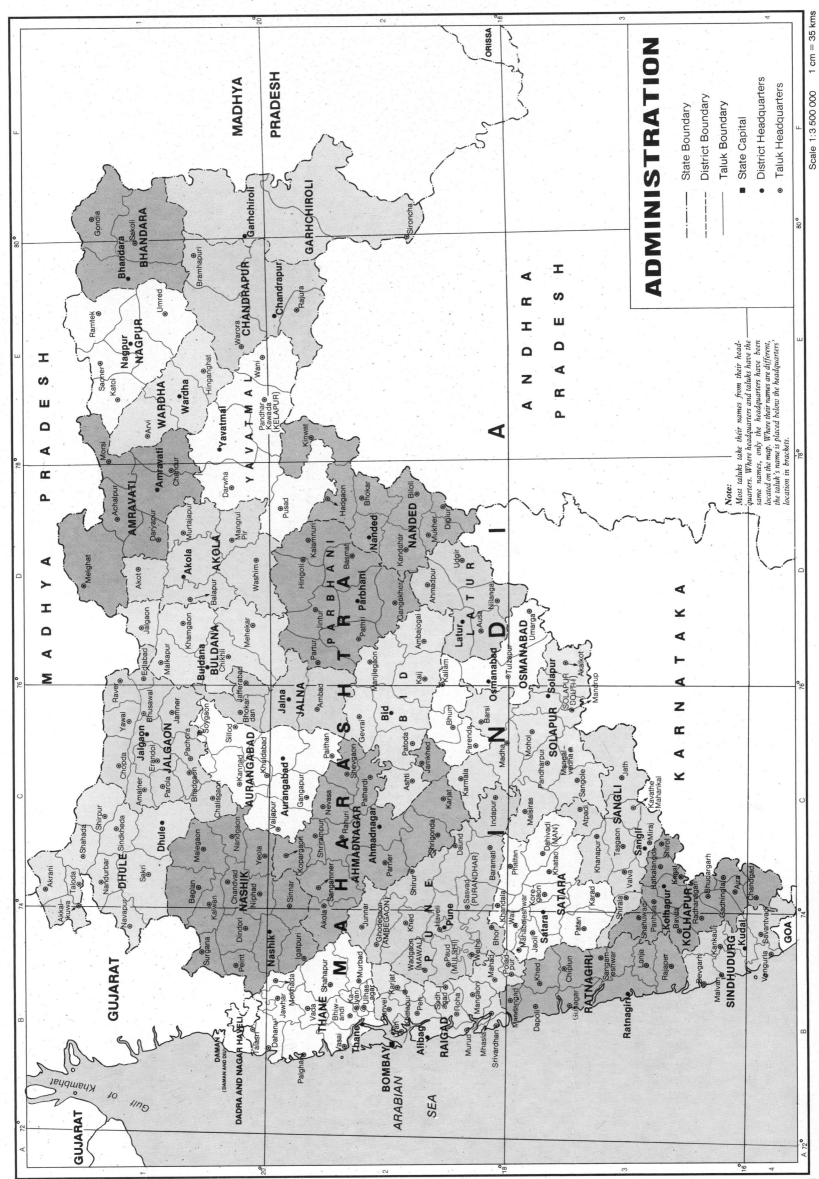

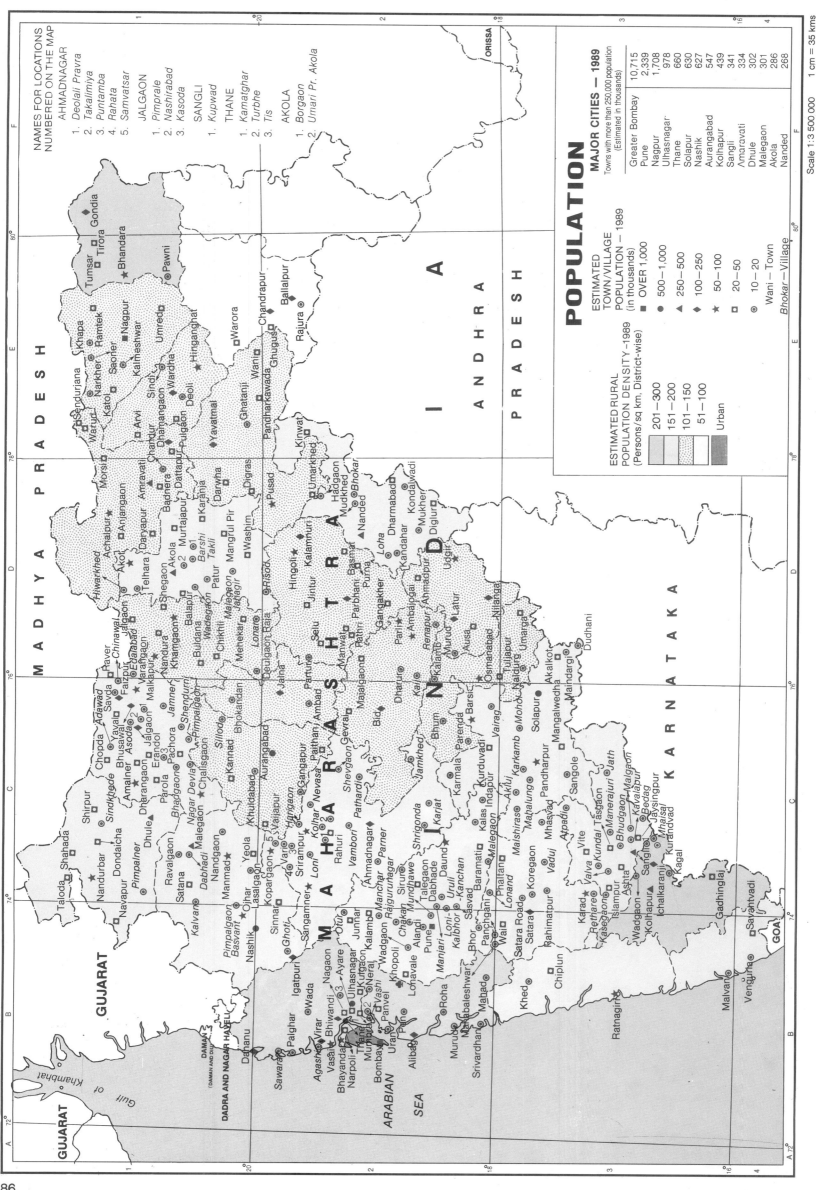

NAMES FOR LOCATIONS
NUMBERED ON THE MAP

AHMADNAGAR
1. Deolali Pravra
2. Takalimiya
3. Puntamba
4. Rahata
5. Samvatsar

JALGAON
1. Pimprale
2. Nashirabad
3. Kasoda

SANGLI
1. Kupwad

THANE
1. Kamatghar
2. Turbhe
3. Tis

AKOLA
1. Borgaon
2. Umari Pr. Akola

POPULATION

MAJOR CITIES — 1989

Towns with more than 250,000 population
(Estimated in thousands)

Greater Bombay	10,715
Pune	2,339
Nagpur	1,708
Ulhasnagar	978
Thane	660
Solapur	630
Nashik	627
Aurangabad	547
Kolhapur	439
Sangli	341
Amaravati	334
Dhule	302
Malegaon	301
Akola	286
Nanded	268

ESTIMATED
TOWN/VILLAGE
POPULATION — 1989
(in thousands)

■ OVER 1,000
● 500—1,000
▲ 250—500
◆ 100—250
★ 50—100
▢ 20—50
◉ 10—20
□ Urban

Wani — Town
Bhokar — Village

ESTIMATED RURAL
POPULATION DENSITY—1989
(Persons/sq. km. District-wise)

201—300
151—200
101—150
51—100

Urban

Scale 1:3 500 000   1 cm = 35 kms

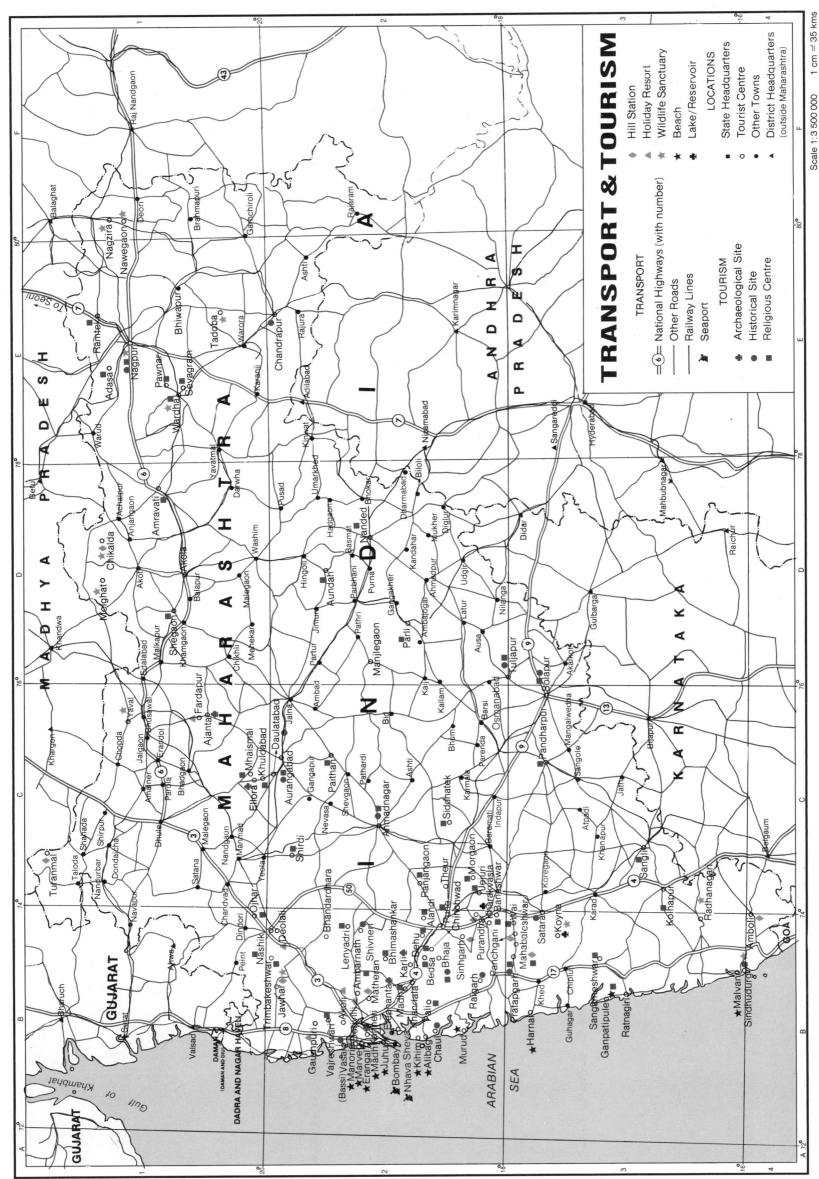

TRANSPORT & TOURISM

TRANSPORT

=6=	National Highways (with number)
	Other Roads
	Railway Lines
⚓	Seaport

TOURISM

♦	Hill Station
▲	Holiday Resort
✦	Wildlife Sanctuary
★	Beach
⚑	Lake/Reservoir

LOCATIONS

⚑	Archaeological Site
●	Historical Site
■	Religious Centre
■	State Headquarters
○	Tourist Centre
●	Other Towns
▲	District Headquarters (outside Maharashtra)

Scale 1:3 500 000   1 cm = 35 kms

87

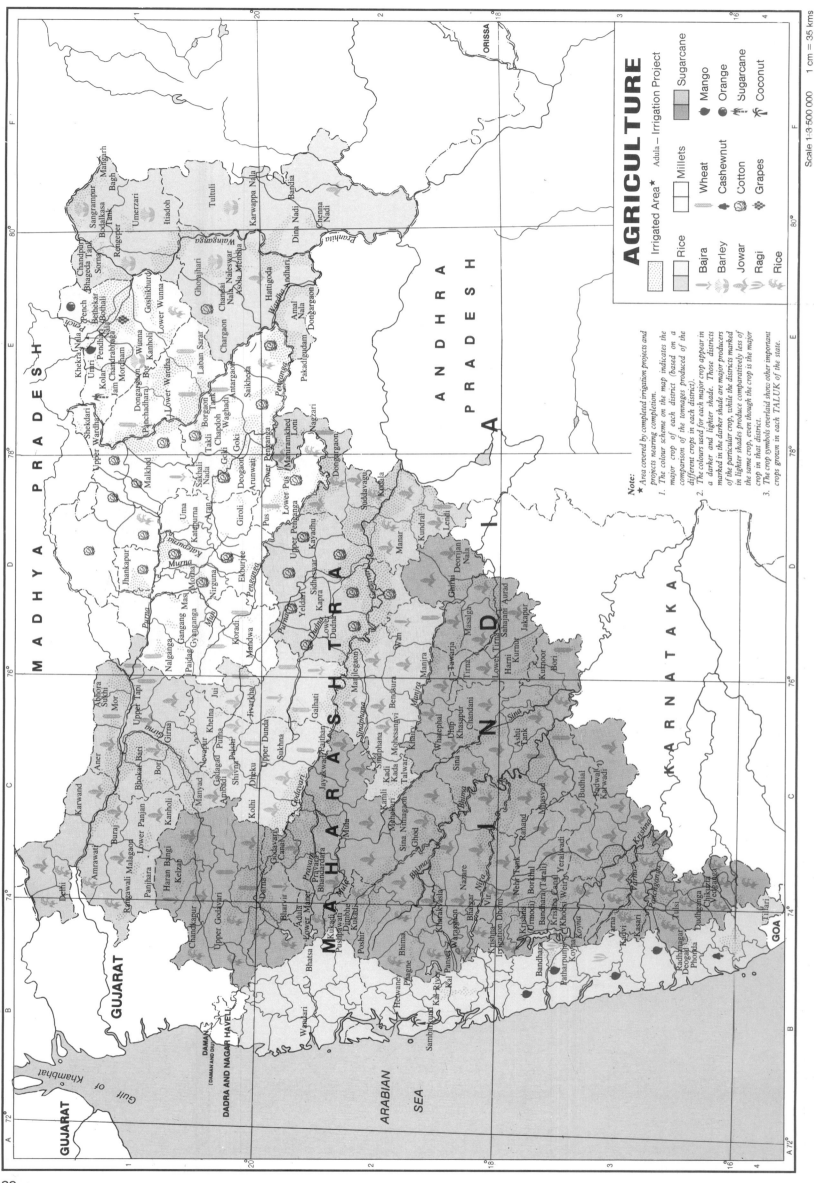

AGRICULTURE

Scale 1:3 500 000   1 cm = 35 kms

Irrigated Area · Adula – Irrigation Project

Irrigated Area	Millets	Sugarcane
Rice	Wheat	Mango
Bajra	Barley	Cashewnut	Orange
	Jowar	Cotton	Sugarcane
	Ragi	Grapes	Coconut
	Rice		

Note:

★ Area covered by completed irrigation projects and projects nearing completion.

1. The colour scheme on the map indicates the major crop of each district (based on a comparison of the tonnages produced of the different crops in each district).

2. The colours used for each major crop appear in a darker and lighter shade. Those districts marked in the darker shade are major producers of the particular crop, while the districts marked in lighter shades produce comparatively less of the same crop, even though the crop is the major crop in that district.

3. The crop symbols overlaid show other important crops grown in each TALUK of the state.

88

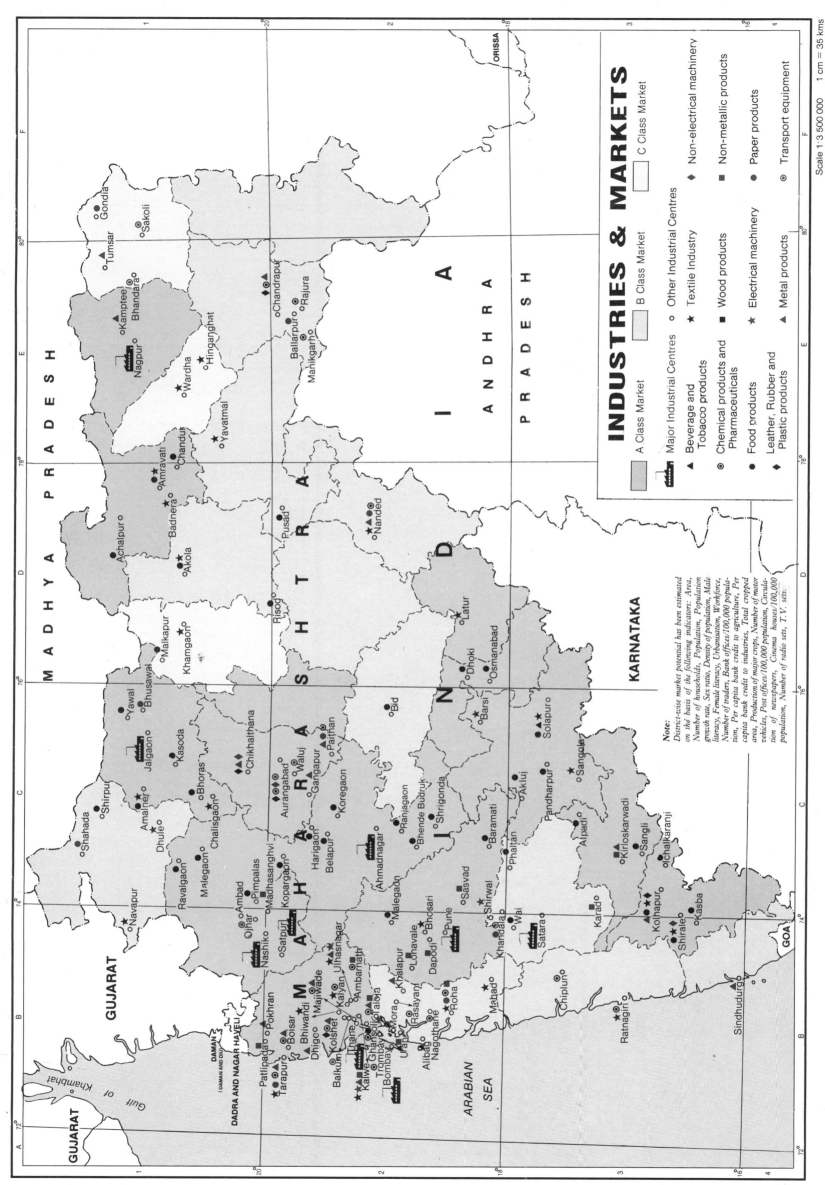

INDUSTRIES & MARKETS

A Class Market
Major Industrial Centres
B Class Market
C Class Market
Other Industrial Centres

○ Textile Industry
▲ Beverage and Tobacco products
◉ Chemical products and Pharmaceuticals
● Food products
◆ Leather, Rubber and Plastic products
◆ Non-electrical machinery
■ Wood products
★ Electrical machinery
● Metal products
■ Non-metallic products
● Paper products
◉ Transport equipment

Scale 1:3 500 000   1 cm = 35 kms

Note:
District-wise market potential has been estimated on the basis of the following indicators: Area, Number of households, Population, Population growth rate, Sex ratio, Density of population, Male literacy, Female literacy, Urbanisation, Workforce, Number of traders, Bank offices/100,000 population, Per capita bank credit to agriculture, Per capita bank credit to industries, Total cropped area, Production of major crops, Number of motor vehicles, Post offices/100,000 population, Circulation of newspapers, Cinema houses/100,000 population, Number of radio sets, T.V. sets.

GUJARAT

MADHYA PRADESH

ANDHRA PRADESH

ORISSA

KARNATAKA

GOA

ARABIAN SEA

Gulf of Khambhat

M A H A R A S H T R A   I N D I A

DAMAN
(DAMAN AND DIU)
DADRA AND NAGAR HAVELI

89

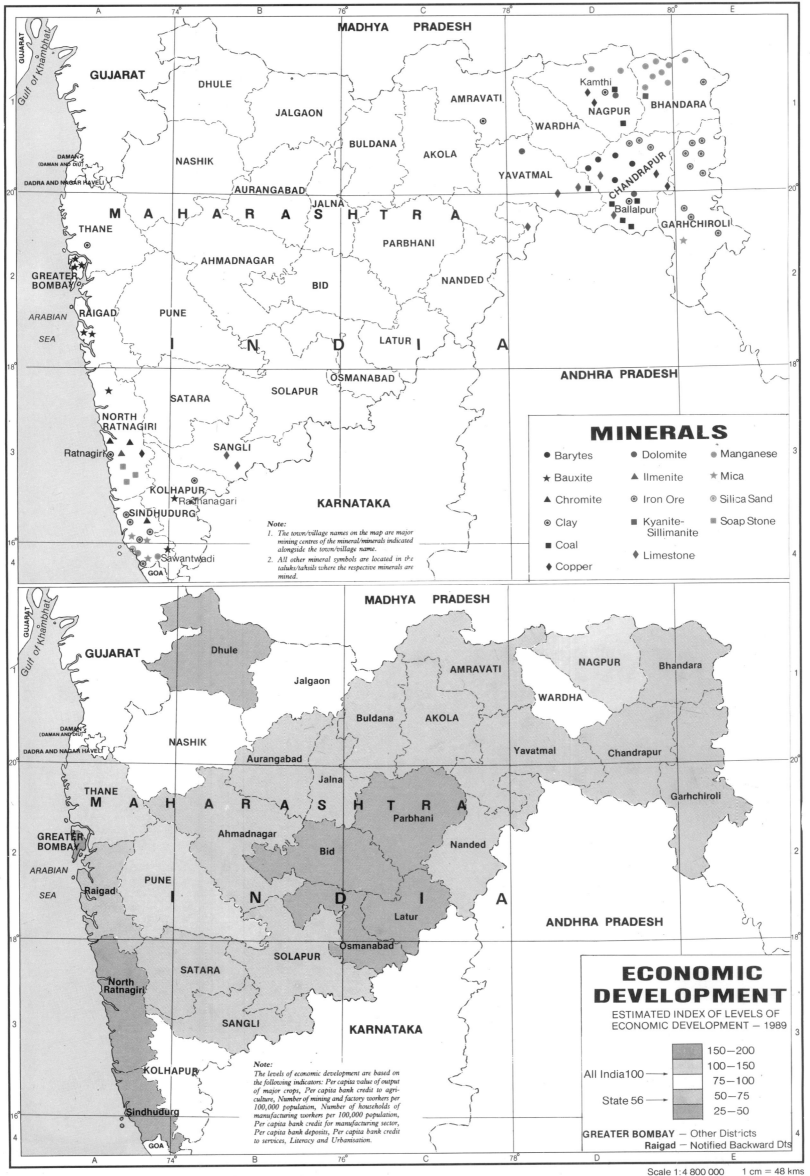

MINERALS

- ● Barytes
- ★ Bauxite
- ▲ Chromite
- ◉ Clay
- ■ Coal
- ◆ Copper
- ● Dolomite
- ▲ Ilmenite
- ◎ Iron Ore
- ■ Kyanite-Sillimanite
- ◆ Limestone
- ● Manganese
- ★ Mica
- ◉ Silica Sand
- ■ Soap Stone

Note:
1. *The town/village names on the map are major mining centres of the mineral/minerals indicated alongside the town/village name.*
2. *All other mineral symbols are located in the taluks/tahsils where the respective minerals are mined.*

ECONOMIC DEVELOPMENT

ESTIMATED INDEX OF LEVELS OF ECONOMIC DEVELOPMENT – 1989

All India 100 →
State 56 →

- 150 – 200
- 100 – 150
- 75 – 100
- 50 – 75
- 25 – 50

GREATER BOMBAY — Other Districts
Raigad — Notified Backward Dts

Note:
The levels of economic development are based on the following indicators: Per capita value of output of major crops, Per capita bank credit to agriculture, Number of mining and factory workers per 100,000 population, Number of households of manufacturing workers per 100,000 population, Per capita bank credit for manufacturing sector, Per capita bank deposits, Per capita bank credit to services, Literacy and Urbanisation.

Scale 1:4 800 000   1 cm = 48 kms

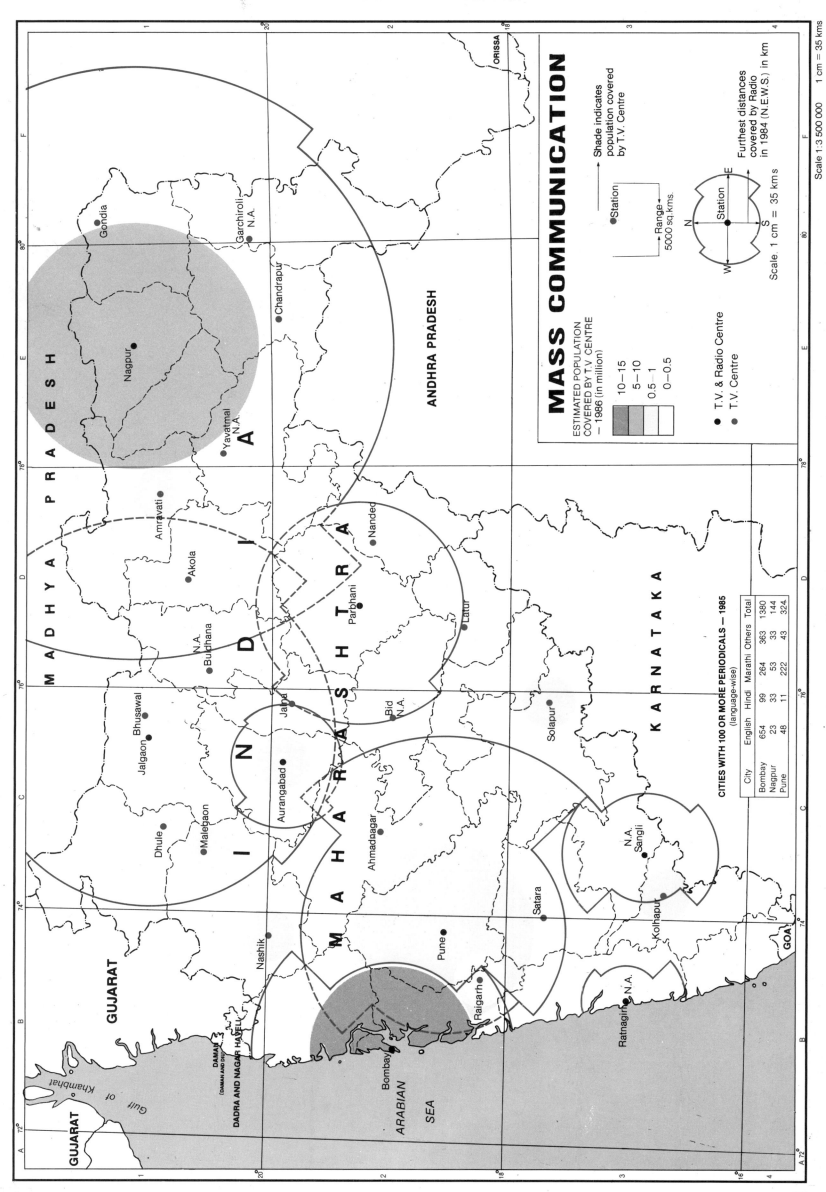

MASS COMMUNICATION

ESTIMATED POPULATION
COVERED BY T.V. CENTRE
— 1986 (in million)

▓	10–15
▒	5–10
░	0.5–1
☐	0–0.5

● T.V. & Radio Centre
● T.V. Centre

→ Station
Shade indicates
population covered
by T.V. Centre

Station
→ Range
5000 sq.kms.

Scale. 1 cm = 35 kms

Furthest distances
covered by Radio
in 1984 (N.E.W.S.) in km

Scale 1:3 500 000   1 cm = 35 kms

CITIES WITH 100 OR MORE PERIODICALS — 1985
(language-wise)

City	English	Hindi	Marathi	Others	Total
Bombay	654	99	264	363	1380
Nagpur	23	33	53	33	144
Pune	48	11	222	43	324

GUJARAT

Gulf of Khambhat

DAMAN
DAMAN AND DIU

DADRA AND NAGAR HAVELI

MADHYA PRADESH

ANDHRA PRADESH

KARNATAKA

GOA

ORISSA

ARABIAN SEA

M A H A R A S H T R A

I N D I A

Gondia
Garchiroli N.A.
Nagpur
Yavatmal N.A.
Chandrapur
Amravati
Akola
Nanded
Buldhana N.A.
Parbhani
Latur
Jalgaon
Bhusawal
Jalna
Bid N.A.
Solapur
Dhule
Malegaon
Aurangabad
Ahmadnagar
Sangli N.A.
Satara
Kolhapur
Nashik
Pune
Raigarh
Ratnagiri N.A.
Bombay

MANIPUR

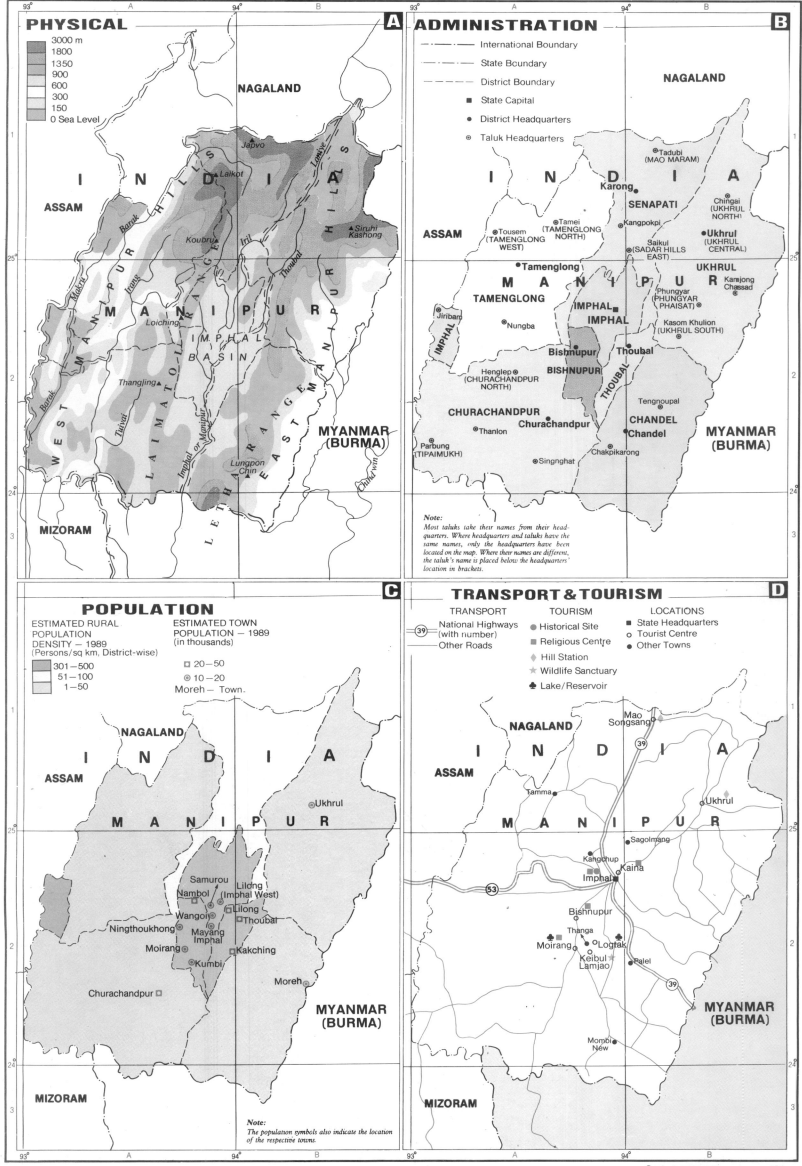

A — PHYSICAL

3000 m
1800
1350
900
600
300
150
0 Sea Level

NAGALAND

ASSAM

INDIA

Japvo

Laikot

WEST MANIPUR HILLS

Barak

Koubru

Siruhi Kashong

Iril

MANIPUR

LAIMATOL RANGE

Loiching

IMPHAL BASIN

Mokru

Irang

Tuivai

Thoubal

EAST MANIPUR HILLS

Thangling

Barak

Imphal or Manipur

Lungpon Chin

Chindwin

MYANMAR (BURMA)

MIZORAM

B — ADMINISTRATION

- – – – International Boundary
- – · – State Boundary
- – – District Boundary
- ■ State Capital
- ● District Headquarters
- ◎ Taluk Headquarters

NAGALAND

INDIA

Tadubi (MAO MARAM)

Karong

SENAPATI

Chingai (UKHRUL NORTH)

Tousem (TAMENGLONG WEST)

Tamei (TAMENGLONG NORTH)

Kangpokpi

Ukhrul (UKHRUL CENTRAL)

ASSAM

Saikul (SADAR HILLS EAST)

UKHRUL

Kamjong Chassad

Tamenglong

MANIPUR

TAMENGLONG

Phungyar (PHUNGYAR PHAISAT)

Jiribam

IMPHAL

IMPHAL

Kasom Khulion (UKHRUL SOUTH)

Nungba

Bishnupur

Thoubal

BISHNUPUR

THOUBAL

Henglep (CHURACHANDPUR NORTH)

Tengnoupal

CHURACHANDPUR

Thanlon

Churachandpur

CHANDEL

Chandel

MYANMAR (BURMA)

Parbung (TIPAIMUKH)

Chakpikarong

Singnghat

Note:

Most taluks take their names from their head-quarters. Where headquarters and taluks have the same names, only the headquarters have been located on the map. Where their names are different, the taluk's name is placed below the headquarters' location in brackets.

C — POPULATION

ESTIMATED RURAL POPULATION DENSITY — 1989
(Persons/sq km, District-wise)

- 301–500
- 51–100
- 1–50

ESTIMATED TOWN POPULATION — 1989
(in thousands)

- ☐ 20–50
- ◎ 10–20
- Moreh — Town.

NAGALAND

INDIA

ASSAM

MANIPUR

Ukhrul

Samurou

Nambol

Lilong (Imphal West)

Wangoi

Lilong

Ningthoukhong

Mayang Imphal

Thoubal

Moirang

Kakching

Kumbi

Moreh

Churachandpur

MYANMAR (BURMA)

MIZORAM

Note:

The population symbols also indicate the location of the respective towns.

D — TRANSPORT & TOURISM

TRANSPORT
- ⑳ National Highways (with number)
- Other Roads

TOURISM
- ● Historical Site
- ■ Religious Centre
- ◆ Hill Station
- ★ Wildlife Sanctuary
- ♣ Lake/Reservoir

LOCATIONS
- ■ State Headquarters
- ○ Tourist Centre
- ● Other Towns

NAGALAND

Mao Songsang

INDIA

ASSAM

39

Tamma

Ukhrul

MANIPUR

Sagolmang

Kangchup

Kaina

Imphal

53

Bishnupur

Thanga

Moirang

Logtak

Keibul Lamjao

Palel

39

MYANMAR (BURMA)

Mombi New

MIZORAM

Scale 1:1 800 000   1 cm = 18 kms

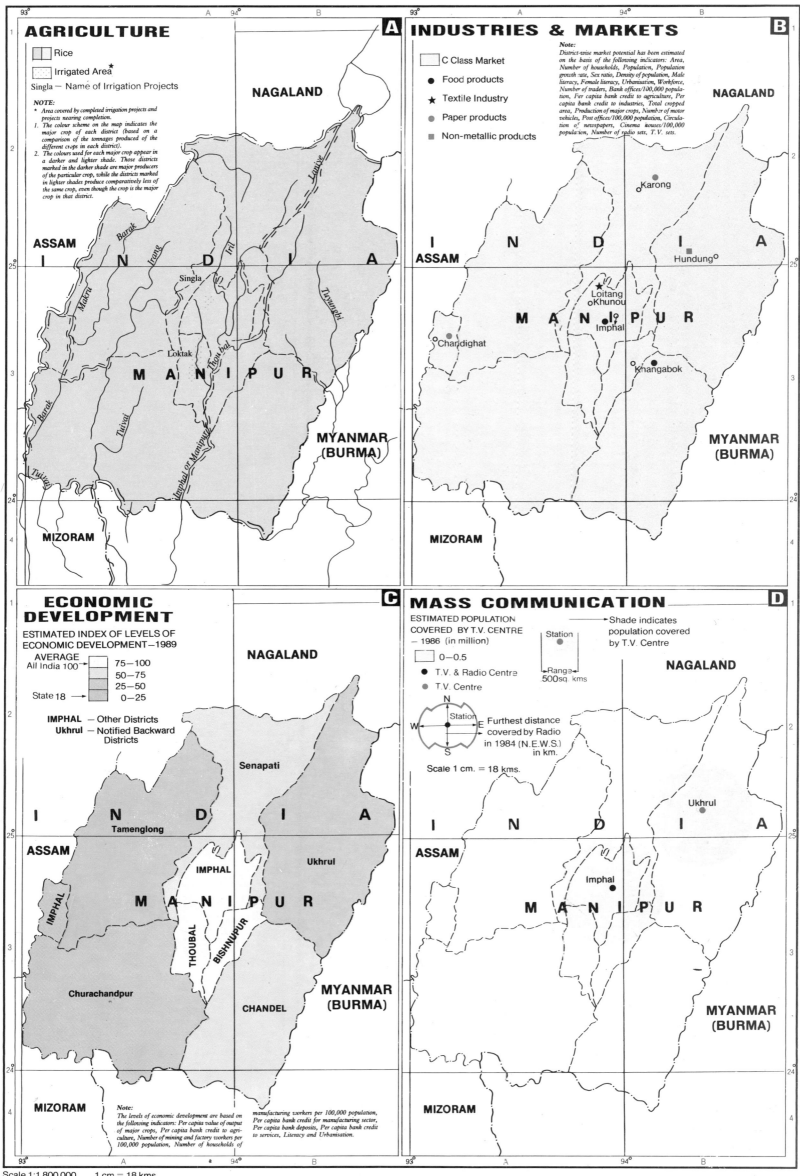

AGRICULTURE

Rice

Irrigated Area ★

Singla — Name of Irrigation Projects

NOTE:
* Area covered by completed irrigation projects and projects nearing completion.
1. The colour scheme on the map indicates the major crop of each district (based on a comparison of the tonnages produced of the different crops in each district).
2. The colours used for each major crop appear in a darker and lighter shade. Those districts marked in the darker shade are major producers of the particular crop, while the districts marked in lighter shades produce comparatively less of the same crop, even though the crop is the major crop in that district.

ASSAM

NAGALAND

I N D I A

Barak
Irang
Iril

Singla

Makru

Loktak

Barak

M A N I P U R

Tuivai

Imphal or Manipur

Thoubal

Tuivai

Lanive

Tizungbi

MYANMAR
(BURMA)

MIZORAM

INDUSTRIES & MARKETS

C Class Market

● Food products

★ Textile Industry

● Paper products

■ Non-metallic products

Note:
District-wise market potential has been estimated on the basis of the following indicators: Area, Number of households, Population, Population growth rate, Sex ratio, Density of population, Male literacy, Female literacy, Urbanisation, Workforce, Number of traders, Bank offices/100,000 population, Per capita bank credit to agriculture, Per capita bank credit to industries, Total cropped area, Production of major crops, Number of motor vehicles, Post offices/100,000 population, Circulation of newspapers, Cinema houses/100,000 population, Number of radio sets, T.V. sets.

NAGALAND

I N D I A

ASSAM

Karong

Hundung

M A N I P U R

Loitang
Khunou

Chandighat

Imphal

Khangabok

MYANMAR
(BURMA)

MIZORAM

ECONOMIC DEVELOPMENT

ESTIMATED INDEX OF LEVELS OF ECONOMIC DEVELOPMENT—1989

AVERAGE
All India 100

75—100
50—75
25—50
0—25

State 18

IMPHAL — Other Districts
Ukhrul — Notified Backward Districts

NAGALAND

I N D I A

Senapati

Tamenglong

ASSAM

IMPHAL

IMPHAL

Ukhrul

M A N I P U R

THOUBAL

BISHNUPUR

Churachandpur

CHANDEL

MYANMAR
(BURMA)

MIZORAM

Note:
The levels of economic development are based on the following indicators: Per capita value of output of major crops, Per capita bank credit to agriculture, Number of mining and factory workers per 100,000 population, Number of households of manufacturing workers per 100,000 population, Per capita bank credit for manufacturing sector, Per capita bank deposits, Per capita bank credit to services, Literacy and Urbanisation.

MASS COMMUNICATION

ESTIMATED POPULATION COVERED BY T.V. CENTRE — 1986 (in million)

Shade indicates population covered by T.V. Centre

Station

0—0.5

● T.V. & Radio Centre

● T.V. Centre

Range
500 sq. kms

N
W   S   E
Station

Furthest distance covered by Radio in 1984 (N.E.W.S.) in km.

Scale 1 cm. = 18 kms.

NAGALAND

I N D I A

Ukhrul

ASSAM

Imphal

M A N I P U R

MYANMAR
(BURMA)

MIZORAM

MEGHALAYA

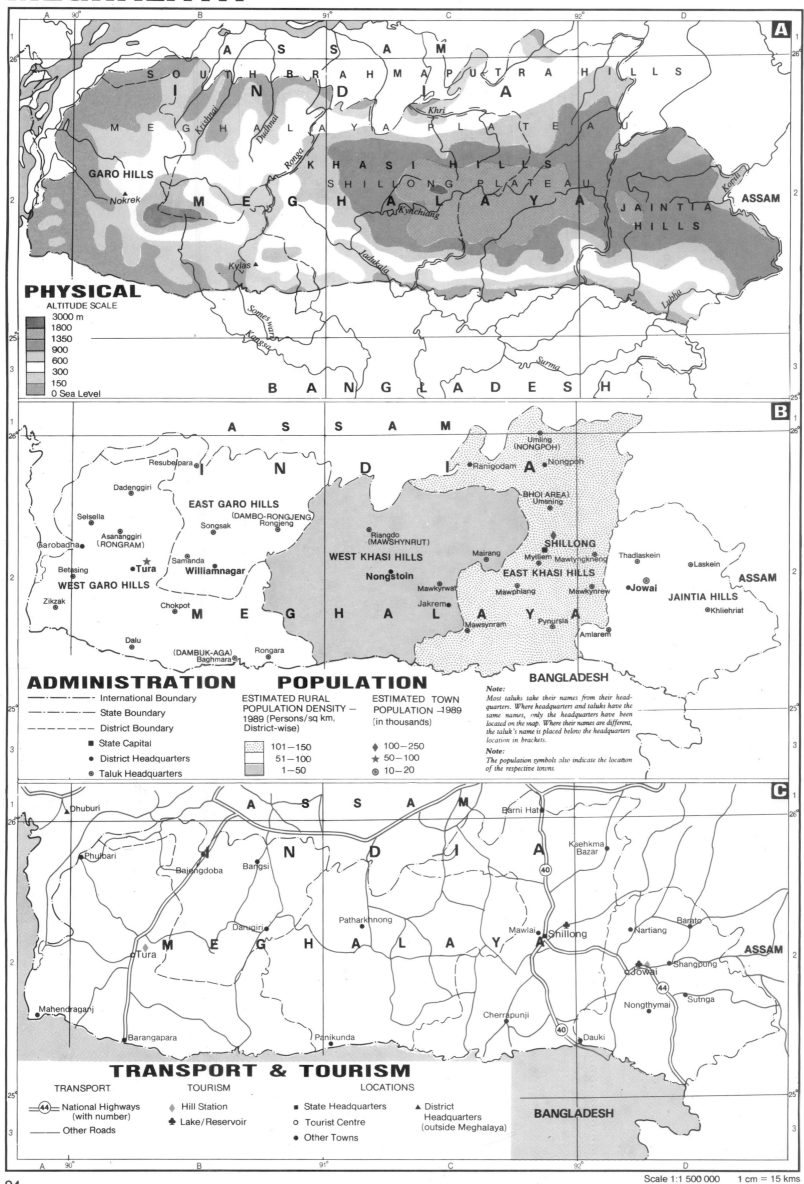

A

ASSAM

SOUTH BRAHMAPUTRA HILLS

INDIA

Khri

S I N G

M E G H A L A Y A P L A T E A U

Krishnai

Dudhnai

Ronga

KHASI HILLS

GARO HILLS

SHILLONG PLATEAU

Koruti

Nokrek

M E G H A L A Y A

Kynchiang

JAINTIA HILLS

ASSAM

Kylas

Tadukala

Lubha

PHYSICAL

ALTITUDE SCALE

	3000 m
	1800
	1350
	900
	600
	300
	150
	0 Sea Level

Someswarn

Kangsa

Surma

B A N G L A D E S H

B

I N D I A   A S S A M   M

Resubelpara

Umling (NONGPOH)

Dadenggiri

Ranigodam   Nongpoh

EAST GARO HILLS

BHOI AREA)

Selsella

(DAMBO-RONGJENG)

Umsning

Asananggiri (RONGRAM)

Songsak   Rongjeng

Riangdo (MAWSHYNRUT)

Garobadha

SHILLONG

Thadlaskein

Samanda

Mairang

Mylliem   Mawlyngkneng

Laskein

Betasing   ★ Tura

Williamnagar

WEST KHASI HILLS

EAST KHASI HILLS

ASSAM

WEST GARO HILLS

Nongstoin

Mawkyrwat   Mawphlang   Mawkynrew   Jowai

JAINTIA HILLS

Zikzak

Jakrem

Khliehriat

M E G H A   L   A   Y   A

Chokpot

Mawsynram   Pynursla

Dalu   Rongara   Amlarem

(DAMBUK-AGA)   Baghmara

BANGLADESH

ADMINISTRATION   POPULATION

ADMINISTRATION	
————	International Boundary
—·—·—	State Boundary
— — —	District Boundary
■	State Capital
●	District Headquarters
◉	Taluk Headquarters

ESTIMATED RURAL POPULATION DENSITY — 1989 (Persons/sq km, District-wise)

	101—150
	51—100
	1—50

ESTIMATED TOWN POPULATION —1989 (in thousands)

◆	100—250
★	50—100
◉	10—20

Note:
Most taluks take their names from their head-quarters. Where headquarters and taluks have the same names, only the headquarters have been located on the map. Where their names are different, the taluk's name is placed below the headquarters location in brackets.

Note:
The population symbols also indicate the location of the respective towns.

C

Dhuburi   A S S A M   M   Barni Hat

Phulbari

Ksehkma Bazar

Bajengdoba   Bangsi

40

Patharkhnong

Barato

Darugiri

Mawlai   Shillong   Nartiang

◆ Tura

M E G H A L A Y A

Jowai   Shangpung

44

Mahendraganj

Nongthymai   Sutnga

Barangapara   Panikunda

Cherrapunji

40   Dauki

BANGLADESH

TRANSPORT & TOURISM

TRANSPORT	TOURISM	LOCATIONS	
〓44〓 National Highways (with number)	◆ Hill Station	■ State Headquarters	▲ District Headquarters (outside Meghalaya)
——— Other Roads	♣ Lake/Reservoir	○ Tourist Centre	
		● Other Towns	

Scale 1:1 500 000   1 cm = 15 kms

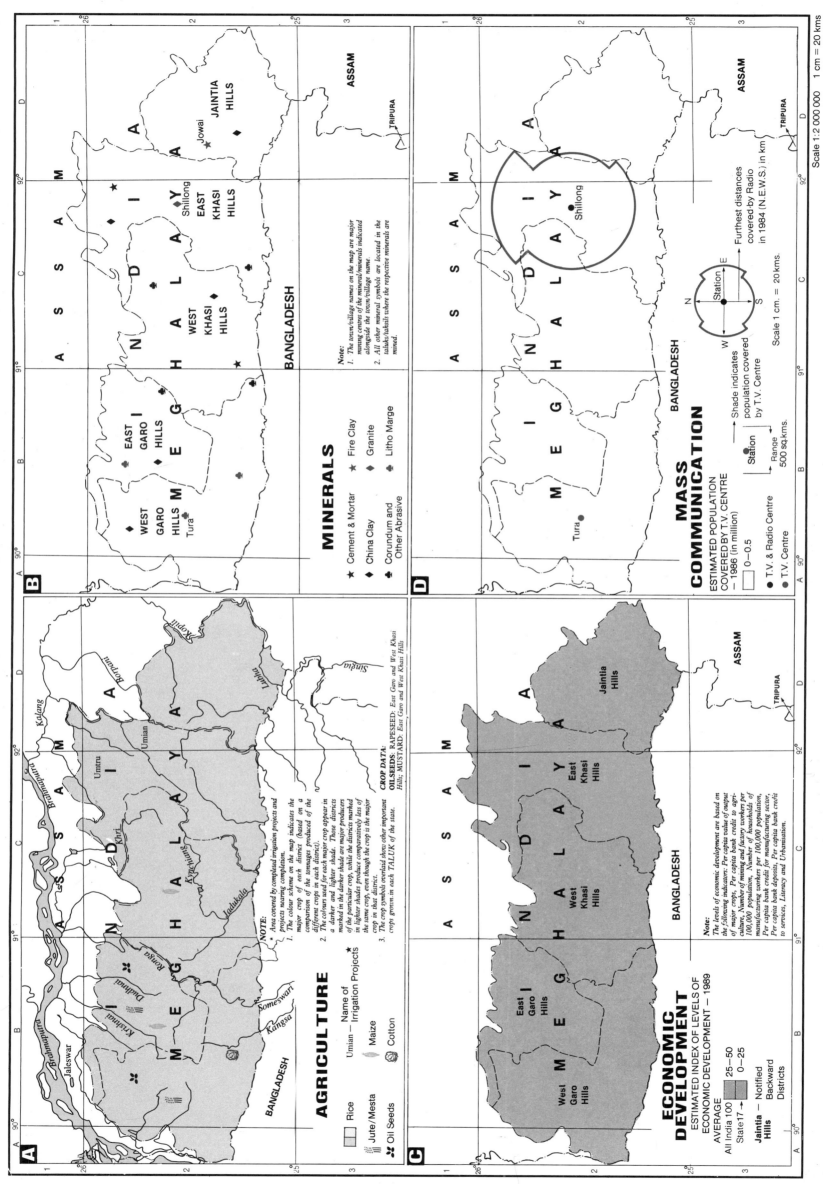

A

AGRICULTURE

Umian — Name of Irrigation Projects

Rice

Jute/Mesta

Oil Seeds

Maize

Cotton

NOTE:

★ Area covered by completed irrigation projects and projects nearing completion.

1. The colour scheme on the map indicates the major crop of each district (based on a comparison of the tonnages produced of the different crops in each district).

2. The colours used for each major crop appear in a darker and lighter shade. Those districts marked in the darker shade are major producers of the particular crop, while the districts marked in lighter shades produce comparatively less of the same crop, even though the crop is the major crop in that district.

3. The crop symbols overlaid show other important crops grown in each TALUK of the state.

CROP DATA:
OILSEEDS: RAPESEED: *East Garo and West Khasi Hills;* MUSTARD: *East Garo and West Khasi Hills*

B

MINERALS

★ Cement & Mortar

◆ China Clay

♣ Corundum and Other Abrasive

★ Fire Clay

◆ Granite

♣ Litho Marge

Note:

1. The town/village names on the map are major mining centres of the mineral/minerals indicated alongside the town/village name.

2. All other mineral symbols are located in the taluks/tahsils where the respective minerals are mined.

C

ECONOMIC DEVELOPMENT

ESTIMATED INDEX OF LEVELS OF ECONOMIC DEVELOPMENT – 1989

AVERAGE
All India 100
State 17

Jaintia Hills — Notified Backward Districts

25–50

0–25

Note:

The levels of economic development are based on the following indicators: Per capita value of output of major crops, Per capita bank credit to agriculture, Number of mining and factory workers per 100,000 population, Number of households of manufacturing workers per 100,000 population, Per capita bank credit for manufacturing sector, Per capita bank deposits, Per capita bank credit to services, Literacy and Urbanisation.

D

MASS COMMUNICATION

ESTIMATED POPULATION COVERED BY T.V. CENTRE – 1986 (in million)

0–0.5

● T.V. & Radio Centre

● T.V. Centre

Station | Range 500 sq.kms.

→ Shade indicates population covered by T.V. Centre

Furthest distances covered by Radio in 1984 (N.E.W.S.) in km

Scale 1 cm. = 20 kms.

Scale 1:2 000 000   1 cm = 20 kms

MIZORAM

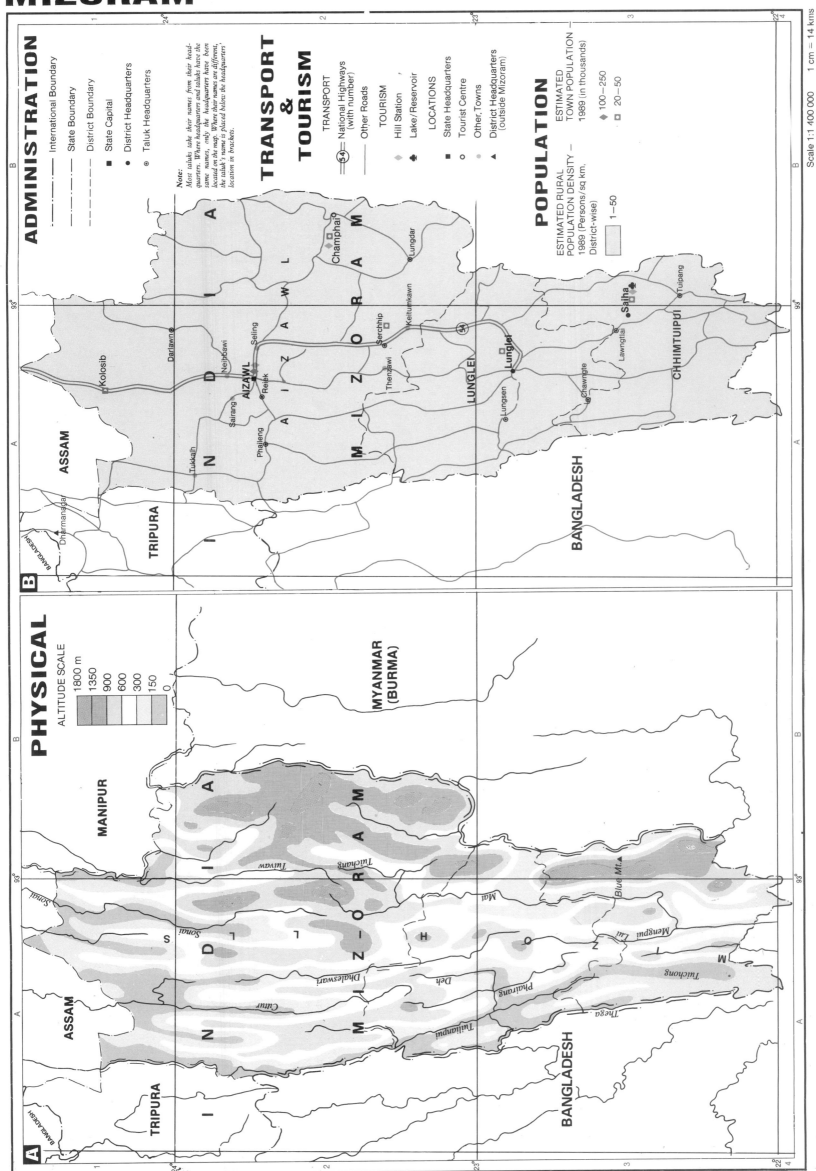

ADMINISTRATION

	International Boundary
	State Boundary
	District Boundary
■ | State Capital |
● | District Headquarters |
◉ | Taluk Headquarters |

Note:
Most taluks take their names from their head-quarters. Where headquarters and taluks have the same names, only the headquarters have been located on the map. Where their names are different, the taluk's name is placed below the headquarters' location in brackets.

TRANSPORT & TOURISM

TRANSPORT
⑤④ National Highways (with number)
Other Roads

TOURISM
◆ Hill Station
♣ Lake/Reservoir

LOCATIONS
■ State Headquarters
○ Tourist Centre
∘ Other Towns
▲ District Headquarters (outside Mizoram)

POPULATION

ESTIMATED RURAL POPULATION DENSITY – 1989 (Persons/sq km, District-wise)

ESTIMATED TOWN POPULATION 1989 (in thousands)
◆ 100–250
□ 20–50
1–50

Scale 1:1 400 000   1 cm = 14 kms

Administration / Transport & Tourism map (B)

ASSAM

INDIA

MIZORAM

AIZAWL

Kolosib
Darlawn
Neihbawi
Seling
Reiek
Sairang
Tukkah
Phaileng

Champhai
Lungdar
Serchhip
Thenzawi
Keitumkawn

LUNGLEI
Lunglei
Lungsen
Chawngte

Saiha
CHHIMTUIPUI
Tuipang
Lawngtlai

TRIPURA

Dharmanagar
BANGLADESH

PHYSICAL map (A)

PHYSICAL

ALTITUDE SCALE
1800 m
1350
900
600
300
150
0

ASSAM
MANIPUR

INDIA

MIZORAM

Sonai
Sonai
Tuivaw
Tuichang
Dhaleswar
Cutur
Deh
Tlailiampui
Phairang
Thega
Tuichong
Mengpui
Lui
Mat
Blue Mt. ▲

MYANMAR (BURMA)

TRIPURA
BANGLADESH

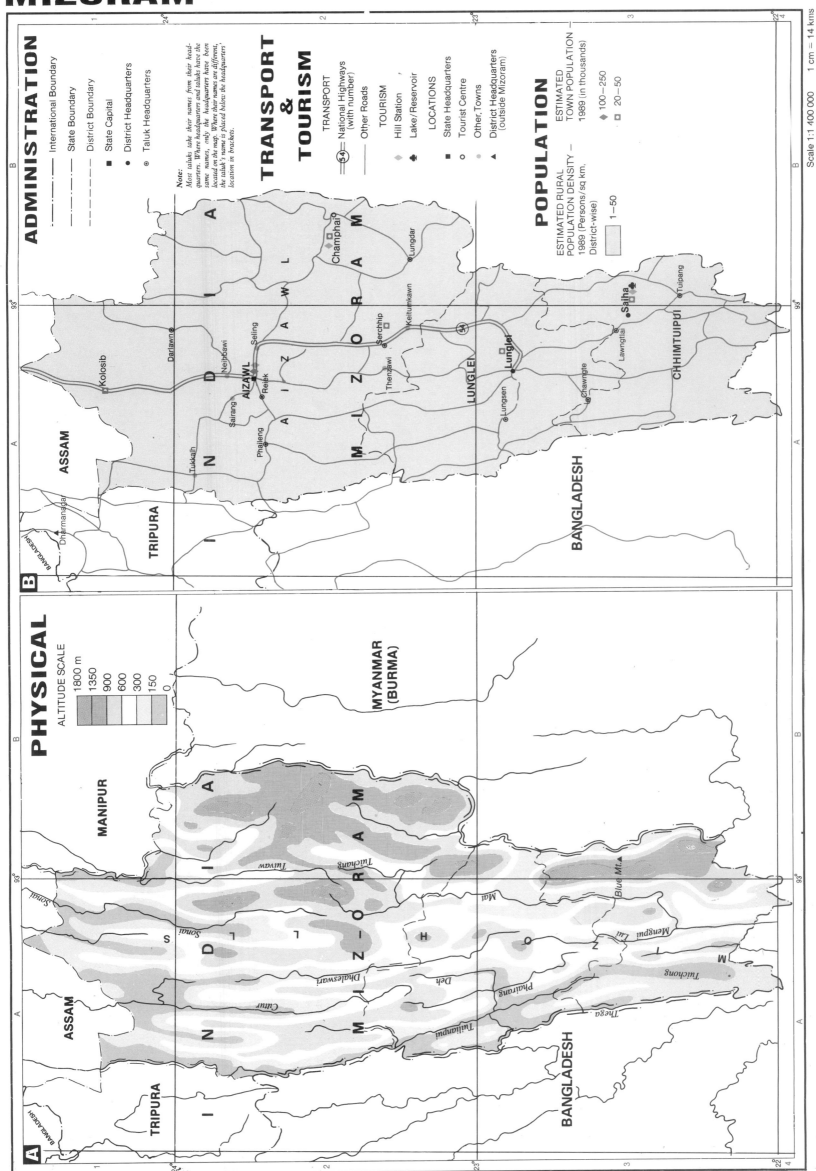

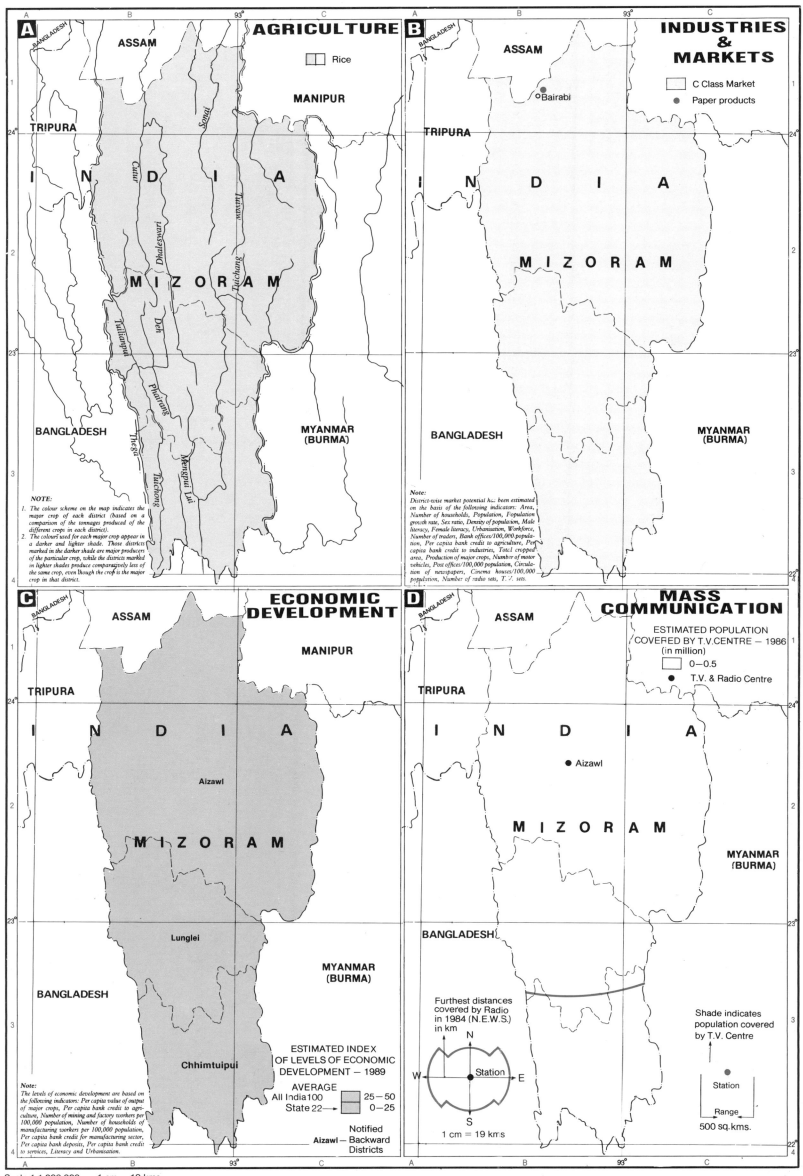

A — AGRICULTURE

Rice

NOTE:
1. The colour scheme on the map indicates the major crop of each district (based on a comparison of the tonnages produced of the different crops in each district).
2. The colour used for each major crop appear in a darker and lighter shade. Those districts marked in the darker shade are major producers of the particular crop, while the districts marked in lighter shades produce comparatively less of the same crop, even though the crop is the major crop in that district.

ASSAM
MANIPUR
TRIPURA
INDIA
MIZORAM
BANGLADESH
MYANMAR (BURMA)

Cutar
Sonai
Tuirini
Dhaleswari
Tuichang
Tuilianpui
Deh
Phairong
Thega
Tuichong
Mengpui Lui

B — INDUSTRIES & MARKETS

C Class Market
Paper products

Note:
District-wise market potential has been estimated on the basis of the following indicators: Area, Number of households, Population, Population growth rate, Sex ratio, Density of population, Male literacy, Female literacy, Urbanisation, Workforce, Number of traders, Bank offices/100,000 population, Per capita bank credit to agriculture, Per capita bank credit to industries, Total cropped area, Production of major crops, Number of motor vehicles, Post offices/100,000 population, Circulation of newspapers, Cinema houses/100,000 population, Number of radio sets, T.V. sets.

ASSAM
Bairabi
TRIPURA
INDIA
MIZORAM
BANGLADESH
MYANMAR (BURMA)

C — ECONOMIC DEVELOPMENT

Note:
The levels of economic development are based on the following indicators: Per capita value of output of major crops, Per capita bank credit to agriculture, Number of mining and factory workers per 100,000 population, Number of households of manufacturing workers per 100,000 population, Per capita bank credit for manufacturing sector, Per capita bank deposits, Per capita bank credit to services, Literacy and Urbanisation.

ESTIMATED INDEX OF LEVELS OF ECONOMIC DEVELOPMENT — 1989
AVERAGE
All India 100
State 22
25—50
0—25
Notified
Aizawl — Backward Districts

ASSAM
MANIPUR
TRIPURA
INDIA
MIZORAM
Aizawl
Lunglei
Chhimtuipui
BANGLADESH
MYANMAR (BURMA)

D — MASS COMMUNICATION

ESTIMATED POPULATION COVERED BY T.V. CENTRE — 1986 (in million)
0—0.5
T.V. & Radio Centre

ASSAM
TRIPURA
INDIA
MIZORAM
Aizawl
BANGLADESH
MYANMAR (BURMA)

Furthest distances covered by Radio in 1984 (N.E.W.S.) in km
N
W — Station — E
S
1 cm = 19 kms

Shade indicates population covered by T.V. Centre
Station
Range
500 sq. kms.

Scale 1:1 900 000   1 cm = 19 kms

NAGALAND

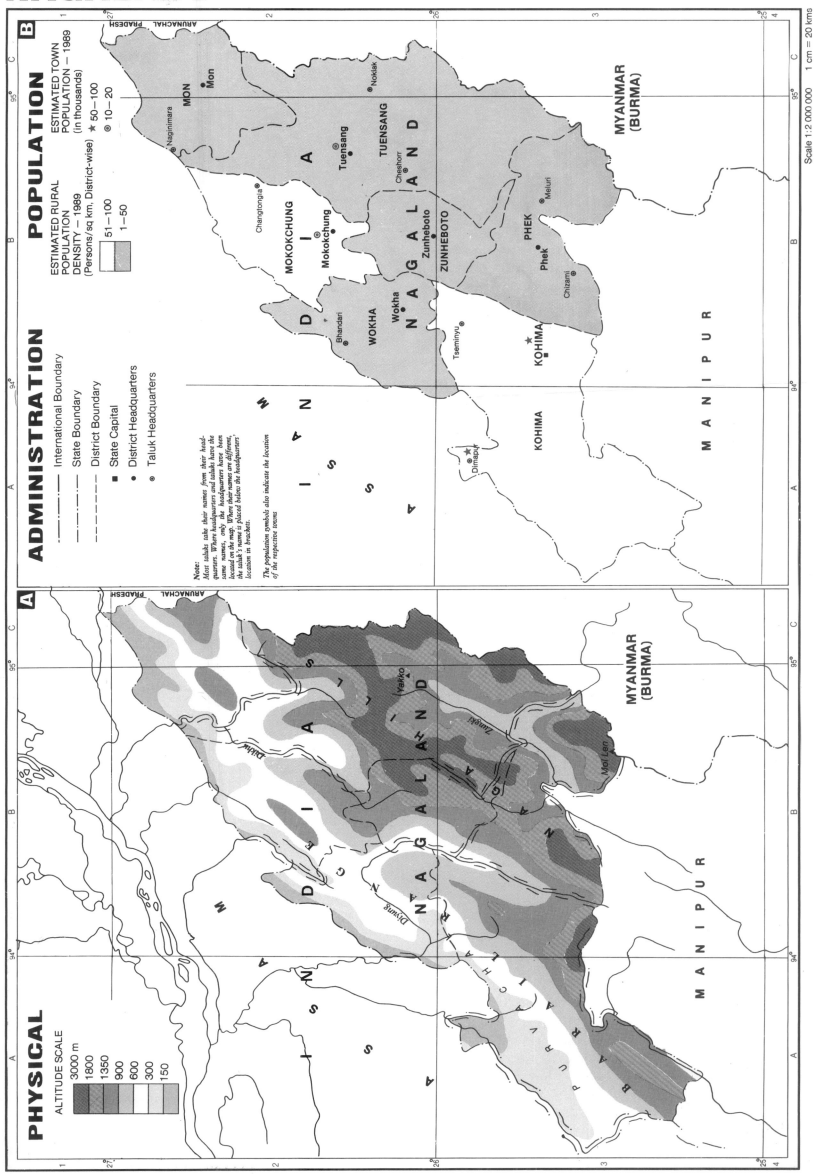

PHYSICAL

ALTITUDE SCALE

3000 m
1800
1350
900
600
300
150

ADMINISTRATION

International Boundary
State Boundary
District Boundary

■ State Capital
● District Headquarters
◎ Taluk Headquarters

Note:
Most taluks take their names from their head-quarters. Where headquarters and taluks have the same names, only the headquarters have been located on the map. Where their names are different, the taluk's name is placed below the headquarters location in brackets.

The population symbols also indicate the location of the respective towns.

POPULATION

ESTIMATED TOWN POPULATION — 1989
(in thousands)
★ 50—100
◎ 10—20

ESTIMATED RURAL POPULATION DENSITY — 1989
(Persons/sq km, District-wise)
51—100
1—50

MYANMAR (BURMA)

MANIPUR

ARUNACHAL PRADESH

KOHIMA

Map A labels: ASSAM, Dikhu, Dising, NAGALAND, Yakko, Zungki, Mai Len, BARAK, PURVACHAL, MANIPUR, MYANMAR (BURMA)

Map B labels: ASSAM, MON, Mon, Naginimara, Nokiak, Changtongia, TUENSANG, Tuensang, Cheshorr, MOKOKCHUNG, Mokokchung, ZUNHEBOTO, Zunheboto, NAGALAND, WOKHA, Wokha, Bhandari, PHEK, Phek, Meluri, Chizami, Tseminyu, KOHIMA, Dimapur, MANIPUR, MYANMAR (BURMA)

Scale 1:2 000 000   1 cm = 20 kms

AGRICULTURE

Rice
Other Crops

NOTE:
1. The colour scheme on the map indicates the major crop of each district (based on a comparison of the tonnages produced of the different crops in each district).
2. The colours used for each major crop appear in a darker and lighter shade. Those districts marked in the darker shade are major producers of the particular crop, while the districts marked in lighter shades produce comparatively less of the same crop, even though the crop is the major crop in that district.
3. The crop symbols overlaid show other important crops grown in each TALUK of the state.

ARUNACHAL PRADESH

N.A.

INDIA

ASSAM

Dikhu

Diyung

N.A.

Zungki

NAGALAND

N.A.

MANIPUR

MYANMAR (BURMA)

TRANSPORT & TOURISM

TRANSPORT
39 National Highways (with number)
Other Roads

TOURISM
★ Wildlife Sanctuary

LOCATIONS
■ State Headquarters
○ Tourist Centre
● Other Towns
▲ District Headquarters (outside Nagaland)

MARKETS
C Class Market

ARUNACHAL PRADESH

Lapa
Naginimara
Borjan   Nyasia
Tamlu   Lungha   Mon
Merangkong   Chen
Chonghyimsen Chantongia Longching
Golaghat   Yonghong

ASSAM

INDIA

Bhandari   Mokokchung   Tuensang
Sanis   Ungma
37   Lungsa   Noklak
Wokha   Cheahorr   Panso
NAGALAND
36   Zunheboto
Dimapur Rengazumi Tseminyu   Thonoknyu
Shedhumi   Sirire
Nichuguard Lazami Kilomi Ghaspani   Sampurre
Chimakudi Sathazuma Kohima
Khonoma   Phek   Laruri
Lakema   Chizami
Peremi   39
Henima
Meehangbung
Intanki

Ukhrul

MANIPUR

MYANMAR (BURMA)

ECONOMIC DEVELOPMENT

ESTIMATED INDEX OF LEVELS OF ECONOMIC DEVELOPMENT — 1989
AVERAGE
All India 100   50—75
                       25—50
State 16 →   0—25

WOKHA – Other Districts
Kohima – Notified Backward Districts

ASSAM

INDIA

MON

Mokokchung

WOKHA

ZUNHEBOTO   Tuensang

NAGALAND

Kohima

PHEK

MANIPUR

MYANMAR (BURMA)

Note:
The levels of economic development are based on the following indicators; Per capita value of output of major crops, Per capita bank credit to agriculture, Number of mining and factory workers per 100,000 population, Number of households of manufacturing workers per 100,000 population, Per capita bank credit for manufacturing sector, Per capita bank deposits, Per capita bank credit to services, Literacy and Urbanisation.

MASS COMMUNICATION

ESTIMATED POPULATION COVERED BY T.V. CENTRE — 1986 (in million)

0—0.5
● T.V. & Radio Centre

Station
Range 500 sq. kms

→ Shade indicates population covered by T.V. Centre

N
W Station E
S
Furthest distance covered by Radio in 1984 (N.E.W.S.) in km
Scale 1 cm. = 20 kms.

ASSAM

INDIA

NAGALAND

Kohima

MANIPUR

MYANMAR (BURMA)

ORISSA

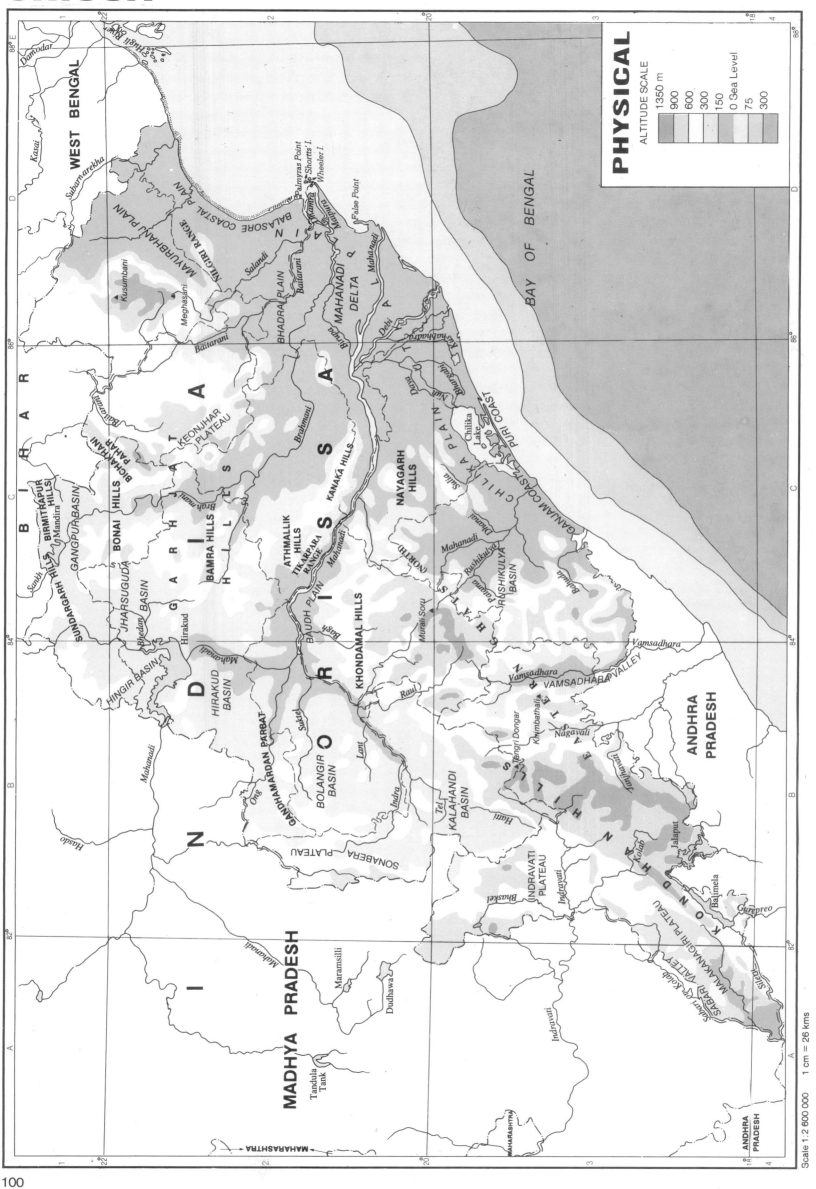

PHYSICAL

ALTITUDE SCALE

1350 m
900
600
300
150
0 Sea Level
75
300

WEST BENGAL

BAY OF BENGAL

B I H A R

O R I S S A

I N D I A

MADHYA PRADESH

ANDHRA PRADESH

MAHARASHTRA

BALASORE COASTAL PLAIN

MAYURBHANJ PLAIN

NILGIRI RANGE

BHADRA PLAIN

MAHANADI DELTA

KEONJHAR PLATEAU

GARHJAT HILLS

BONAI HILLS

BICHAKHANI PAHAR

BIRMITRAPUR HILLS

SUNDARGARH HILLS

GANGPUR BASIN

JHARSUGUDA

BHEDAN BASIN

HINGIR BASIN

HIRAKUD BASIN

GANDHAMARDAN PARBAT

BOLANGIR BASIN

SONABERA PLATEAU

BAMRA HILLS

ATHMALLIK HILLS

TIKARPARA RANGE

BAUDH PLAIN

KHONDAMAL HILLS

NAYAGARH HILLS

KANAKA HILLS

CHILKA PLAIN

GANJAM COAST

PURI COAST

EASTERN GHATS

(R)SHIKULYA BASIN

VAMSADHARA VALLEY

VAMSADHARA

KALAHANDI BASIN

INDRAVATI PLATEAU

KANDH

MALAKANAGIRI PLATEAU

SABARI VALLEY

Damodar
Kasai
Subarnarekha
Hugli River
Palmyras Point
Shortts I.
Wheeler I.
False Point
Kusumbani
Meghasani
Salandi
Baitarani
Baitarani
Brahmani
Burha
Mahanadi
L. Mahanadi
Debi
Kushbhadra
Daya
Nun
Bhargavi
Chilka Lake
Bahuda
Sundar
Kolab
Sileru
Indravati
Maramsilli
Dudhawa
Tandula Tank
Mahanadi
Hasdo
Ong
Suktel
Lant
Indra
Tel
Haiti
Bhaskel
Jalaput
Balimela
Gurepreo
Bagh
Raul
Murali Soru
Sabut
Dhanui
Bahuda
Rushikulya
Vamsadhara
Nagavali
Janhavati
Khimbathali
Tangiri Dongar
Mahanadi
Hirakud
Bolangir
Mandira
Sankh
Ib
Bhedan
Mahanadi

100

Scale 1:2 600 000   1 cm = 26 kms

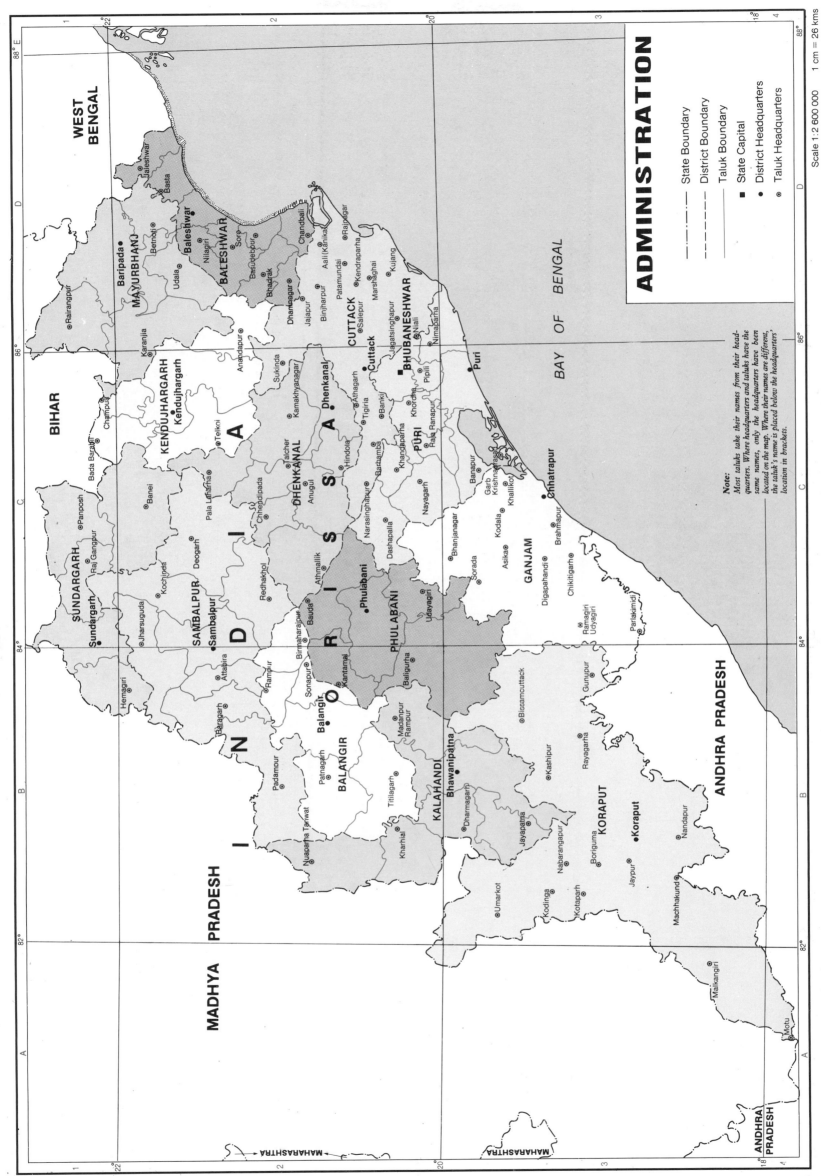

ADMINISTRATION

	State Boundary
	District Boundary
	Taluk Boundary
■	State Capital
●	District Headquarters
⊙	Taluk Headquarters

Note:
*Most taluks take their names from their head-
quarters. Where headquarters and taluks have the
same names, only the headquarters have been
located on the map. Where their names are different,
the taluk's name is placed below the headquarters'
location in brackets.*

Scale 1:2 600 000   1 cm = 26 kms

WEST BENGAL

BIHAR

MADHYA PRADESH

ANDHRA PRADESH

BAY OF BENGAL

MAYURBHANJ

BALESHWAR

KENDUJHARGARH

SUNDARGARH

SAMBALPUR

DHENKANAL

CUTTACK

BHUBANESHWAR

PURI

PHULABANI

BALANGIR

KALAHANDI

GANJAM

KORAPUT

ORISSA

INDIA

← MAHARASHTRA

ANDHRA PRADESH

MAHARASHTRA →

101

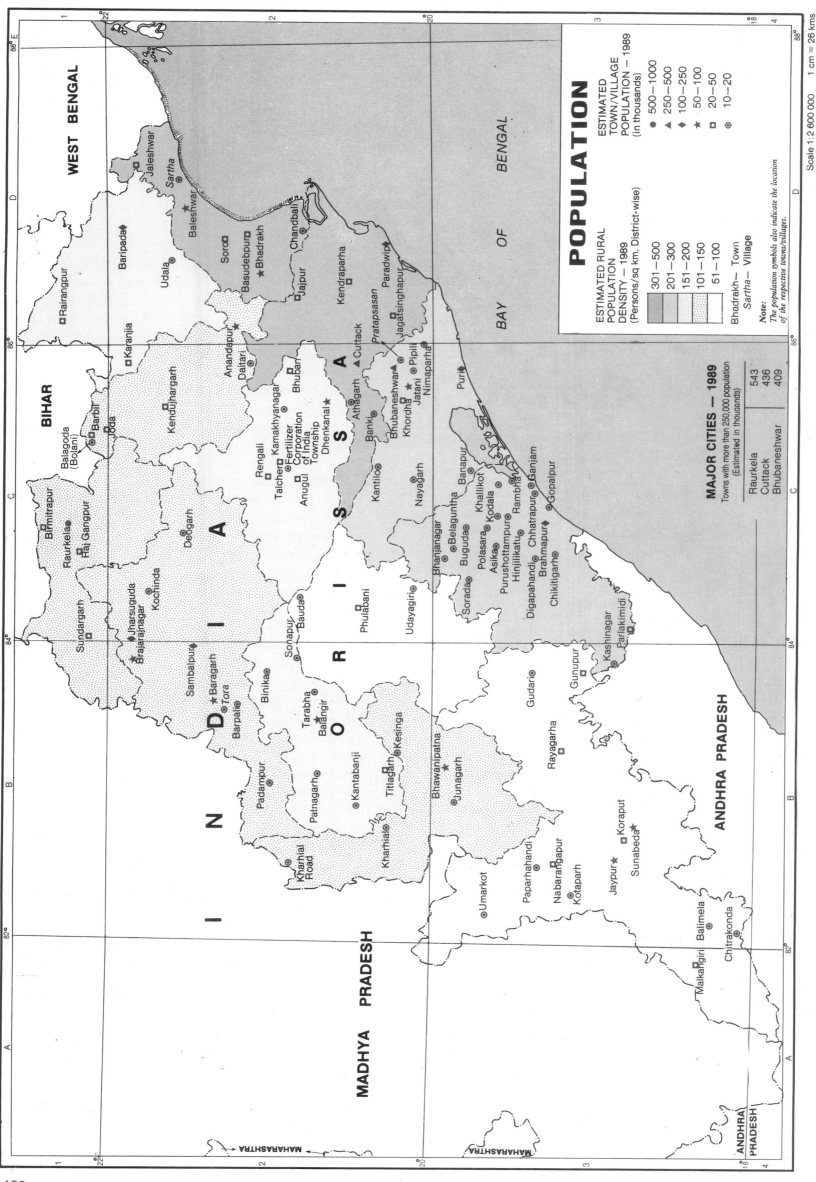

POPULATION

ESTIMATED RURAL POPULATION DENSITY — 1989
(Persons/sq km, District-wise)

- 301—500
- 201—300
- 151—200
- 101—150
- 51—100

ESTIMATED TOWN/VILLAGE POPULATION — 1989
(in thousands)

- ● 500—1000
- ▲ 250—500
- ◆ 100—250
- ★ 50—100
- ▫ 20—50
- ⊙ 10—20

Bhadrakh— Town
Sartha— Village

Note:
The population symbols also indicate the location of the respective towns/villages.

MAJOR CITIES — 1989
Towns with more than 250,000 population
(Estimated in thousands)

Raurkela	543
Cuttack	436
Bhubaneshwar	409

Scale 1:2 600 000   1 cm = 26 kms

BAY OF BENGAL

WEST BENGAL

BIHAR

ORISSA

MADHYA PRADESH

ANDHRA PRADESH

MAHARASHTRA

Jaleshwar
Sartha
Baleshwar
Baripada
Soro
Basudebpura
Bhadrakh
Chandbali
Rairangpur
Udala
Jajpur
Karanjia
Kendraparha
Pratapsasan
Paradwip
Jagatsinghapur
Anandapur
Daitari
Bhuban
Kamakhyanagar
Cuttack
Athagarh
Kendujhargarh
Rengali
Talchera
Fertilizer Corporation of India Township
Anugul
Dhenkanal
Banki
Bhubaneshwar
Khordha
Jatani
Pipili
Nimaparha
Balagoda (Bolani)
Barbil
Joda
Kantilo
Nayagarh
Puri
Birmitrapur
Raurkela
Rai Gangpur
Deogarh
Udayagiri
Phulabani
Sundargarh
Jharsuguda
Kochinda
Brajarajnagar
Bhanjanagar
Belaguntha
Bugudа
Banapur
Khallikot
Kodala
Asika
Polasara
Purushottampur
Hinjilikatu
Chhatrapur
Rambha
Ganjam
Gopalpur
Brahmapur
Chikitigarh
Sambalpur
Baragarh
Tora
Sonapur
Bauda
Sorada
Gudari
Rayagarha
Gunupur
Kashinagar
Parlakimidi
Barpali
Tarabha
Balangir
Binika
Kantabanji
Titlagarh
Kesinga
Padampur
Patnagarh
Bhawanipatna
Junagarh
Kharhiala
Kharhiala Road
Umarkot
Paparhahandi
Nabarangapur
Kotaparh
Jaypur
Koraput
Sunabeda
Balimela
Malkangiri
Chitrakonda

108

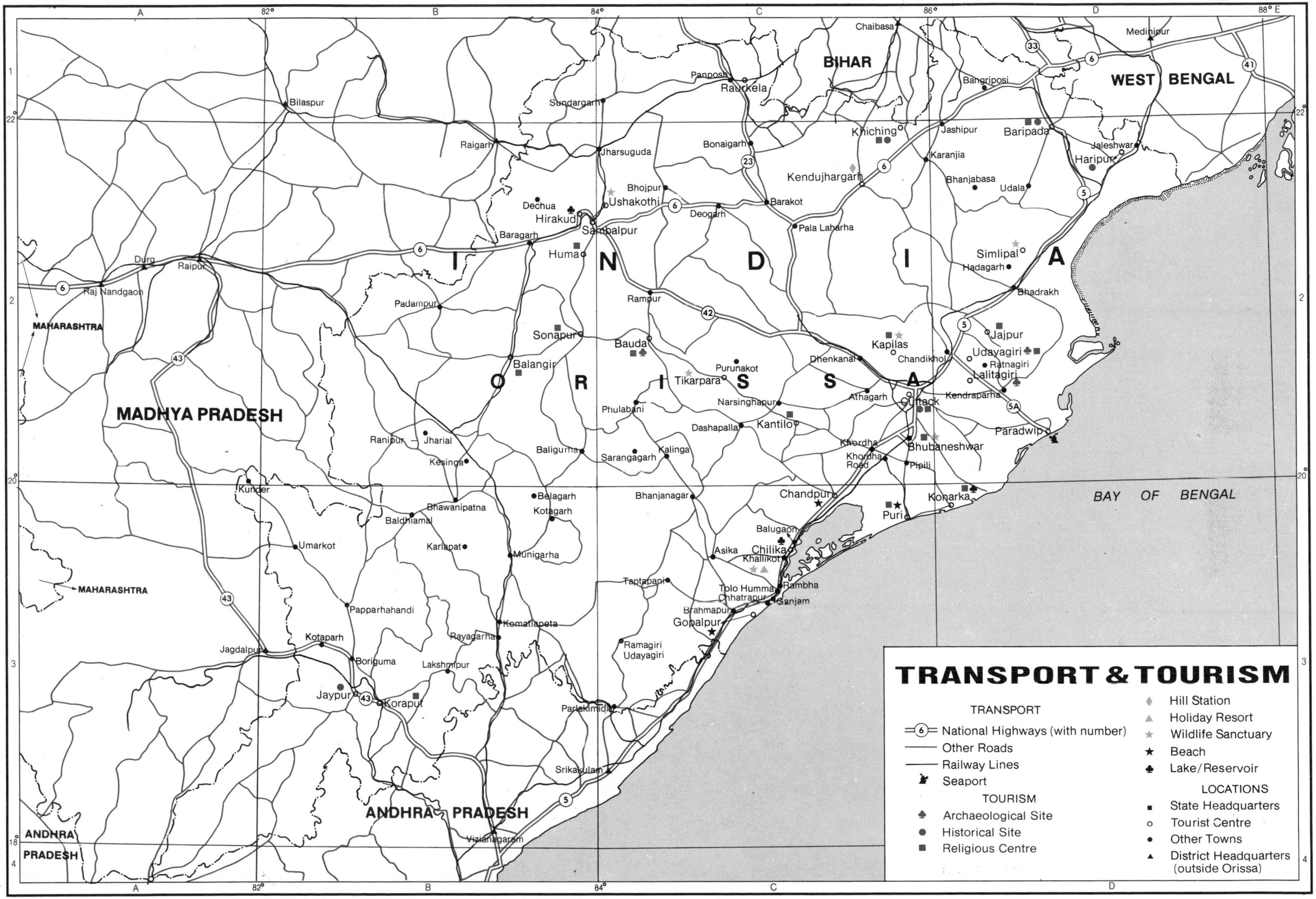

TRANSPORT & TOURISM

Scale 1:2 600 000   1 cm = 26 kms

TRANSPORT

6	National Highways (with number)
	Other Roads
	Railway Lines
	Seaport

TOURISM

	Archaeological Site
	Historical Site
	Religious Centre

	Hill Station
	Holiday Resort
	Wildlife Sanctuary
	Beach
	Lake/Reservoir

LOCATIONS

	State Headquarters
	Tourist Centre
	Other Towns
	District Headquarters (outside Orissa)

WEST BENGAL

BIHAR

ORISSA

MADHYA PRADESH

ANDHRA PRADESH

MAHARASHTRA

BAY OF BENGAL

Medinipur
Jaleshwar
Haripur
Balgiposi
Baripada
Bhanjabasa
Udala
Simlipal
Hadagarh
Bhadrakh
Jajpur
Udayagiri
Ratnagiri
Lalitgiri
Kendrapara
Paradwip
Jashipur
Karanjia
Chandikhol
Chaibasa
Khiching
Kapilas
Chandikhol
Cuttack
Bhubaneshwar
Phulli
Konarka
Puri
Kendujhargarh
Pala Lahara
Athagarh
Khurda
Khurda Road
Barakot
Dhenkanal
Raurkela
Paposi
Bonaigarh
Deogarh
Purunakot
Narsinghpur
Kantilo
Chandpur
Balugaon
Chilika
Khallikot
Tolo Humma
Chhatrapur
Ganjam
Jharsuguda
Bhojpur
Ushakothi
Sambalpur
Rampur
Tikarpara
Dashapalla
Kalinga
Asika
Brahmapur
Gopalpur
Sadargarh
Dechua
Hirakud
Huma
Sonapur
Bauda
Phulabani
Sarangagarh
Bhanjanagar
Ramagiri
Udayagiri
Taptapani
Shbgargarh
Baragarh
Balangir
Baligurha
Balagarh
Kotagarh
Munigarha
Rayagarha
Paralakimidi
Srikakulam
Raigarh
Padampur
Jharial
Kesinga
Bhawanipatna
Karlapat
Lakshmipur
Rayagarha
Kamakhyapeta
Bilaspur
Ranipur
Baldhamal
Papparhahandi
Borjjuma
Koraput
Vizianagaram
Umarkot
Papparhahandi
Jaypur
Kunder
Kotaparh
Jagdalpur
Durg
Raj Nandgaon
Raipur

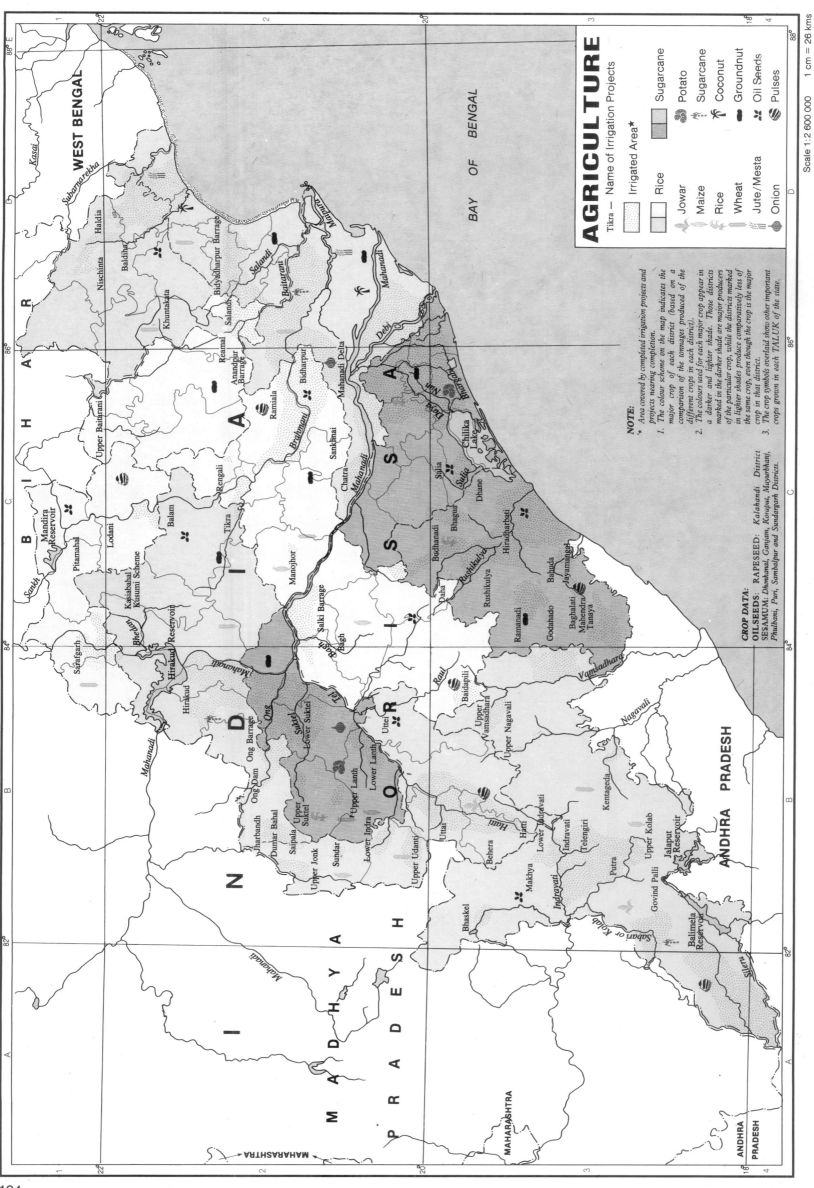

AGRICULTURE

Tikra — Name of Irrigation Projects

Irrigated Area★		Sugarcane	Potato
Rice		Jowar	Maize
		Sugarcane	Rice
		Coconut	Wheat
		Groundnut	Jute/Mesta
		Oil Seeds	Onion
		Pulses	

Scale 1 : 2 600 000   1 cm = 26 kms

NOTE:

★ Area covered by completed irrigation projects and projects nearing completion.

1. The colour scheme on the map indicates the major crop of each district (based on a comparison of the tonnages produced of the different crops in each district).

2. The colours used for each major crop appear in a darker and lighter shade. Those districts marked in the darker shade are major producers of the particular crop, while the districts marked in lighter shades produce comparatively less of the same crop, even though the crop is the major crop in that district.

3. The crop symbols overlaid show other important crops grown in each TALUK of the state.

CROP DATA:

OILSEEDS: RAPESEED: *Kalahandi District*
SESAMUM: *Dhenkanal, Ganjam, Koraput, Mayurbhanj, Phulbani, Puri, Sambalpur and Sundargarh Districts.*

BAY OF BENGAL

WEST BENGAL

B I H A R

O R I S S A

M A D H Y A   P R A D E S H

ANDHRA PRADESH

MAHARASHTRA

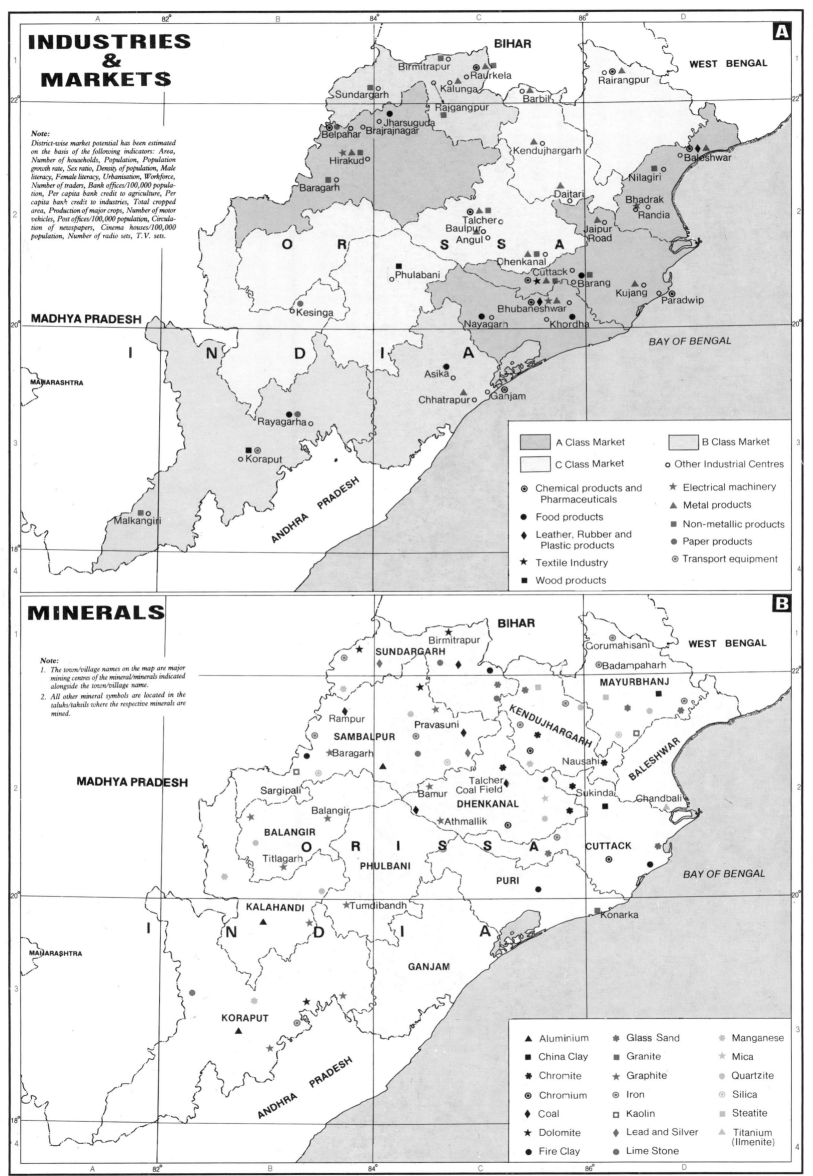

INDUSTRIES & MARKETS

A

BIHAR
WEST BENGAL

Birmitrapur
Raurkela
Sundargarh
Kalunga
Rajgangpur
Barbil
Rairangpur

Jharsuguda
Belpahar Brajrajnagar
Kendujhargarh
Baleshwar
Hirakud
Nilagiri
Baragarh
Daitari
Bhadrak
Randia

O R I S S A

Talcher
Baulpur
Angul
Jaipur Road
Dhenkanal
Cuttack
Barang
Phulabani
Kujang
Paradwip
Bhubaneshwar
Nayagarh
Khordha

MADHYA PRADESH

I N D I A

MAHARASHTRA

Kesinga

BAY OF BENGAL

Asika
Chhatrapur
Ganjam

Rayagarha

Koraput

ANDHRA PRADESH

Malkangiri

	A Class Market		B Class Market
	C Class Market	◦	Other Industrial Centres

⊙ Chemical products and Pharmaceuticals   ★ Electrical machinery
● Food products   ▲ Metal products
◆ Leather, Rubber and Plastic products   ■ Non-metallic products
★ Textile Industry   ● Paper products
■ Wood products   ◉ Transport equipment

MINERALS

B

BIHAR
Birmitrapur
SUNDARGARH
Gorumahisani
Badampaharh
WEST BENGAL

MAYURBHANJ

Rampur
Pravasuni
KENDUJHARGARH
SAMBALPUR
Nausahi
BALESHWAR

MADHYA PRADESH

Baragarh
Bamur
Talcher Coal Field
Sukinda
Chandbali

Sargipali
DHENKANAL
Balangir
Athmallik

BALANGIR
O R I S S A
CUTTACK
Titlagarh
PHULBANI

PURI
Konarka

KALAHANDI
Tumdibandh

I N D I A

MAHARASHTRA

GANJAM

BAY OF BENGAL

KORAPUT

ANDHRA PRADESH

▲ Aluminium	✹ Glass Sand	✶ Manganese
■ China Clay	■ Granite	★ Mica
✲ Chromite	★ Graphite	● Quartzite
◉ Chromium	◉ Iron	◉ Silica
◆ Coal	□ Kaolin	■ Steatite
★ Dolomite	◆ Lead and Silver	▲ Titanium (Ilmenite)
● Fire Clay	● Lime Stone	

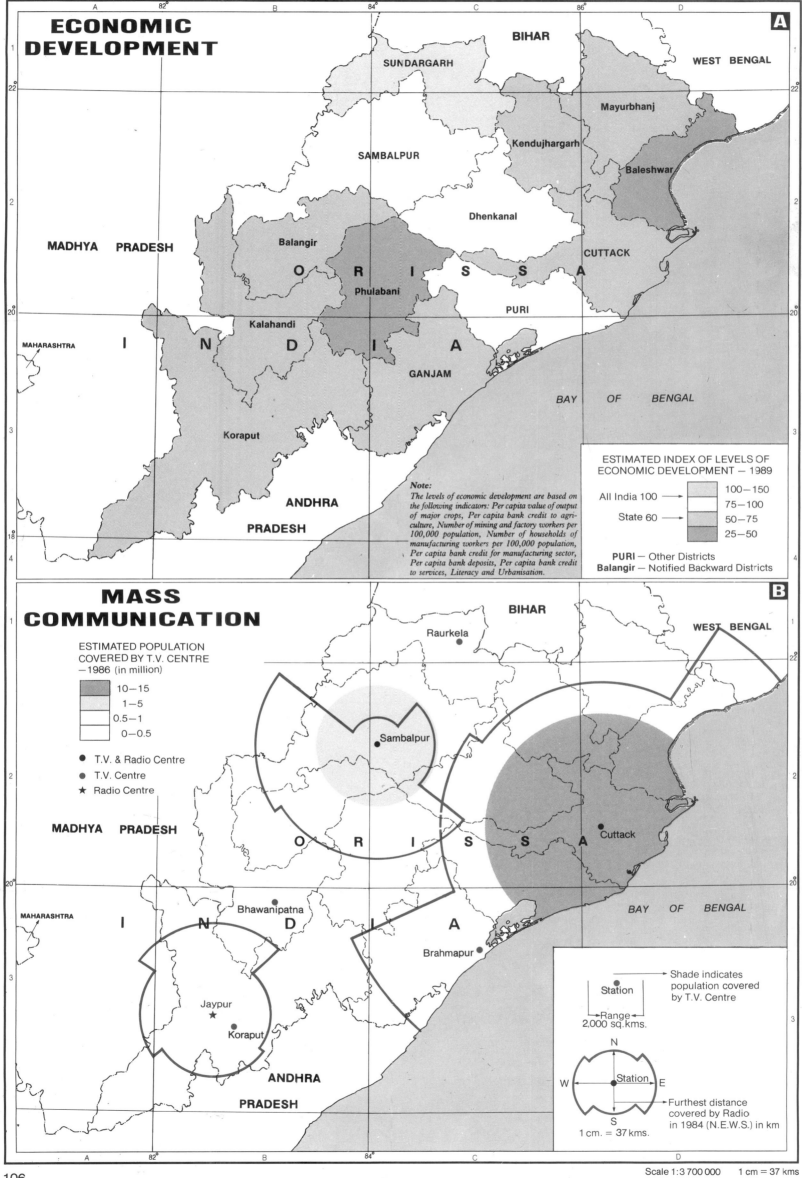

ECONOMIC DEVELOPMENT

BIHAR

WEST BENGAL

SUNDARGARH

Mayurbhanj

SAMBALPUR

Kendujhargarh

Baleshwar

Dhenkanal

MADHYA PRADESH

Balangir

O   R   I   S   S   A

CUTTACK

Phulabani

PURI

Kalahandi

I   N   D   I   A

MAHARASHTRA

GANJAM

Koraput

BAY   OF   BENGAL

ANDHRA

PRADESH

Note:
The levels of economic development are based on the following indicators: Per capita value of output of major crops, Per capita bank credit to agriculture, Number of mining and factory workers per 100,000 population, Number of households of manufacturing workers per 100,000 population, Per capita bank credit for manufacturing sector, Per capita bank deposits, Per capita bank credit to services, Literacy and Urbanisation.

ESTIMATED INDEX OF LEVELS OF ECONOMIC DEVELOPMENT — 1989

All India 100 →
State 60 →

100—150
75—100
50—75
25—50

PURI — Other Districts
Balangir — Notified Backward Districts

MASS COMMUNICATION

ESTIMATED POPULATION COVERED BY T.V. CENTRE —1986 (in million)

10—15
1—5
0.5—1
0—0.5

● T.V. & Radio Centre
● T.V. Centre
★ Radio Centre

BIHAR

Raurkela

WEST BENGAL

MADHYA PRADESH

Sambalpur

O   R   I   S   S   A

Cuttack

I   N   D   I   A

MAHARASHTRA

Bhawanipatna

BAY   OF   BENGAL

Brahmapur

Jaypur

Koraput

ANDHRA

PRADESH

Station
→ Shade indicates population covered by T.V. Centre
Range
2,000 sq.kms.

N
W — Station — E
S

→ Furthest distance covered by Radio in 1984 (N.E.W.S.) in km

1 cm. = 37 kms.

Scale 1:3 700 000   1 cm = 37 kms.

PUNJAB

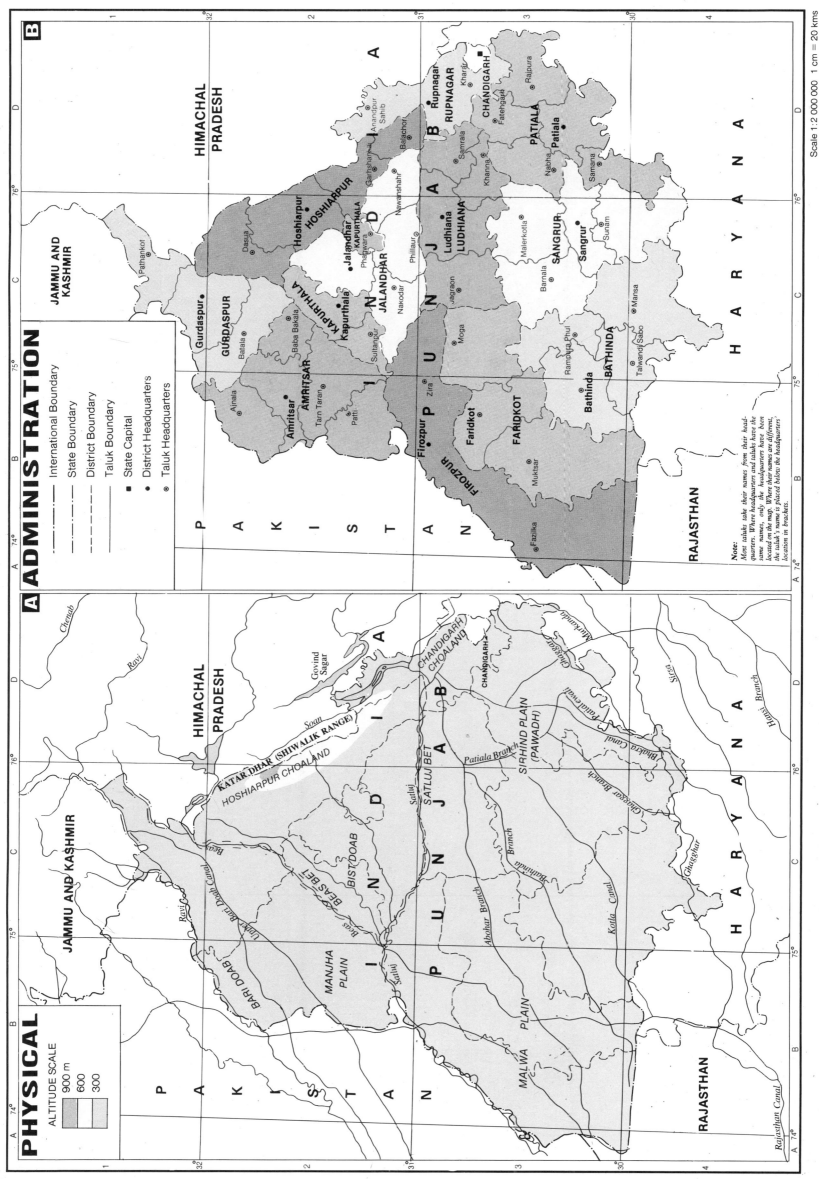

B ADMINISTRATION

———	International Boundary
— · —	State Boundary
— — —	District Boundary
- - - -	Taluk Boundary
■	State Capital
●	District Headquarters
⊛	Taluk Headquarters

Note:
Most taluks take their names from their headquarters. Where headquarters and taluks have the same names, only the headquarters have been located on the map. Where their names are different, the taluk's name is placed below the headquarters' location in brackets.

HIMACHAL PRADESH

JAMMU AND KASHMIR

H A R Y A N A

P A K I S T A N

RAJASTHAN

Pathankot • Gurdaspur
GURDASPUR
Batala ⊛ Baba Bakala
Ajnala ⊛
Amritsar • AMRITSAR
Tarn Taran ⊛
Patti ⊛
Sultanpur ⊛
Kapurthala • KAPURTHALA
Phagwara ⊛
Jalandhar • JALANDHAR
Nakodar ⊛
Phillaur ⊛
Nawanshahr ⊛
Dasua ⊛
Hoshiarpur • HOSHIARPUR
Garhshankar ⊛
Anandpur Sahib ⊛
Balachor ⊛
Rupnagar • RUPNAGAR
Kharar ⊛
CHANDIGARH ■
Fatehgarh ⊛
Rajpura ⊛
PATIALA • Patiala
Nabha ⊛
Samana ⊛
Ludhiana • LUDHIANA
Khanna ⊛
Samrala ⊛
Jagraon ⊛
Moga ⊛
Malerkotla ⊛
SANGRUR • Sangrur
Barnala ⊛
Sunam ⊛
Mansa ⊛
Rampura Phul ⊛
BATHINDA Bathinda •
Talwandi Sabo ⊛
Faridkot • FARIDKOT
Muktsar ⊛
Firozpur • FIROZPUR (Zira)
Fazilka ⊛

A PHYSICAL

ALTITUDE SCALE

	900 m
	600
	300

HIMACHAL PRADESH

JAMMU AND KASHMIR

H A R Y A N A

P A K I S T A N

RAJASTHAN

Chenab
Ravi
Ravi
Beas
Govind Sagar
Soan
KATAR DHAR (SHIWALIK RANGE)
HOSHIARPUR CHOALAND
CHANDIGARH CHOALAND
CHANDIGARH
BARI DOAB
Upper Bari Doab Canal
Satluj
BEAS BET
BIST DOAB
MANJHA PLAIN
SATLUJ BET
Satluj
SIRHIND PLAIN (PAWADH)
Patiala Branch
Abohar Branch
Bhakra Canal
Ghaggar Branch
Kotla Canal
Patiala Branch
Bhakra
Ghaggar
Sirsa
Hansi Branch
Markanda
MALWA PLAIN
Rajasthan Canal

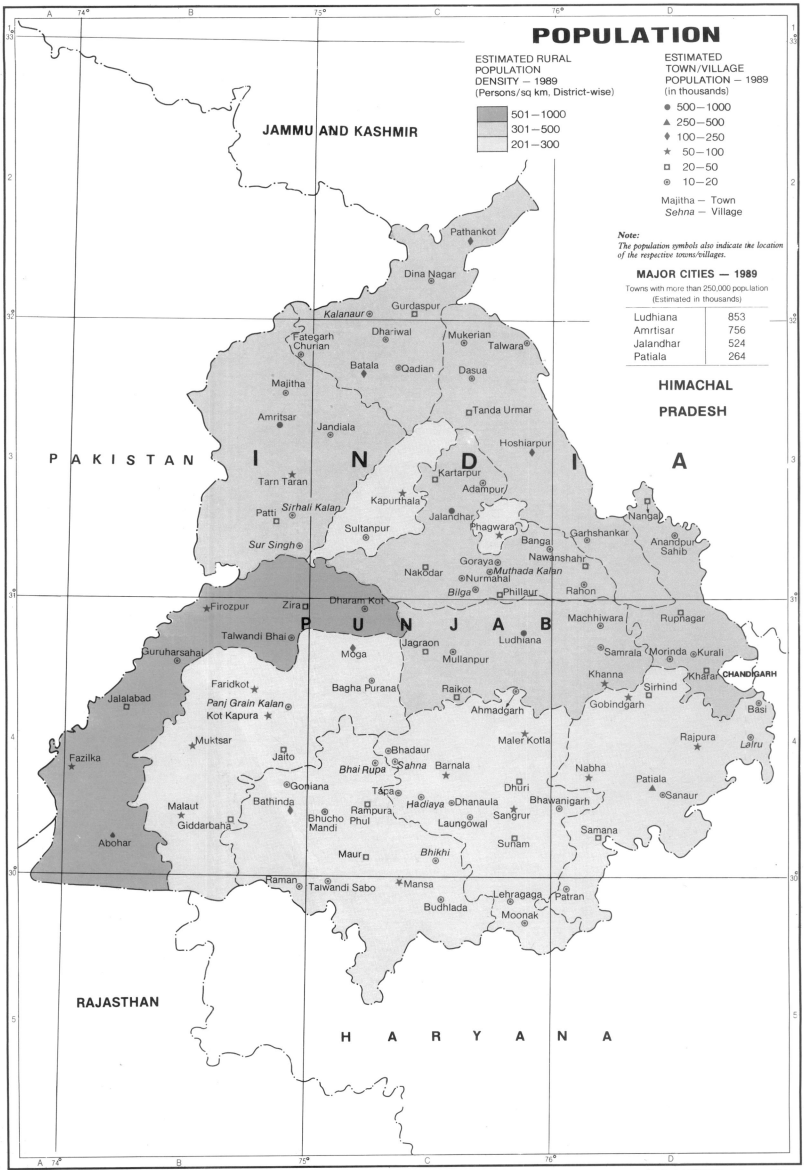

POPULATION

ESTIMATED RURAL POPULATION DENSITY — 1989
(Persons/sq km, District-wise)

- 501—1000
- 301—500
- 201—300

ESTIMATED TOWN/VILLAGE POPULATION — 1989
(in thousands)

- ● 500—1000
- ▲ 250—500
- ◆ 100—250
- ★ 50—100
- □ 20—50
- ◉ 10—20

Majitha — Town
Sehna — Village

Note:
The population symbols also indicate the location of the respective towns/villages.

MAJOR CITIES — 1989
Towns with more than 250,000 population
(Estimated in thousands)

Ludhiana	853
Amrtisar	756
Jalandhar	524
Patiala	264

JAMMU AND KASHMIR

PAKISTAN

HIMACHAL PRADESH

Pathankot

Dina Nagar

Gurdaspur
Kalanaur
Fategarh Churian
Dhariwal
Mukerian
Talwara

Majitha
Batala
Qadian
Dasua

Amritsar
Jandiala
Tanda Urmar

Tarn Taran
Kartarpur
Hoshiarpur

Patti
Kapurthala
Adampur
Nangal

Sirhali Kalan
Sultanpur
Jalandhar
Phagwara
Banga
Garhshankar
Anandpur Sahib

Sur Singh
Nakodar
Goraya
Muthada Kalan
Nawanshahr
Nurmahal
Rahon

Bilga
Phillaur

I N D I A

Firozpur
Zira
Dharam Kot
Machhiwara
Rupnagar

P U N J A B
Talwandi Bhai
Jagraon
Ludhiana
Samrala
Morinda
Kurali

Guruharsahai
Moga
Mullanpur
Khanna
Sirhind
Kharar
CHANDIGARH

Faridkot
Bagha Purana
Raikot
Gobindgarh
Basi

Jalalabad
Panj Grain Kalan
Ahmadgarh

Kot Kapura
Maler Kotla
Rajpura
Lalru

Muktsar
Jaito
Bhadaur
Nabha
Patiala

Fazilka
Bhai Rupa
Sahna
Barnala
Dhuri
Sanaur

Goniana
Tapa
Dhanaula
Bhawanigarh

Malaut
Bathinda
Hadiaya
Sangrur
Samana

Giddarbaha
Bhucho Mandi
Rampura Phul
Laungowal
Sunam

Abohar
Maur
Bhikhi

Raman
Talwandi Sabo
Mansa
Lehragaga
Patran

Budhlada
Moonak

RAJASTHAN

H A R Y A N A

Scale 1:1 500 000   1 cm = 15 kms

B

AGRICULTURE

Harike — Name of Irrigation Projects

Irrigated Area*

- Wheat
- Sugarcane

Crop symbols:
- Barley
- Maize
- Rice
- Wheat
- Chilli
- Cotton
- Potato
- Sugarcane
- Tobacco
- Groundnut
- Oil Seeds
- Pulses

HIMACHAL PRADESH

JAMMU AND KASHMIR

PAKISTAN

PUNJAB

INDIA

HARYANA

Rivers/places: Basantpur, Soan, Nangal, Beas, Ravi, Sutlej, Harike, Ferozpur Feeder Works, Rajasthan, Roper, Panala Branch, Kotla Branch, Abohar Branch, Bathinda Branch, Bhakra Canal, Ghaggar, Saraswati

CROP DATA:
OILSEEDS: MUSTARD: *Moga, Muktsar Taluks (Faridkot Dt.), Barnala (Sangrur), Talwandi Sabo (Bathinda).*
PULSES: GRAM: *Muktsar Taluk (Faridkot Dt.), Ludhiana (Ludhiana), Sunam (Sangrur), Amritsar (Amritsar), Talwandi Sabo (Bathinda).*

Note:

* Area covered by completed irrigation projects and projects nearing completion.

1. The colour scheme on the map indicates the major crop of each district (based on a comparison of the tonnages produced of the different crops in each district).

2. The colours used for each major crop appear in a darker and lighter shade. Those districts marked in the darker shade are major producers of the particular crop, while the districts marked in lighter shades produce comparatively less of the same crop, even though the crop is the major crop in that district.

3. The crop symbols overlaid show other important crops grown in each TALUK of the state.

Scale 1:2 000 000   1 cm = 20 kms

A

TRANSPORT & TOURISM

TRANSPORT

- (15) National Highways (with number)
- Other Roads
- Railway Lines

TOURISM

- ♣ Archaeological Site
- ● Historical Site
- ■ Religious Centre
- ◄ Holiday Resort
- ★ Wildlife Sanctuary
- Lake/Reservoir

LOCATIONS

- ♣ State Headquarters
- ■ Tourist Centre
- ○ Other Towns
- ● District Headquarters
- ▲ (outside Punjab)

HIMACHAL PRADESH

JAMMU AND KASHMIR

PUNJAB

INDIA

HARYANA

PAKISTAN

RAJASTHAN

Places: Chamba, Dharmsala, Mandi, Bilaspur, Nangal, Hamirpur, Una, Kiratpur, Anandpur, Nawanshahr, Rahon, Ropar, Kurali, Kurali, Chandigarh, Fatehgarh, Dera Bassi, Rajpura, Ambala, Kurukshetra, Shahpur, Pathankot, Dholbaha, Dasuya, Mukerian, Tanda, Hoshiarpur, Phagwara, Phillaur, Machhiwara, Neelon, Samrala, Khanna, Patiala, Nabha, Bhawanigarh, Doraha, Ind, Gurdaspur, Qadian, Dhariwal, Batala, Dera Baba Nanak, Govindwal, Dera Baba Jaimal Singh, Sultanpur, Kapurthala, Jalandhar, Nakodar, Ludhiana, Jagraon, Dhuri, Maler Kotla, Barnala, Sangrur, Sunam, Budhlada, Dogal, Amritsar, Atari, Tarn Taran, Ajnala, Patti, Harike, Husainiwala, Zira, Moga, Nihal Singhwala, Nathana, Bhikhi, Mansa, Sirsa, Firozpur, Faridkot, Kot Kapura, Muktsar, Baja Khana, Ballinda, Lambi, Jalalabad, Malaut, Fazilka, Abohar, Ganganagar, Sirhind

109

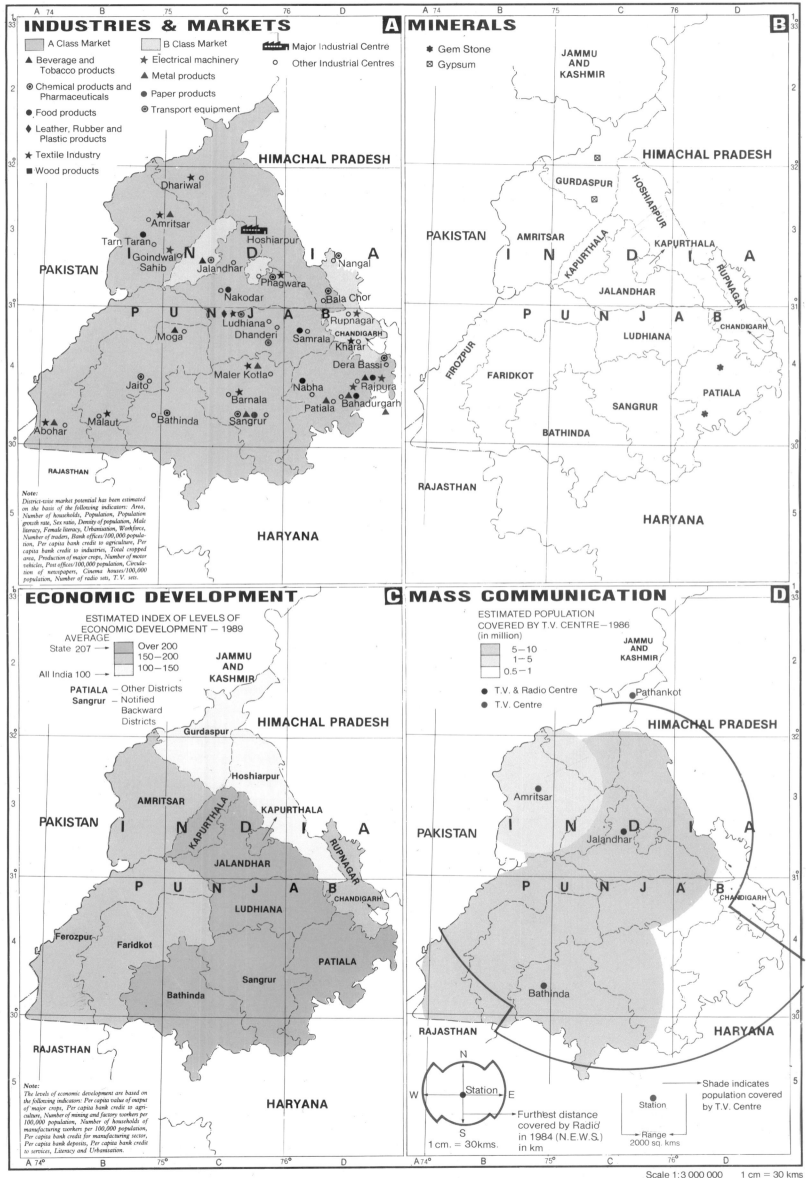

INDUSTRIES & MARKETS

- ▨ A Class Market
- ▨ B Class Market
- 🏭 Major Industrial Centre
- ▲ Beverage and Tobacco products
- ★ Electrical machinery
- ○ Other Industrial Centres
- ▲ Metal products
- ◉ Chemical products and Pharmaceuticals
- ● Paper products
- ● Food products
- ◉ Transport equipment
- ◆ Leather, Rubber and Plastic products
- ★ Textile Industry
- ■ Wood products

HIMACHAL PRADESH

PAKISTAN

I N D I A

P U N J A B

Dhariwal
Amritsar
Tarn Taran
Goindwal Sahib
Jalandhar
Hoshiarpur
Phagwara
Nangal
Nakodar
Bala Chor
Moga
Ludhiana
Dhanderi
Rupnagar
Samrala
CHANDIGARH
Kharar
Dera Bassi
Maler Kotla
Nabha
Rajpura
Jaito
Barnala
Patiala
Bahadurgarh
Malaut
Bathinda
Sangrur
Abohar

RAJASTHAN

HARYANA

Note:
District-wise market potential has been estimated on the basis of the following indicators: Area, Number of households, Population, Population growth rate, Sex ratio, Density of population, Male literacy, Female literacy, Urbanisation, Workforce, Number of traders, Bank offices/100,000 population, Per capita bank credit to agriculture, Per capita bank credit to industries, Total cropped area, Production of major crops, Number of motor vehicles, Post offices/100,000 population, Circulation of newspapers, Cinema houses/100,000 population, Number of radio sets, T.V. sets.

MINERALS

- ✱ Gem Stone
- ⊠ Gypsum

JAMMU AND KASHMIR

HIMACHAL PRADESH

PAKISTAN

GURDASPUR
HOSHIARPUR
AMRITSAR
KAPURTHALA
KAPURTHALA
RUPNAGAR
JALANDHAR
CHANDIGARH
I N D I A
P U N J A B
FIROZPUR
LUDHIANA
FARIDKOT
PATIALA
SANGRUR
BATHINDA

RAJASTHAN

HARYANA

ECONOMIC DEVELOPMENT

ESTIMATED INDEX OF LEVELS OF ECONOMIC DEVELOPMENT — 1989

AVERAGE
State 207 →
All India 100 →

- ▨ Over 200
- ▨ 150–200
- ▨ 100–150

PATIALA – Other Districts
Sangrur – Notified Backward Districts

JAMMU AND KASHMIR

HIMACHAL PRADESH

PAKISTAN

Gurdaspur
Hoshiarpur
AMRITSAR
KAPURTHALA
KAPURTHALA
RUPNAGAR
I N D I A
JALANDHAR
CHANDIGARH
P U N J A B
LUDHIANA
Ferozpur
Faridkot
PATIALA
Sangrur
Bathinda

RAJASTHAN

HARYANA

Note:
The levels of economic development are based on the following indicators: Per capita value of output of major crops, Per capita bank credit to agriculture, Number of mining and factory workers per 100,000 population, Number of households of manufacturing workers per 100,000 population, Per capita bank credit for manufacturing sector, Per capita bank deposits, Per capita bank credit to services, Literacy and Urbanisation.

MASS COMMUNICATION

ESTIMATED POPULATION COVERED BY T.V. CENTRE–1986 (in million)

- ▨ 5–10
- ▨ 1–5
- ▨ 0.5–1

- ● T.V. & Radio Centre
- ● T.V. Centre

JAMMU AND KASHMIR

HIMACHAL PRADESH

PAKISTAN

Pathankot
Amritsar
I N D I A
Jalandhar
P U N J A B
CHANDIGARH
Bathinda

RAJASTHAN

HARYANA

N
W — Station — E
S

1 cm. = 30 kms.

Furthest distance covered by Radio in 1984 (N.E.W.S.) in km

Shade indicates population covered by T.V. Centre

● Station

Range 2000 sq. kms

Scale 1:3 000 000   1 cm = 30 kms

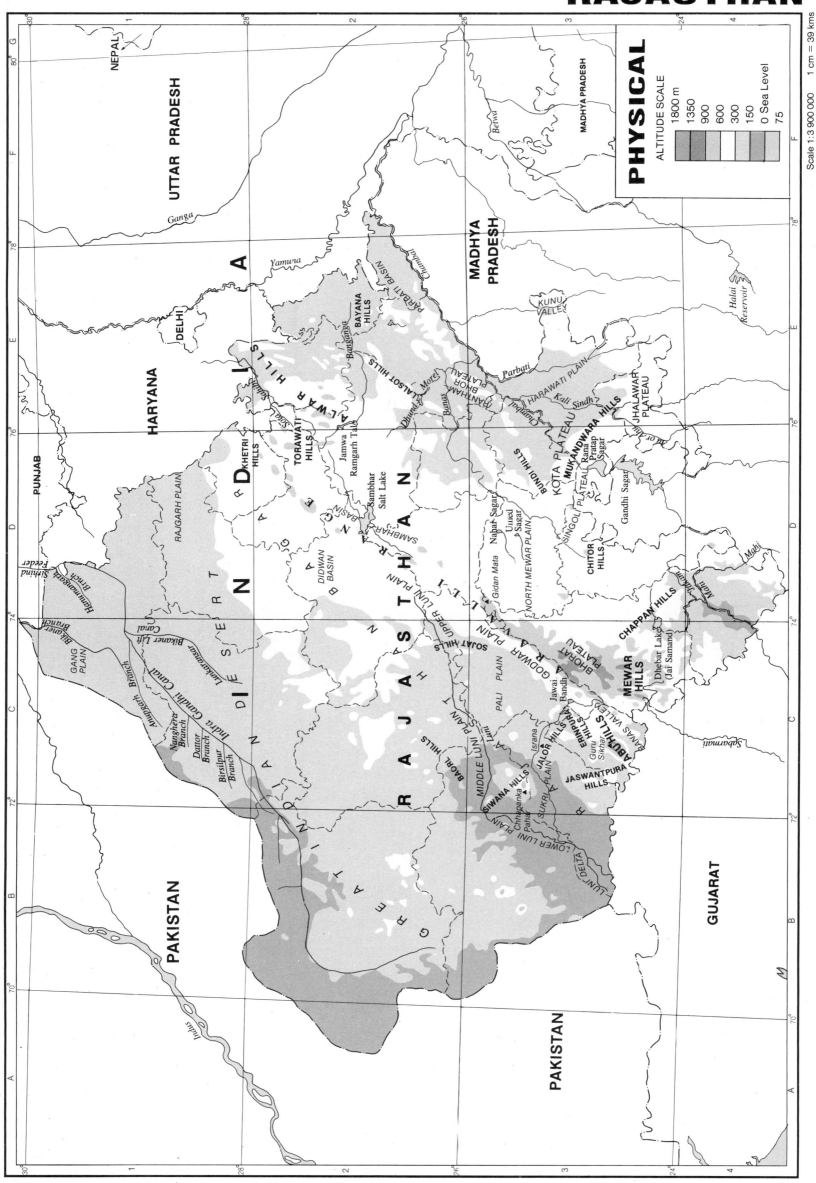

RAJASTHAN

PHYSICAL

ALTITUDE SCALE

	1800 m
	1350
	900
	600
	300
	150
	0 Sea Level
	75

NEPAL

UTTAR PRADESH

HARYANA

PUNJAB

DELHI

PAKISTAN

GUJARAT

MADHYA PRADESH

Ganga

Yamuna

Banganga

BAYANA HILLS

PARBATI BASIN

KUNU VALLEY

Halai Reservoir

Benwa

MADHYA PRADESH

Chambal

ALWAR HILLS

KHETRI HILLS

TORAWATI HILLS

LALSOT HILLS

Sota

Sabi

Sabi

Morel

Dhund

Banas

RANTHAMBHOR PLATEAU

BHOR PLATEAU

HARAWATI PLAIN

Parbati

Kali Sindh

Chambal

JHALAWAR PLATEAU

MUKANDWARA HILLS

KOTA PLATEAU

Rana Pratap Sagar

Jamwa Ramgarh Tal

Sambhar Salt Lake

Sambhar

SAMBHAR BASIN

DIDWAN BASIN

BUNDI HILLS

SINGOLI PLATEAU

Uniel Sagar

Nahar Sagar

Gandhi Sagar

CHITOR HILLS

NORTH MEWAR PLAIN

Gidan Mata

RAIGARH PLAIN

THAR DESERT

GREAT INDIAN DESERT

ARAVALLI RANGE

Sirhind Feeder

Hanumangarh Birch

Bikaner Branch

GANG PLAIN

Amarpura Branch

Nanghera Branch

Dattor Branch

Birsilpur Branch

Lunkaransar Lift Canal

Indira Gandhi Canal

Bikaner Lift Canal

UPPER LUNI PLAIN

SOJAT HILLS

GODWAR PLAIN

PALI PLAIN

Luni

Jawai

MIDDLE LUNI

BAORI HILLS

SIWANA HILLS

JALOR HILLS

Israna

Chhapanka Pahar

SUKRI PLAIN

JASWANTPURA HILLS

Guru Sikhar

ERINPURA HILLS

ABU HILLS

MEWAR HILLS

BHORAT PLATEAU

CHAPPAN HILLS

Dhebar Lake (Jai Samand)

BANAS VALLEY

Mahi

Sabarmati

Mahi

LOWER LUNI PLAIN

LUNI DELTA

Indus

A M A B C D E F G

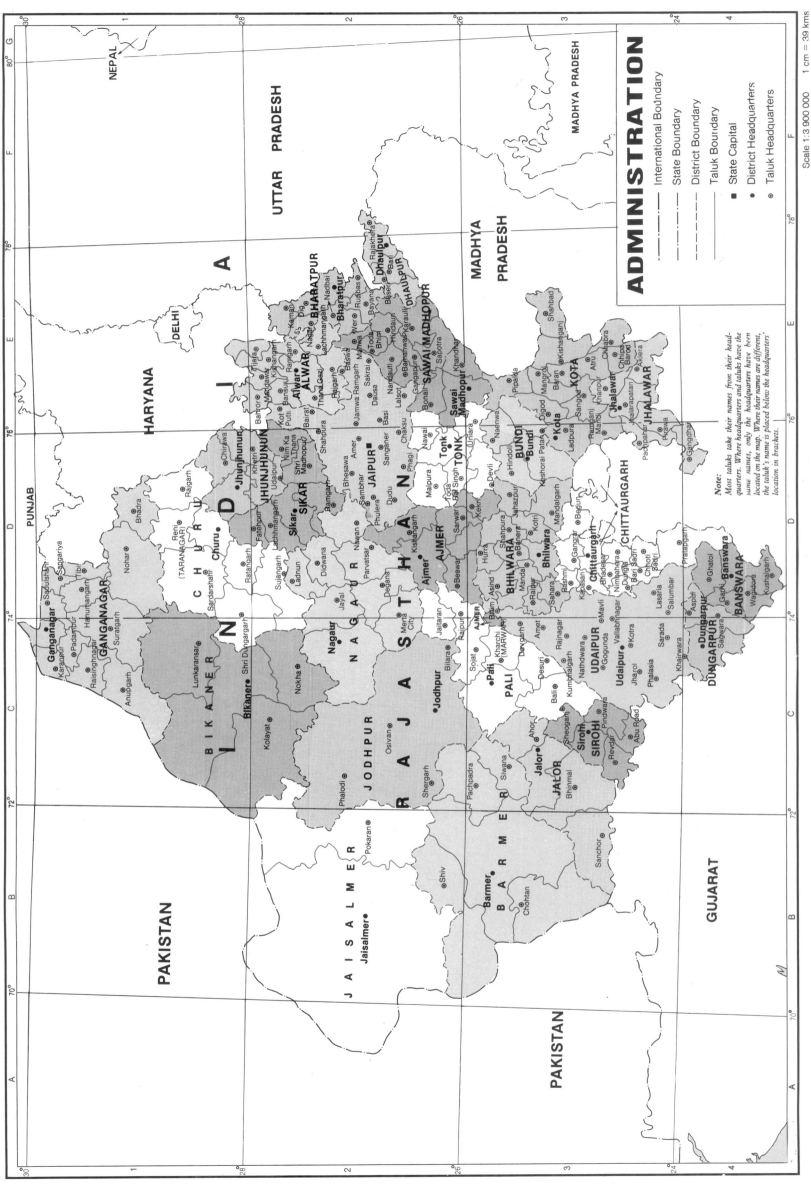

ADMINISTRATION

	International Boundary
	State Boundary
	District Boundary
	Taluk Boundary
■	State Capital
●	District Headquarters
⊚	Taluk Headquarters

Scale 1:3 900 000   1 cm = 39 kms

Note:
Most taluks take their names from their headquarters. Where headquarters and taluks have the
same names, only the headquarters have been
located on the map. Where their names are different,
the taluk's name is placed below the headquarters'
location in brackets.

PAKISTAN

PAKISTAN

NEPAL

PUNJAB

HARYANA

DELHI

UTTAR   PRADESH

MADHYA
PRADESH

MADHYA PRADESH

GUJARAT

R A J A S T H A N

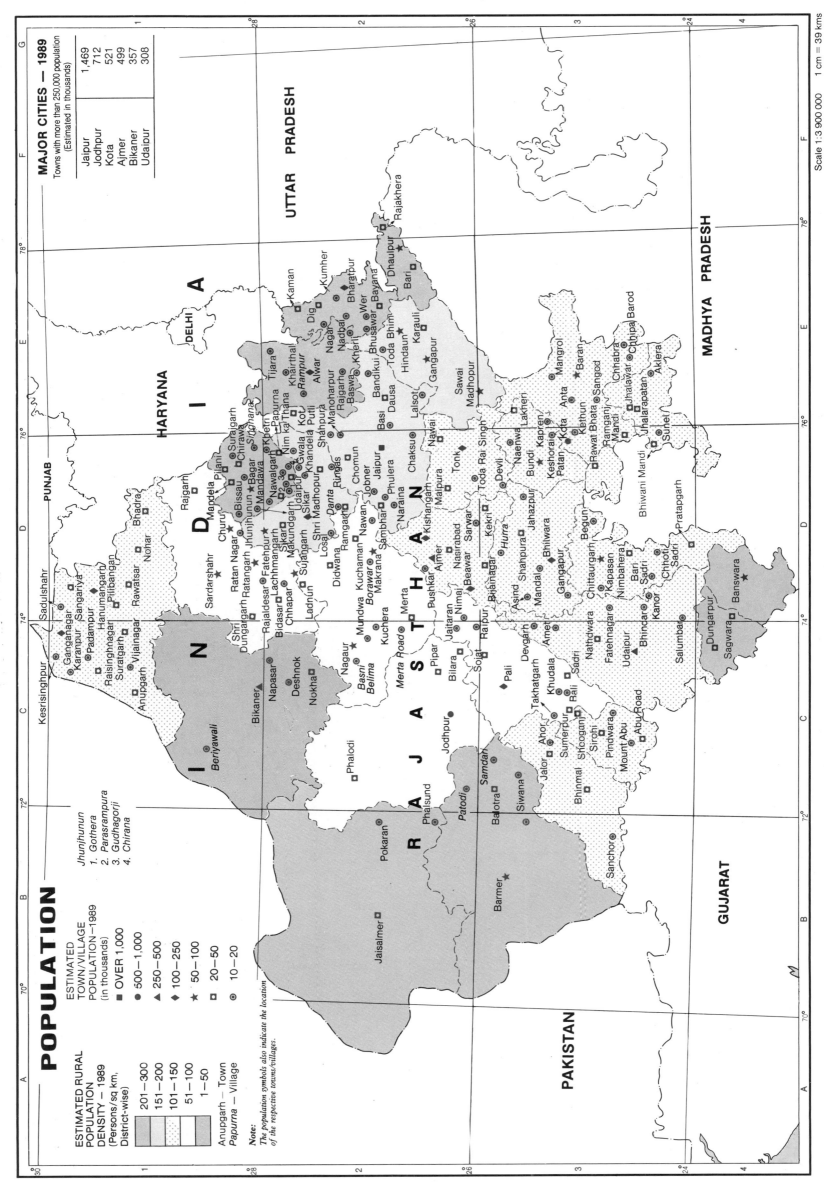

POPULATION

ESTIMATED RURAL
POPULATION
DENSITY – 1989
(Persons/sq km,
District-wise)

201–300
151–200
101–150
51–100
1–50

ESTIMATED
TOWN/VILLAGE
POPULATION–1989
(in thousands)

■ OVER 1,000
● 500–1,000
▲ 250–500
◆ 100–250
★ 50–100
□ 20–50
◎ 10–20

Anupgarh – Town
Papurna – Village

Note:
*The population symbols also indicate the location
of the respective towns/villages.*

Jhunjhunun
1. *Gothera*
2. *Parasrampura*
3. *Gudhagorji*
4. *Chirana*

MAJOR CITIES — 1989

Towns with more than 250,000 population
(Estimated in thousands)

Jaipur	1,469
Jodhpur	712
Kota	521
Ajmer	499
Bikaner	357
Udaipur	308

Scale 1:3 900 000   1 cm = 39 kms

PAKISTAN

PUNJAB

HARYANA

DELHI

UTTAR PRADESH

MADHYA PRADESH

GUJARAT

I N D I A

R A J A S T H A N

Kesrisinghpur
Sadulshahr
Ganganagar
Karanpur
Sangariya
Raisinghnagar
Padampur
Hanumangarh
Pilibangan
Suratgarh
Anupgarh
Vijainagar
Rawatsar
Bhadra
Nohar
Rajgarh
Sardarshahr
Shri Dungargarh
Ratan Nagar
Churu
Mandela
Pilani
Surajgarh
Chirawa
Bissau
Bagar
Sidmana
Mandawa
Nawalgarh
Jhunjhunun
Khetri
Papurna
Tijara
Kaman
Dig
Kumher
Bharatpur
Nadbai
Wer
Bayana
Rupbas
Dhaulpur
Bari
Rajakhera
Nagar
Kherli
Alwar
Rampur
Khairthal
Manoharpur
Bandikui
Bhusawar
Baswa
Rajgarh
Dausa
Basi
Toda Bhim
Hindaun
Karauli
Gangapur
Sawai Madhopur
Kotputli
Shahpura
Khandela
Gwala Putli
Sikar
Udaipurwati
Ramgarh
Losal
Danta
Sri Madhopur
Rungas
Chomu
Jobner
Jaipur
Phulera
Chaksu
Lalsot
Nawai
Tonk
Toda Rai Singh
Malpura
Deoli
Newai
Lakheri
Bundi
Kapren
Keshoraipatan
Kota
Anta
Baran
Chhabra
Mangrol
Chhipa Barod
Aklera
Jhalarapatan
Jhalawar
Mandi
Chhabra
Ramganj Mandi
Rawat Bhata
Sangod
Bhinmal
Mardawa
Makundgarh
Shri Mangal
Ratangarh
Lachhmangarh
Rajaldesar
Fatehpur
Bidasar
Chhapar
Ladnun
Didwana
Kuchaman
Nawan
Makrana
Borawar
Sambhar
Naraina
Kishangarh
Pushkar
Ajmer
Nasirabad
Beawar
Sarwar
Bilainagar
Asind
Shahpura
Jahazpur
Kekri
Hurra
Nimaj
Vijainagar
Jaitaran
Raipur
Sojat
Bilara
Pipar
Merta Road
Merta
Mundwa
Kuchera
Nagaur
Basni
Belima
Deshnok
Nokha
Napasar
Bikaner
Beriyawali
Phalodi
Jaisalmer
Pokaran
Phalsund
Barmer
Balotra
Siwana
Samdari
Patodi
Sanchor
Bhinmal
Jalor
Ahor
Sumerpur
Sheoganj
Sirohi
Pindwara
Mount Abu
Abu Road
Takhatgarh
Khudala
Bali
Raipur
Sumerpur
Devgarh
Kumbhalgarh
Amet
Nathdwara
Fatehnagar
Udaipur
Bhinder
Kanor
Salumbar
Dungarpur
Sagwara
Banswara
Pratapgarh
Pipar
Sadri
Nimbahera
Kapasan
Chittaurgarh
Gangapur
Mandal
Bhilwara
Begun
Jodhpur

TRANSPORT & TOURISM

TRANSPORT
- ══9══ National Highways (with number)
- ────── Other Roads
- ────── Railway Lines

TOURISM
- ✠ Archaeological Site
- ● Historical Site
- ■ Religious Centre
- ◆ Hill Station
- ◀ Holiday Resort
- ★ Wildlife Sanctuary
- ✤ Lake/Reservoir

LOCATIONS
- ■ State Headquarters
- ○ Tourist Centre
- ● Other Towns
- ◀ District Headquarters (outside Rajasthan)

Scale 1:3 900 000   1 cm = 39 kms

RAJASTHAN

PAKISTAN

PUNJAB

HARYANA

UTTAR PRADESH

MADHYA PRADESH

GUJARAT

NEPAL

DELHI

Selected place names: Ganganagar, Hanumangarh, Suratgarh, Anupgarh, Sarupsar, Bikaner, Gajner, Kolayat, Nokhra, Pugal, Ranjitpura, Naukh, Chhattargarh, Hanseran, Sahawa, Nohar, Bhadra, Sirsa, Taranagar (Reni), Churu, Sardarshahr, Ratangarh, Sujangarh, Chhapar, Sri Dungargarh, Deshnok, Nokha, Mandi, Phalodi, Lohawat, Osiyan, Shaitrawa, Shergarh, Nagaur, Deh, Didwana, Degana, Merta, Tarnau, Jayal, Ramgarh, Parvatsar, Kishangarh, Ajmer, Pushkar, Nasirabad, Beawar, Bar, Ras, Sojat, Marwar, Bilara, Rohat, Pipar, Jodhpur, Mandor, Bhopalgarh, Khangar, Pokaran, Ramdevra, Bandevra, Baniyana, Phalsund, Kanasar, Shiv, Bharka, Myajlar, Harsani, Ramsar, Barmer, Sindari, Gurha, Siwana, Balotra, Jalor, Bhinmal, Ramsin, Sirohi, Sanchor, Chitalwana, Jaisalmer, Mandha, Hamira, Lodurva, Sam, Devikot, Bhadasar, Ramgarh, Ghotaru, Sadhewala Tar, Tanot, Sarkari Tala, Chinnu, Nachna, Ghantiyali, Mount Abu, Abu Road, Pindwara, Sheoganj, Ranakpur, Kumbhalgarh, Nathdwara, Haldi Ghat, Udaipur, Nai, Nimbahera, Chittaurgarh, Bhilwara, Bundi, Kota, Baran, Jhalawar, Jhalrapatan, Ak.lera, Rajgarh, Pratapgarh, Dungarpur, Banswara, Kushalgarh, Pratapnur, Dhariawad, Som, Mandalgarh, Shahpura, Kekri, Devli, Tonk, Malpura, Phagi, Phuda, Chomun, Sambhar, Salt Lake, Sikar, Ringas, Udaipur, Shahpura, Neem-Ka-Thana, Khetri, Jhunjhunun, Nawalgarh, Pilani, Narnaul, Bahror, Kot Putli, Siliserh, Sariska, Alwar, Bayana, Bharatpur, Dig, Kaman, Deeg, Dhaulpur, Karauli, Mahwa, Hindaun, Gangapur, Bharoti, Sawai Madhopur, Ranthambhor, Lalsot, Dausa, Jaipur, Amer, Ramgarh, Jamwa Ramgarh

UTTAR PRADESH towns: Sitapur, Bareilly, Rampur, Moradabad, Etawah, Agra, Shivpuri, Guna, Bhopal

HARYANA towns: Kurukshetra, Karnal, Sonipat, Jind, Hisar, Bhiwani, Rohtak, Ghaziabad, Faridabad, Gurgaon

National Highways: 1, 2, 3, 8, 10, 11, 12, 15, 24

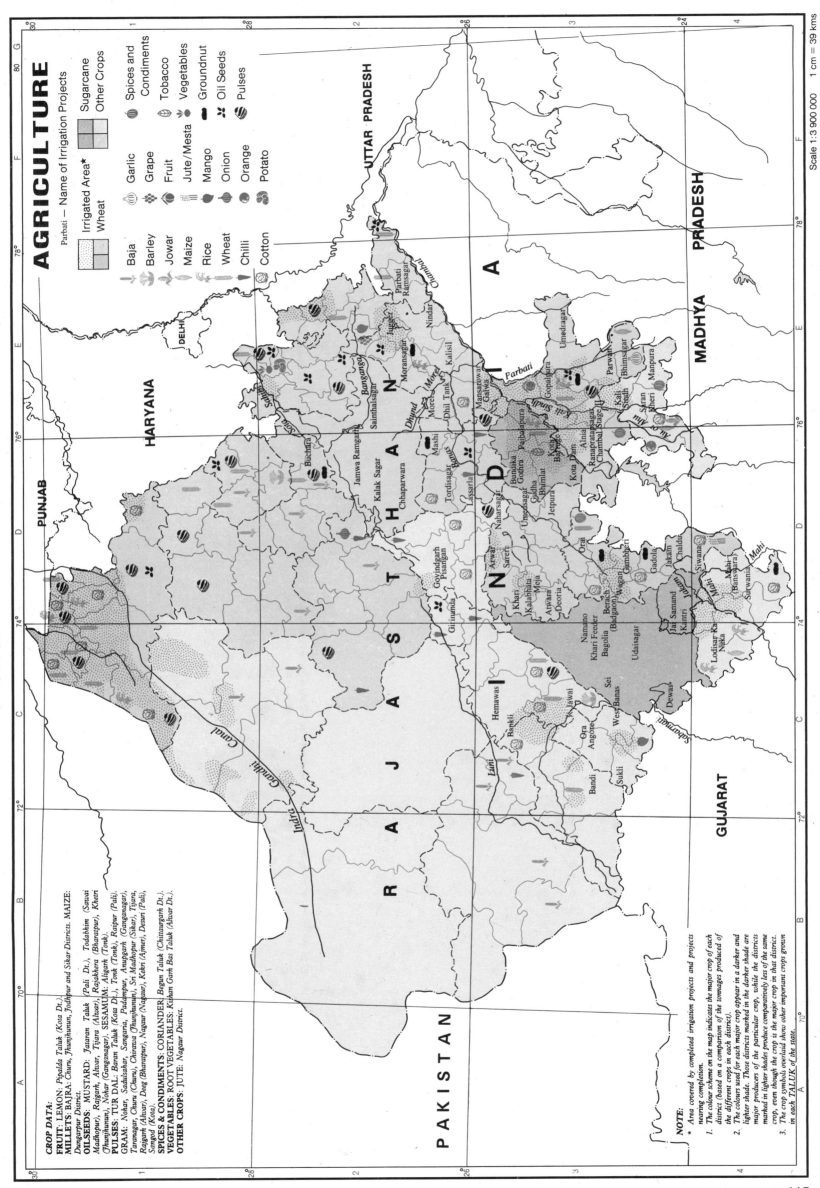

AGRICULTURE

Parbati – Name of Irrigation Projects

Legend

Irrigated Area*
Wheat
Sugarcane
Other Crops

Garlic   Spices and Condiments
Grape   Tobacco
Fruit   Vegetables
Jute/Mesta   Groundnut
Mango   Oil Seeds
Onion   Pulses
Orange
Potato

Baja   Barley   Jowar   Maize   Rice   Wheat   Chilli   Cotton

CROP DATA:
FRUIT: LEMON: *Pipalda Taluk (Kota Dt.).*
MILLETS: BAJRA: *Churu, Jhunjhunun, Jodhpur and Sikar Districts.* MAIZE: *Dungarpur District.*
OILSEEDS: MUSTARD: *Jaitaran Taluk (Pali Dt.), Todabhim (Sawai Madhopur), Rajgarh, Alwar, Tijara (Alwar), Rajakhera (Bharatpur), Khetri (Jhunjhunun), Nohar (Ganganagar).* SESAMUM: *Aligarh (Tonk).*
PULSES: TUR DAL: *Baran Taluk (Koa Dt.), Tonk (Tonk), Raipur (Pali).*
GRAM: *Nohar, Sadulshahar, Sangaria, Padampur, Anupgarh (Ganganagar), Taranagar, Churu (Churu), Chirawa (Jhunjhunun), Sri Madhopur (Sikar), Tijara, Rajgarh (Alwar), Deeg (Bharatpur), Nagaur (Nagaur), Kekri (Ajmeri), Desuri (Pali), Songod (Kota).*
SPICES & CONDIMENTS: CORIANDER: *Begun Taluk (Chittaurgarh Dt.).*
VEGETABLES: ROOT VEGETABLES: *Kishan Garh Bas Taluk (Alwar Dt.).*
OTHER CROPS: JUTE: *Nagaur District.*

NOTE:
* Area covered by completed irrigation projects and projects nearing completion.
1. The colour scheme on the map indicates the major crop of each district (based on a comparison of the tonnages produced of the different crops in each district).
2. The colours used for each major crop appear in a darker and lighter shade. Those districts marked in the darker shade are major producers of the particular crop, while the districts marked in lighter shades produce comparatively less of the same crop, even though the crop is the major crop in that district.
3. The crop symbols overlaid show other important crops grown in each TALUK of the state.

Scale 1:3 900 000   1 cm = 39 kms

115

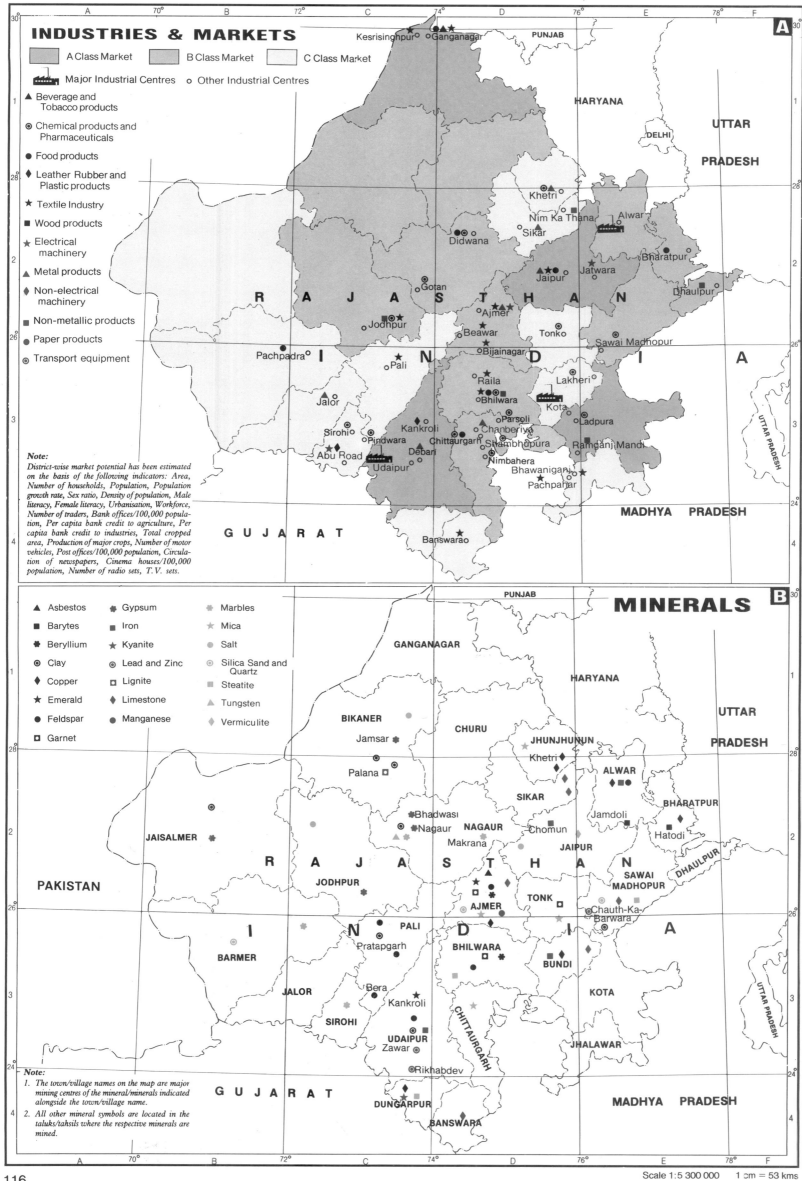

INDUSTRIES & MARKETS

A Class Market   B Class Market   C Class Market

Major Industrial Centres   o Other Industrial Centres

▲ Beverage and Tobacco products
⊙ Chemical products and Pharmaceuticals
● Food products
◆ Leather Rubber and Plastic products
★ Textile Industry
■ Wood products
★ Electrical machinery
▲ Metal products
◆ Non-electrical machinery
■ Non-metallic products
● Paper products
⊙ Transport equipment

Note:
District-wise market potential has been estimated on the basis of the following indicators: Area, Number of households, Population, Population growth rate, Sex ratio, Density of population, Male literacy, Female literacy, Urbanisation, Workforce, Number of traders, Bank offices/100,000 population, Per capita bank credit to agriculture, Per capita bank credit to industries, Total cropped area, Production of major crops, Number of motor vehicles, Post offices/100,000 population, Circulation of newspapers, Cinema houses/100,000 population, Number of radio sets, T.V. sets.

Kesrisinghpur   Ganganagar   PUNJAB

HARYANA

DELHI   UTTAR PRADESH

Khetri
Nim Ka Thana   Alwar
Sikar
Didwana
Bharatpur

RAJASTHAN INDIA

Gotan
Jaipur   Jatwara
Dhaulpur
Jodhpur   Ajmer
Beawar   Tonk
Bijainagar   Sawai Madhopur
Pachpadra   Pali
Raila   Lakheri
Bhilwara   Kota
Jalor   Parsoli   Ladpura
Sirohi   Kankroli   Chanberiya
Pindwara   Chittaurgarh   Shambhupura   Ramganj Mandi
Abu Road   Debari   Nimbahera
Udaipur   Bhawaniganj
Pachpahar

Banswarao

GUJARAT   MADHYA PRADESH

MINERALS

▲ Asbestos   ✹ Gypsum   ✤ Marbles
■ Barytes   ■ Iron   ★ Mica
✤ Beryllium   ★ Kyanite   ● Salt
⊙ Clay   ⊚ Lead and Zinc   ⊙ Silica Sand and Quartz
◆ Copper   □ Lignite   ■ Steatite
★ Emerald   ◆ Limestone   ▲ Tungsten
● Feldspar   ● Manganese   ◆ Vermiculite
□ Garnet

PUNJAB

GANGANAGAR

HARYANA

CHURU   UTTAR PRADESH

BIKANER
Jamsar
Palana
JHUNJHUNUN
Khetri
ALWAR
SIKAR
BHARATPUR
Jamdoli
Hatodi

RAJASTHAN INDIA

PAKISTAN

JAISALMER

Bhadwasi
Nagaur   NAGAUR
Makrana   Chomun
JAIPUR
DHAULPUR

JODHPUR
Ajmer
TONK   SAWAI MADHOPUR
Chauth-Ka-Barwara

PALI
Pratapgarh   BHILWARA
BARMER
Bera   BUNDI
JALOR   Kankroli
SIROHI   KOTA
UDAIPUR
Zawar   CHITTAURGARH   JHALAWAR
Rikhabdev

DUNGARPUR
BANSWARA

GUJARAT   MADHYA PRADESH

Note:
1. The town/village names on the map are major mining centres of the mineral/minerals indicated alongside the town/village name.
2. All other mineral symbols are located in the taluks/tahsils where the respective minerals are mined.

Scale 1:5 300 000   1 cm = 53 kms

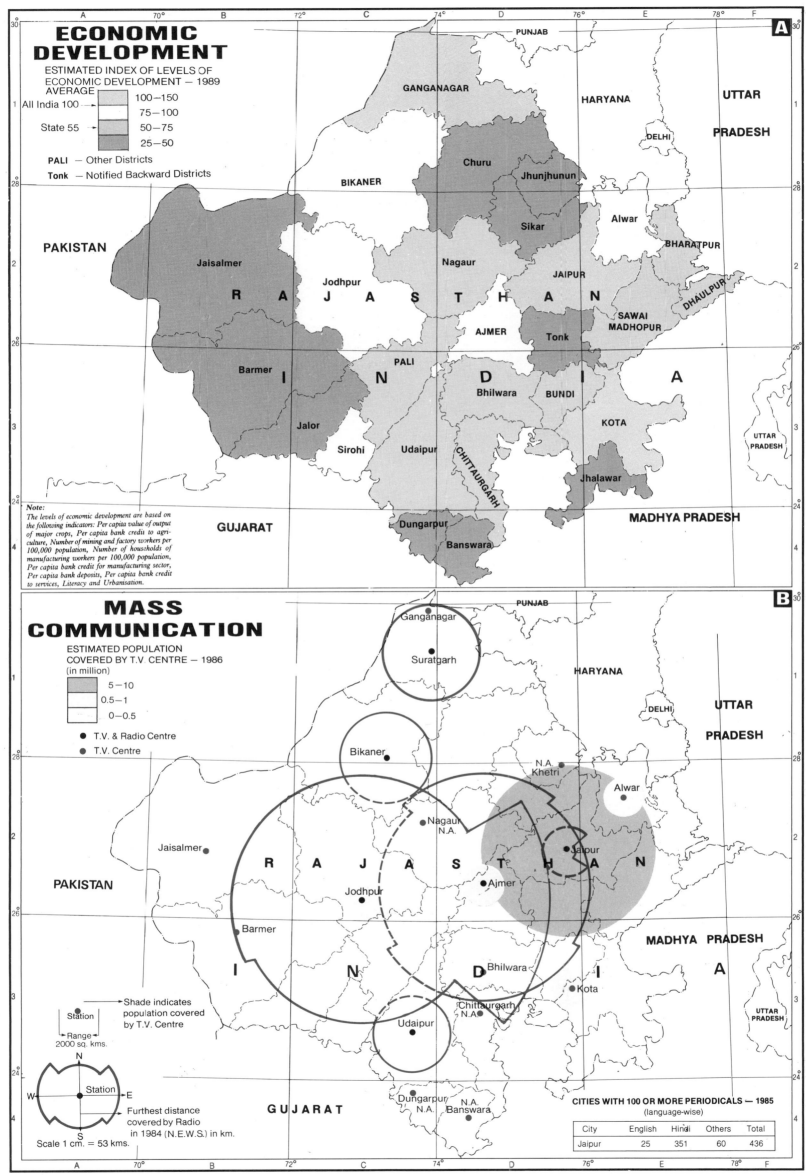

ECONOMIC DEVELOPMENT

ESTIMATED INDEX OF LEVELS OF
ECONOMIC DEVELOPMENT — 1989
AVERAGE

All India 100 → 100—150
75—100
State 55 → 50—75
25—50

PALI — Other Districts
Tonk — Notified Backward Districts

PUNJAB

GANGANAGAR

HARYANA

UTTAR

PRADESH

DELHI

BIKANER

CHURU

JHUNJHUNUN

SIKAR

Alwar

BHARATPUR

PAKISTAN

Jaisalmer

Jodhpur

Nagaur

JAIPUR

DHAULPUR

R A J A S T H A N

SAWAI
MADHOPUR

AJMER

Tonk

I   N

D   I   A

Barmer

PALI

Bhilwara

BUNDI

Jalor

KOTA

UTTAR
PRADESH

Sirohi

Udaipur

CHITTAURGARH

Jhalawar

Dungarpur

Note:
*The levels of economic development are based on
the following indicators: Per capita value of output
of major crops, Per capita bank credit to agriculture, Number of mining and factory workers per
100,000 population, Number of households of
manufacturing workers per 100,000 population,
Per capita bank credit for manufacturing sector,
Per capita bank deposits, Per capita bank credit
to services, Literacy and Urbanisation.*

GUJARAT

Banswara

MADHYA PRADESH

A

MASS
COMMUNICATION

ESTIMATED POPULATION
COVERED BY T.V. CENTRE — 1986
(in million)

5—10
0.5—1
0—0.5

● T.V. & Radio Centre
● T.V. Centre

PUNJAB

Ganganagar

Suratgarh

HARYANA

UTTAR

PRADESH

DELHI

Bikaner

N.A.
Khetri

Alwar

PAKISTAN

Jaisalmer

R A J A S

Nagaur
N.A.

T H A N

Jaipur

Jodhpur

Ajmer

Barmer

I   N

D

A

Bhilwara

MADHYA PRADESH

Kota

Chittaurgarh
N.A.

Station

Range
2000 sq. kms.

→ Shade indicates
population covered
by T.V. Centre

Udaipur

UTTAR
PRADESH

N

W — Station — E

S

→ Furthest distance
covered by Radio
in 1984 (N.E.W.S.) in km.

Scale 1 cm. = 53 kms.

Dungarpur
N.A.

N.A.
Banswara

GUJARAT

CITIES WITH 100 OR MORE PERIODICALS — 1985
(language-wise)

City	English	Hindi	Others	Total
Jaipur	25	351	60	436

B

Scale 1:5 300 000   1 cm = 53 kms

SIKKIM

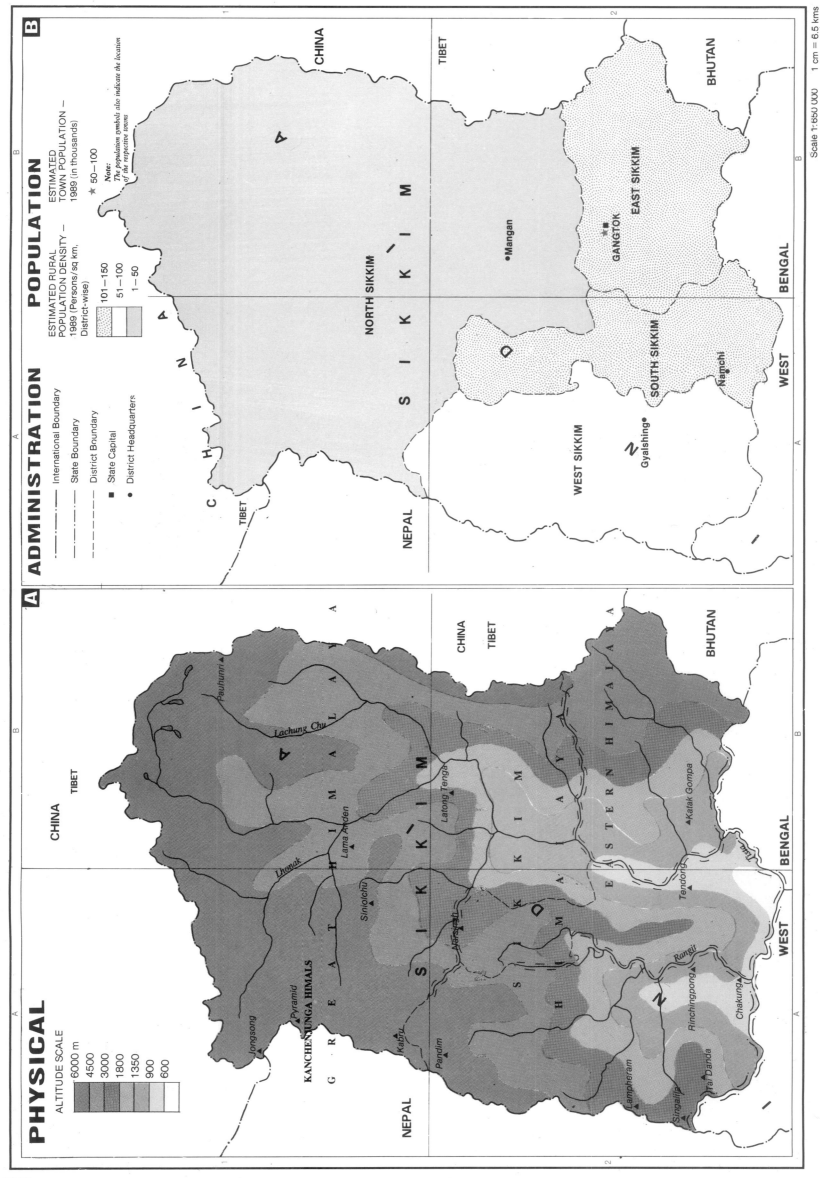

PHYSICAL

ALTITUDE SCALE

| 6000 m |
| 4500 |
| 3000 |
| 1800 |
| 1350 |
| 900 |
| 600 |

A

CHINA
TIBET

NEPAL

KANCHENJUNGA HIMALS

GREAT HIMALAYA

Pauhunri ▲

Lachung Chu

Jongsong ▲
Pyramid ▲
Lhonak
Lama Anden ▲
Siniolchu ▲
Kabru ▲
Pandim ▲

Latong Tenga

SIKKIM HIMALAYA

EASTERN HIMALAYA

Katak Gompa ▲

Tendong ▲

Rangit
Rinchingpong ▲
Chakung
Lampheram ▲
Singalila ▲
Tha Danda ▲

BHUTAN

BENGAL

WEST

ADMINISTRATION

International Boundary
State Boundary
District Boundary
■ State Capital
● District Headquarters

POPULATION

ESTIMATED RURAL
POPULATION DENSITY —
1989 (Persons/sq km,
District-wise)

| 101—150 |
| 51—100 |
| 1— 50 |

ESTIMATED
TOWN POPULATION —
1989 (in thousands)
★ 50—100

Note:
*The population symbols also indicate the location
of the respective towns*

B

CHINA

TIBET

NEPAL

NORTH SIKKIM

•Mangan

WEST SIKKIM

Gyalshing•

EAST SIKKIM

★■ GANGTOK

SOUTH SIKKIM

Namchi•

BHUTAN

TIBET

BENGAL

WEST

Scale 1:650 000   1 cm = 6.5 kms

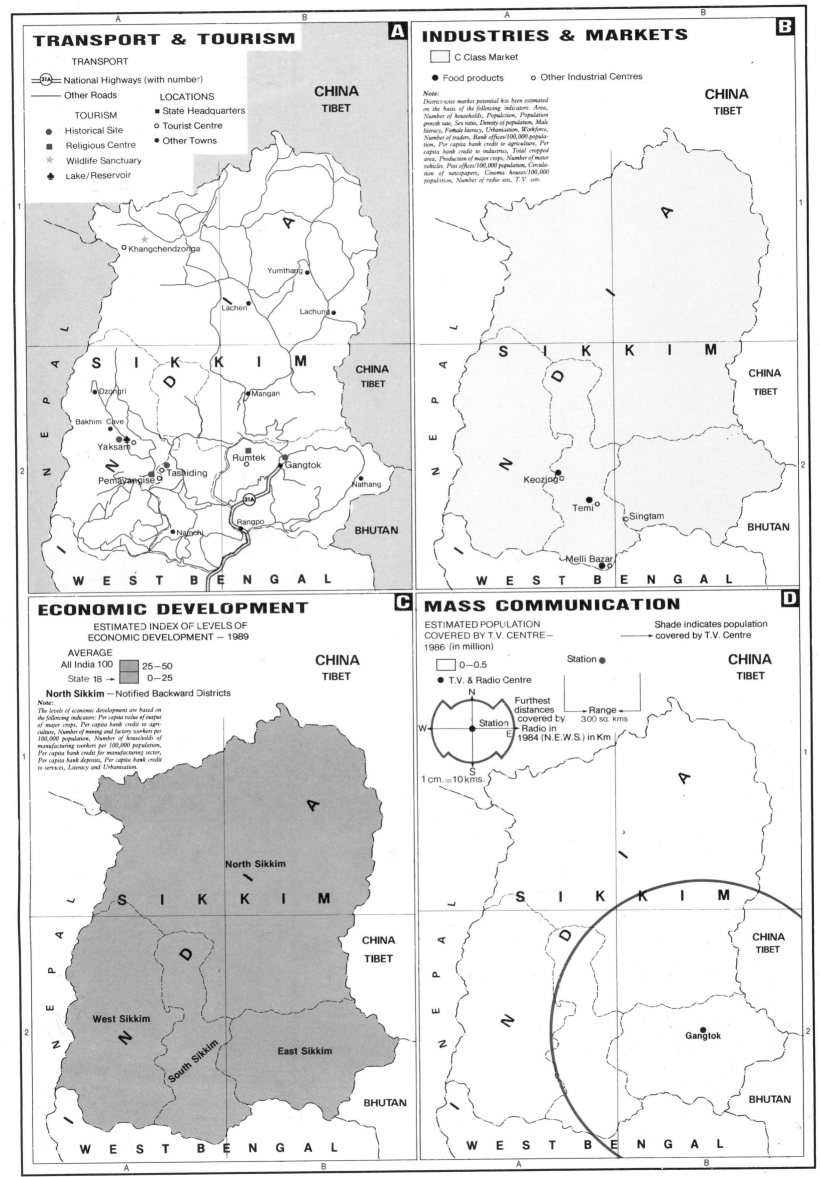

TRANSPORT & TOURISM

A

TRANSPORT

National Highways (with number) 31A

Other Roads

TOURISM
- ● Historical Site
- ■ Religious Centre
- ★ Wildlife Sanctuary
- ♣ Lake/Reservoir

LOCATIONS
- ■ State Headquarters
- ○ Tourist Centre
- ● Other Towns

CHINA
TIBET

○ Khangchendzonga

Yumthang

Lachen

Lachung

N E P A L

S I K K I M

Dzongri

Mangan

Bakhim Cave

Yaksam

Rumtek

Gangtok

Pemayangtse

Tashiding

Nathang

Namchi

Rangpo

CHINA
TIBET

BHUTAN

W E S T   B E N G A L

INDUSTRIES & MARKETS

B

☐ C Class Market

- ● Food products
- ○ Other Industrial Centres

Note:
District-wise market potential has been estimated on the basis of the following indicators: Area, Number of households, Population, Population growth rate, Sex ratio, Density of population, Male literacy, Female literacy, Urbanization, Workforce, Number of traders, Bank offices/100,000 population, Per capita bank credit to agriculture, Per capita bank credit to industries, Total cropped area, Production of major crops, Number of motor vehicles, Post offices/100,000 population, Circulation of newspapers, Cinema houses/100,000 population, Number of radio sets, T.V. sets.

CHINA
TIBET

N E P A L

S I K K I M

Keozing

Temi

Singtam

Melli Bazar

CHINA
TIBET

BHUTAN

W E S T   B E N G A L

ECONOMIC DEVELOPMENT

C

ESTIMATED INDEX OF LEVELS OF
ECONOMIC DEVELOPMENT — 1989

AVERAGE
All India 100
State 18 →

☐ 25—50
☐ 0—25

North Sikkim — Notified Backward Districts

Note:
The levels of economic development are based on the following indicators: Per capita value of output of major crops, Per capita bank credit to agriculture, Number of mining and factory workers per 100,000 population, Number of households of manufacturing workers per 100,000 population, Per capita bank credit for manufacturing sector, Per capita bank deposits, Per capita bank credit to services, Literacy and Urbanisation.

CHINA
TIBET

N E P A L

North Sikkim

S I K K I M

West Sikkim

South Sikkim

East Sikkim

CHINA
TIBET

BHUTAN

W E S T   B E N G A L

MASS COMMUNICATION

D

ESTIMATED POPULATION
COVERED BY T.V. CENTRE—
1986 (in million)

Shade indicates population
covered by T.V. Centre

☐ 0—0.5

● T.V. & Radio Centre

Station ●

Furthest distances covered by Radio in 1984 (N.E.W.S.) in Km

Range
300 sq. kms

1 cm. = 10 kms.

CHINA
TIBET

N E P A L

S I K K I M

Gangtok

CHINA
TIBET

BHUTAN

W E S T   B E N G A L

TAMIL NADU

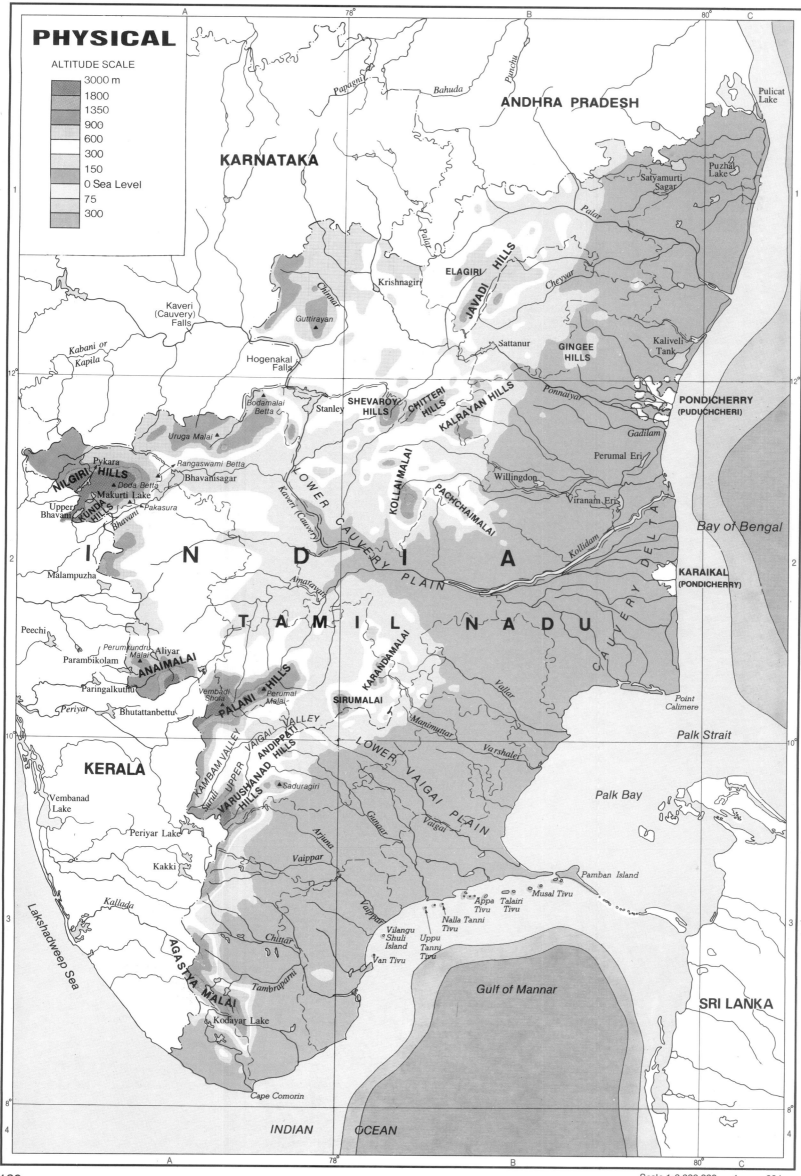

PHYSICAL

ALTITUDE SCALE

3000 m	
1800	
1350	
900	
600	
300	
150	
0 Sea Level	
75	
300	

KARNATAKA

ANDHRA PRADESH

Papagni

Bahuda

Panchu

Pulicat
Lake

Puzhal
Lake

Satyamurti
Sagar

Palar

Krishnagiri

ELAGIRI
HILLS

JAVADI HILLS

Cheyyar

Kaveri
(Cauvery)
Falls

Chinnar

▲ *Guttirayan*

GINGEE
HILLS

Kaliveli
Tank

*Kabani or
Kapila*

Hogenakal
Falls

Sattanur

▲ *Bodamalai
Betta*

Stanley

SHEVAROY
HILLS

CHITTERI
HILLS

KALRAYAN HILLS

Ponnaiyar

PONDICHERRY
(PUDUCHCHERI)

▲ *Uruga Malai*

Gadilam

Pykara

NILGIRI HILLS

▲ *Doda Betta*

Rangaswami Betta ▲

Bhavanisagar

KOLLAI MALAI

PACHCHAIMALAI

Perumal Eri

Makurti Lake

KUNDA
HILLS

▲ Pakasura

Bhavani

Willingdon

Viranam Eri

Kaveri (Cauvery)

LOWER CAUVERY PLAIN

Upper
Bhavani

Kollidam

Bay of Bengal

Malampuzha

I N D I A

Amaravati

KARAIKAL
(PONDICHERRY)

Peechi

T A M I L   N A D U

CAUVERY DELTA

Perumkundru
Malai ▲

Aliyar

ANAIMALAI

KARANDAMALAI

Parambikolam

Vallar

Point
Calimere

Paringalkuthu

▲ *Vembadi
Shola*

PALANI HILLS

▲ *Perumal
Malai*

SIRUMALAI

Manimuttar

Palk Strait

Periyar

Bhutattanbettu

KAMBAM VALLEY

VAIGAI VALLEY

UPPER

ANDIPPATTI
HILLS

VARUSHANAD
HILLS

LOWER VAIGAI PLAIN

Varshalei

KERALA

Vembanad
Lake

Sirull

▲ *Saduragiri*

Gundar

Palk Bay

Periyar Lake

Vaigai

Kakki

Ariuna

Pamban Island

Vaippar

Musal Tivu

Kallada

Vaippar

Appa
Tivu

Talairi
Tivu

Nalla Tanni
Tivu

Vilangu
Shuli
Island

Uppu
Tanni
Tivu

Van Tivu

Lakshadweep Sea

AGASTYA MALAI

Chittar

Tambraparni

Gulf of Mannar

SRI LANKA

Kodayar Lake

Cape Comorin

INDIAN   OCEAN

Scale 1:2 300 000   1 cm = 23 kms

120

ADMINISTRATION

- —·—·— State Boundary
- — — — District Boundary
- —————— Taluk Boundary
- ■ State Capital
- ● District Headquarters
- ⊙ Taluk Headquarters

ANDHRA PRADESH

KARNATAKA

I N D I A

MADRAS
Saidapet

Pallipattu
Uttukkottai
Ponneri
Gummidipundi
Tiruttani
Tiruvallur
Arakkonam
Sriperumbudur
Gudiyattam
Vellore
Walajapet (WALAJA)
CHENGAI ANNA
Kanchipuram
AMBEDKAR
Arcot (VANNUR)
Vaniyambadi
Arni
Chengalpattu
Tiruvettipuram (CHEYYAR)
Uttiramerur
TIRUVANNAMALAI
Polur
Vandavasi
SAMBUVARAYAR
Madurantakam
Hosur
Denkanikota
Krishnagiri
Tiruppattur
Chengam
Tiruvannamalai
Gingee
Tindivanam
DHARMAPURI
Pennagaram
Patakkodu
Uttangarai
Harur
Vanur

Tirukkovilur
Viluppuram
SOUTH ARCOT
Mettur
Yercaud
Kallakkurichchi
Ulundurpettai
Panrut
PONDICHERRY (PUDUCHCHERI)
Omalur
SALEM
Salem
Attur
Vriddhachalam
Cuddalore
Satyamangalam
Sankaridrug
Rasipuram
Tittagudi
Chidambaram
Bay of Bengal
NILGIRI
Gudalur
Kotagiri
Bhavani
Tiruchengodu
Mannarcudi (KATTU MANNAR KOIL)
Sirkazhi
Udagamandalam
Gopichettipalaiyam
Namakkal
Perambalur
Jayamkondacholapuram (UDAIYARPALAIYAM)
Mayiladuturai
Coonoor
Mettuppalaiyam
Perundurai
Erode
Turaiyur
Ariyalur
Tiruvidaimarudur
Tarangambadi
KARAIKAL (PONDICHERRY)
Ayanashi
PERIYAR
Karur
Kumbakonam
Valangiman
Nannilam
Coimbatore
Palladam
Kangayam
Kulittalai
Musiri
Lalgudi
Papanasam
Kudavasal
Nidamangalam
Tiruvarur
Nagappattinam
COIMBATORE
TIRUCHCHIRAPPALLI
Thiruvaiyaru
Pollachi
Dharapuram
T A M I L   N A D U
Tiruchchirappalli
Thanjavur
THANJAVUR
Orattanadu
Mannargudi
Udumalaippettai
Manapparai
Kiranure (KULATTUR)
Gandarvakottai
Pattukkottai
Tiruthuraippundi
Palani
Vedasandur
PUDUKKOTTAI
Valparai
QUAID-E-MILLETH
Pudukkottai
Alangudi
Eravurani
Kodaikanal
Dindigul
Nattam
Tirumayam
Vedaranyam
Arantangi
Periyakulam
Nilakkottai
Tiruppattur
Karaikkudi
Avadaiyarkovil
Melur
Palk Strait
Usilampatti
Madurai
PASUMPON
Devakottai
KERALA
M A D U R A I
THEVAR THIRUMAGAN
Sivaganga
Tirumangalam
Tiruvadanai
Manamadurai
Ilaiyankudi
Virudunagar
Tiruchuli
Palk Bay
Srivilliputtur
Aruppukkottai
Paramakkudi
RAMANATHAPURAM
Rajapalaiyam
K A M A R A J A R
Kamudi
Ramanathapuram
Sattur
Sivagiri
Mudukulattur
Rameswaram
Sankarankovil
Vilettikulam
Kovilpatti
Gulf of Mannar
SRI LANKA
Sengottai
CHIDAMBARANAR
Tenkasi
NELLAI
Ottappidaram
KATTABOMMAN
Tuticorin
Lakshadweep Sea
Tirunelveli
Palayamkottai
Ambasamudram
Srivaikuntam
Nanguneri
Tiruchchendur
Sattankulam
KANNIYAKUMARI
Kuzhitturai (VILAVANKOD)
Bhutapandi (TOVALA)
Radhapuram
Takkalai (KALKULAM)
Nagercoil (AGASTISWARAM)

INDIAN OCEAN

Note:
Most taluks take their names from their head-quarters. Where headquarters and taluks have the same names, only the headquarters have been located on the map. Where their names are different, the taluk's name is placed below the headquarters' location in brackets.

Scale 1:2 300 000   1 cm = 23 kms

121

POPULATION

ESTIMATED RURAL POPULATION DENSITY-1989
(Persons/sq km, District-wise)

Above 1000
501–1000
301–500
201–300
151–200
Urban

Note:
The population symbols also indicate the location of the respective towns/villages.

ESTIMATED TOWN/VILLAGE POPULATION — 1989
(in thousands)

■ OVER 1,000
● 500–1,000
▲ 250–500
♦ 100–250
★ 50–100
□ 20–50
◉ 10–20

Pallapatti — Town
Adavattur — Village

MAJOR CITIES — 1989
Towns with more than 250,000 population
(Estimated in thousands)

Madras	5,471
Madurai	1,101
Coimbatore	1,094
Tiruchchirappalli	758
Salem	613
Erode	412
Tirunelveli	380
Tuticorin	326
Vellore	323
Tiruppur	290

ANDHRA PRADESH

KARNATAKA

I N D I A

T A M I L N A D U

KERALA

Lakshadweep Sea

PONDICHERRY (PUDUCHCHERI)

Bay of Bengal

KARAIKAL (PONDICHERRY)

Palk Strait

Palk Bay

Gulf of Mannar

INDIAN OCEAN

NAMES FOR LOCATIONS NUMBERED ON THE MAP
TIRUCHCHIRAPALLI
1. *Uppidamangalam*
2. *Krishnarayapuram*
3. *Mahadanapuram*
THANJAVUR
4. *Tirukkattupalli*
COIMBATORE
5. *Samalapuram*
KAMÁRAJAR
6. *Sundarapandiyam*

Scale 1:2 300 000   1 cm = 23 kms

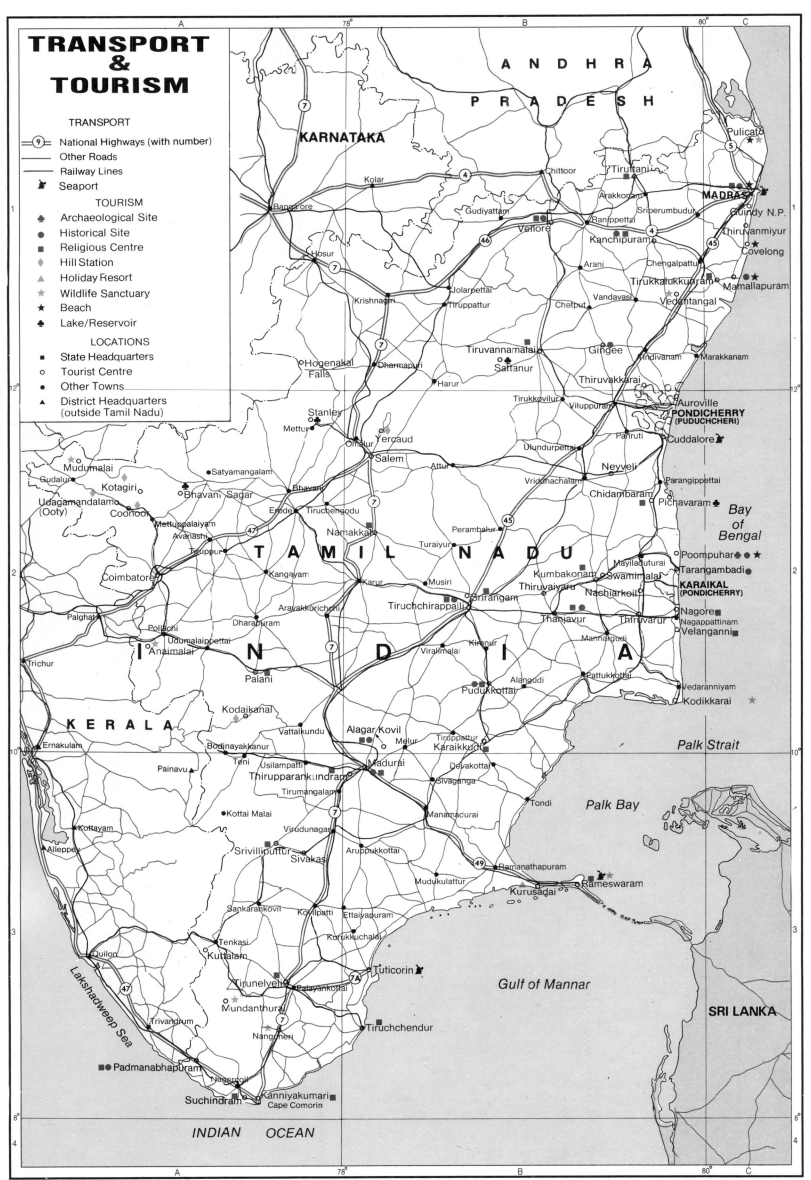

TRANSPORT & TOURISM

TRANSPORT
- ⑨ National Highways (with number)
- Other Roads
- Railway Lines
- ⚓ Seaport

TOURISM
- ♣ Archaeological Site
- ● Historical Site
- ■ Religious Centre
- ♦ Hill Station
- ▲ Holiday Resort
- ★ Wildlife Sanctuary
- ★ Beach
- ♣ Lake/Reservoir

LOCATIONS
- ■ State Headquarters
- ○ Tourist Centre
- ● Other Towns
- ▲ District Headquarters (outside Tamil Nadu)

ANDHRA PRADESH

KARNATAKA

Kolar
Chittoor
Tiruttani
Bangalore
Gudiyattam
Arakkonam
Sriperumbudur
MADRAS
Ranippettai
Guindy N.P.
Hosur
Vellore
Kanchipuram
Thiruvanmiyur
Covelong
Arani
Chengalpattu
Jolarpettai
Tirukkalukkunram
Mamallapuram
Krishnagiri
Tiruppattur
Vandavasi
Vedantangal
Chetput

Hogenakal Falls
Dharmapuri
Tiruvannamalai
Sattanur
Gingee
Tindivanam
Marakkanam
Harur
Thiruvakkarai
Stanley
Tirukkovilur
Viluppuram
Mettur
Yercaud
Auroville
Omalur
Salem
PONDICHERRY (PUDUCHCHERI)
Mudumalai
Satyamangalam
Bhavani
Ulundurpettai
Cuddalore
Gudalur
Kotagiri
Bhavani Sagar
Attur
Neyveli
Udagamandalam (Ooty)
Erode
Tiruchengodu
Vriddhachalam
Parangippettai
Coonoor
Namakkal
Chidambaram
Pichavaram
Mettuppalaiyam
Perambalur
Bay of Bengal
Avanashi
Turaiyur
Tiruppur
TAMIL NADU
Coimbatore
Kangayam
Karur
Musiri
Mayiladuturai
Poompuhar
Palghat
Aravakkurichchi
Srirangam
Kumbakonam
Swamimalai
Tarangambadi
Dharapuram
Tiruchchirappalli
Thiruvaiyaru
Nachiarkoil
KARAIKAL (PONDICHERRY)
Pollachi
Thanjavur
Thiruvarur
Nagore
Trichur
Udumalaippettai
Kiranur
Nagappattinam
Anaimalai
Viralimalai
Mannargudi
Velanganni
INDIA
Palani
Alangudi
Pattukkottai
Pudukkottai
Vedaranniyam
Kodikkarai
KERALA
Kodaikanal
Alagar Kovil
Ernakulam
Vattalkundu
Melur
Bodinayakkanur
Karaikkudi
Teni
Usilampatti
Madurai
Devakottai
Painavu
Thirupparankundram
Sivaganga
Kottayam
Tirumangalam
Tondi
Palk Strait
Kottai Malai
Manamadurai
Palk Bay
Alleppey
Virudunagar
Srivilliputtur
Sivakasi
Aruppukkottai
Ramanathapuram
Mudukulattur
Rameswaram
Kurusadai
Sankarankovil
Kovilpatti
Ettaiyapuram
SRI LANKA
Tenkasi
Kurukkuchalai
Kuttalam
Tuticorin
Gulf of Mannar
Quilon
Tirunelveli
Palayankottai
Mundanthurai
Nangneri
Tiruchchendur
Trivandrum
Padmanabhapuram
Nagercoil
Suchindram
Kanniyakumari
Cape Comorin

INDIAN OCEAN

Lakshadweep Sea

Pulicat

Scale 1:2 300 000   1 cm = 23 kms

123

9

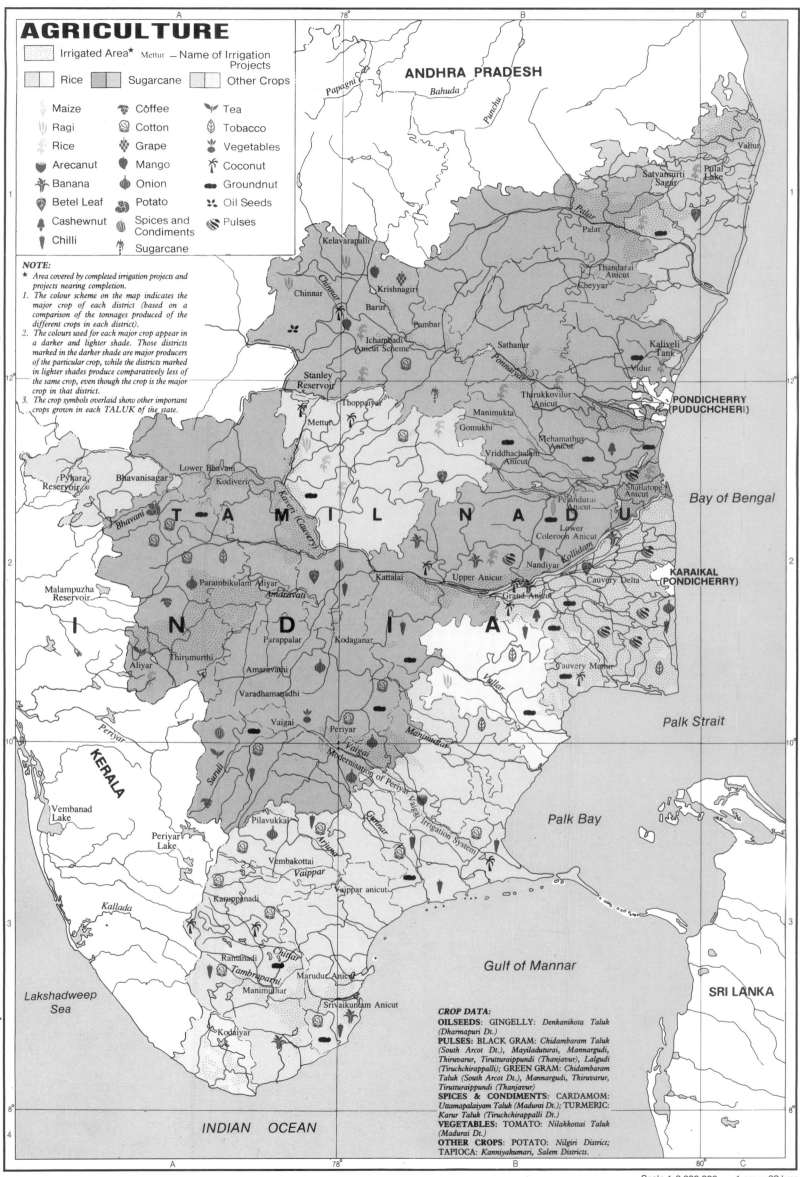

AGRICULTURE

Irrigated Area*   *Mettur* — Name of Irrigation Projects

Rice   Sugarcane   Other Crops

- Maize
- Ragi
- Rice
- Arecanut
- Banana
- Betel Leaf
- Cashewnut
- Chilli
- Coffee
- Cotton
- Grape
- Mango
- Onion
- Potato
- Spices and Condiments
- Sugarcane
- Tea
- Tobacco
- Vegetables
- Coconut
- Groundnut
- Oil Seeds
- Pulses

NOTE:

* Area covered by completed irrigation projects and projects nearing completion.
1. The colour scheme on the map indicates the major crop of each district (based on a comparison of the tonnages produced of the different crops in each district).
2. The colours used for each major crop appear in a darker and lighter shade. Those districts marked in the darker shade are major producers of the particular crop, while the districts marked in lighter shades produce comparatively less of the same crop, even though the crop is the major crop in that district.
3. The crop symbols overlaid show other important crops grown in each TALUK of the state.

ANDHRA PRADESH

KERALA

TAMIL NADU

INDIA

Bay of Bengal

PONDICHERRY (PUDUCHCHERI)

KARAIKAL (PONDICHERRY)

Palk Strait

Palk Bay

Gulf of Mannar

SRI LANKA

Lakshadweep Sea

INDIAN OCEAN

CROP DATA:
OILSEEDS: GINGELLY: *Denkanikota Taluk (Dharmapuri Dt.)*
PULSES: BLACK GRAM: *Chidambaram Taluk (South Arcot Dt.), Mayiladuturai, Mannargudi, Thiruvarur, Tiruthuraippundi (Thanjavur), Lalgudi (Tiruchchirappalli);* GREEN GRAM: *Chidambaram Taluk (South Arcot Dt.), Mannargudi, Thiruvarur, Tirutturaippundi (Thanjavur)*
SPICES & CONDIMENTS: CARDAMOM: *Uttamapalaiyam Taluk (Madurai Dt.);* TURMERIC: *Karur Taluk (Tiruchchirappalli Dt.)*
VEGETABLES: TOMATO: *Nilakkottai Taluk (Madurai Dt.)*
OTHER CROPS: POTATO: *Nilgiri District;* TAPIOCA: *Kanniyakumari, Salem Districts.*

Scale 1:2 300 000   1 cm = 23 kms

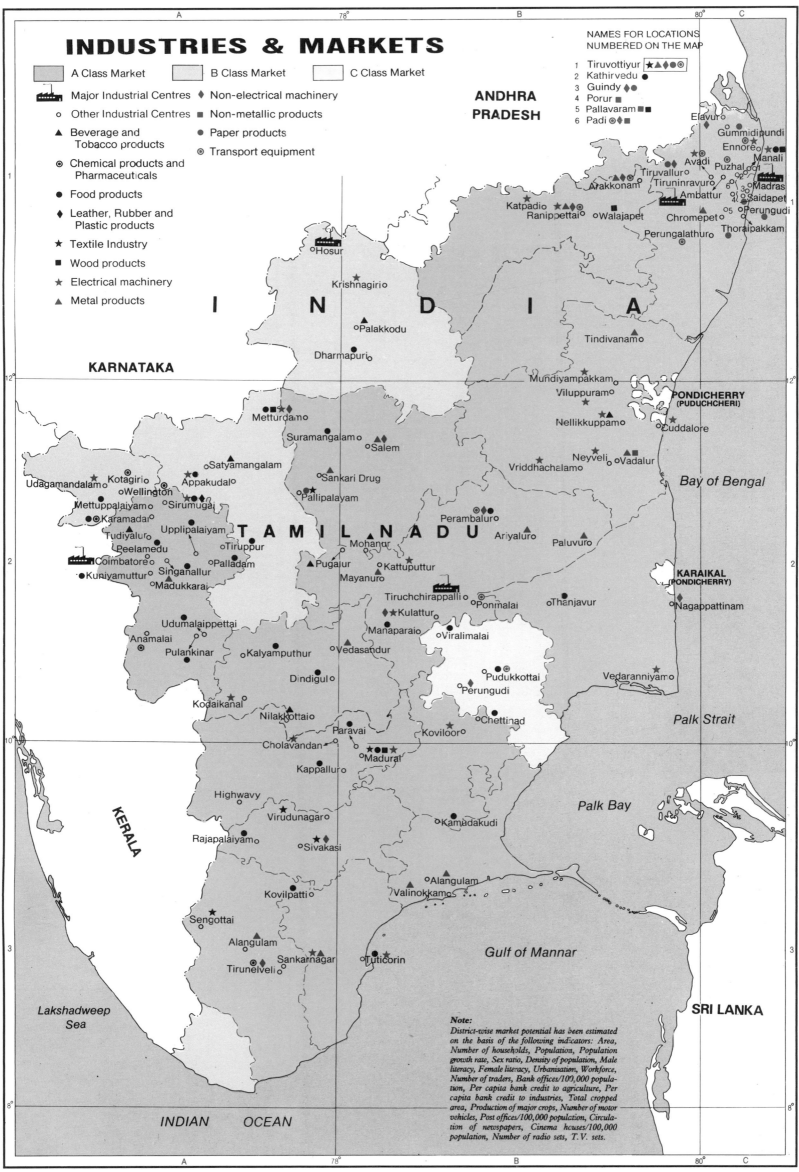

INDUSTRIES & MARKETS

A Class Market   B Class Market   C Class Market

🏭 Major Industrial Centres   ◆ Non-electrical machinery
○ Other Industrial Centres   ■ Non-metallic products
▲ Beverage and Tobacco products   ● Paper products
◎ Chemical products and Pharmaceuticals   ◉ Transport equipment
● Food products
◆ Leather, Rubber and Plastic products
★ Textile Industry
■ Wood products
★ Electrical machinery
▲ Metal products

NAMES FOR LOCATIONS NUMBERED ON THE MAP
1 Tiruvottiyur ★▲◆●◎
2 Kathirvedu ●
3 Guindy ◆●
4 Porur ■
5 Pallavaram ■■
6 Padi ◎◆■

ANDHRA PRADESH

KARNATAKA

I N D I A

T A M I L N A D U

KERALA

Hosur
Krishnagiri○
Palakkodu○
Dharmapuri○
Tindivanam○

Elavur
Gummidipundi
Ennore○
Manali
Avadi
Tiruvallur○
Tiruninravur○
Puzhal○
Madras
Arakkonam
Ambattur
Saidapet
Katpadio
Ranippettai
Walajapet○
Chromepet
Perungudi
Perungalathuro
Thoraipakkam

Mundiyampakkam
Viluppuram○
PONDICHERRY (PUDUCHCHERI)
Metturdam○
Nellikkuppam
Suramangalam○
Salem○
Cuddalore○
Satyamangalam○
Appakudal○
Sankari Drug○
Neyveli
Vadalur○
Udagamandalam
Kotagiri○
Wellington○
Pallipalayam○
Vriddhachalamo
Mettuppalaiyam○
Sirumugai○
Perambaluro
Karamadai○
Ariyaluro
Paluvuro
Tudiyaluro
Upplipalaiyam○
Peelamedu○
Tiruppur○
Mohanur
Pugalur▲
Kattuputtur
Coimbatore
Palladam○
Mayanuro
Singanallur○
Kuniyamuttur○
Madukkarai○
Tiruchchirappalli○
Ponmalai○
Thanjavur○
KARAIKAL (PONDICHERRY)
Kulattur★
Nagappattinam
Udumalaippettai○
Manaparaio
Viralimalai○
Anamalai○
Pulankinar○
Kalyamputhur○
Vedasandur○
Pudukkottai○
Vedaranniyam★
Dindigul○
Perungudi○
Kodaikanal★
Chettinad○
Nilakkottaio
Paravai○
Koviloor○
Cholavandan○
Madurai
Kappalluro
Highwavy○
Kamudakudi●
Palk Bay
Virudunagaro
Rajapalaiyam○
Sivakasi○
Alangulam○
Valinokkamo
Kovilpatti○
Sengottai★
Alangulam○
Sankarnagar★
Tuticorin○
Tirunelveli○

Bay of Bengal

Palk Strait

Gulf of Mannar

Lakshadweep Sea

SRI LANKA

INDIAN OCEAN

Note:
District-wise market potential has been estimated on the basis of the following indicators: Area, Number of households, Population, Population growth rate, Sex ratio, Density of population, Male literacy, Female literacy, Urbanisation, Workforce, Number of traders, Bank offices/100,000 population, Per capita bank credit to agriculture, Per capita bank credit to industries, Total cropped area, Production of major crops, Number of motor vehicles, Post offices/100,000 population, Circulation of newspapers, Cinema houses/100,000 population, Number of radio sets, T.V. sets.

Scale 1:2 300 000   1 cm = 23 kms

125

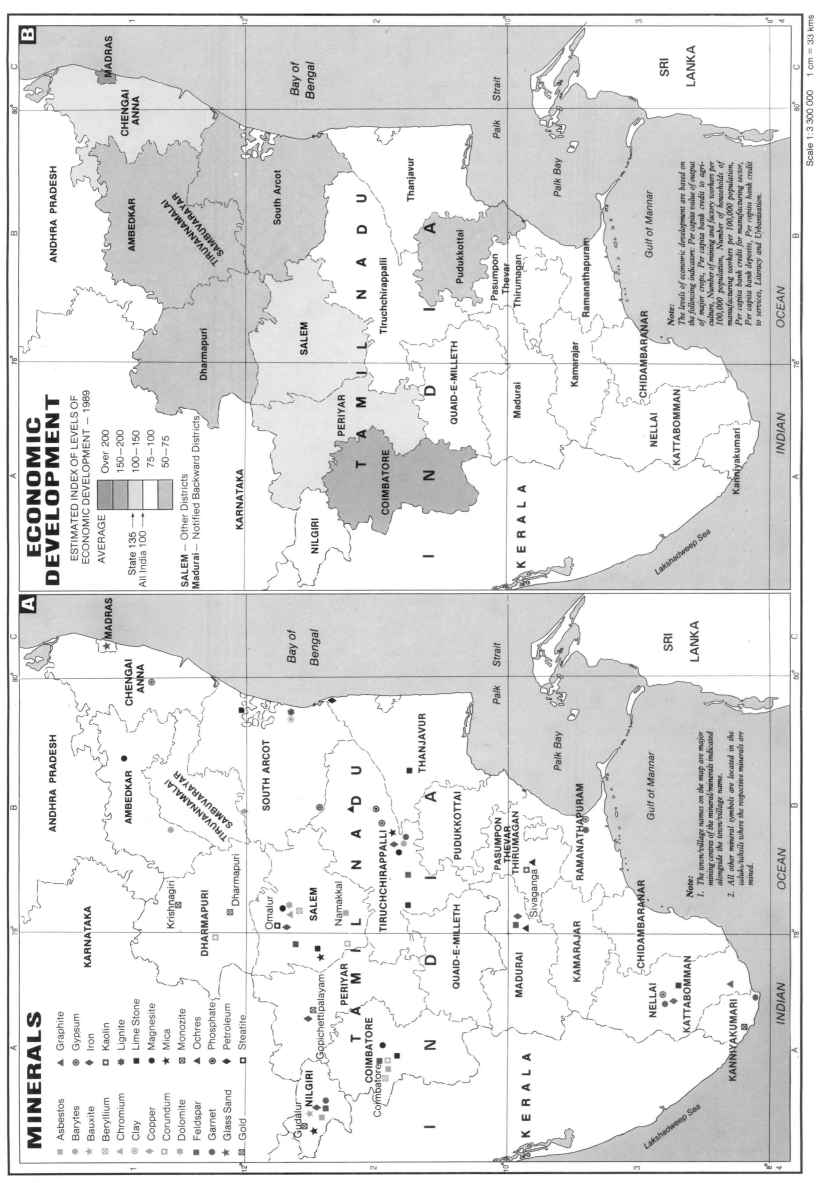

MINERALS

Asbestos
Barytes
Bauxite
Beryllium
Chromium
Clay
Copper
Corundum
Dolomite
Feldspar
Garnet
Glass Sand
Gold

Graphite
Gypsum
Iron
Kaolin
Lignite
Lime Stone
Magnesite
Mica
Monozite
Ochres
Phosphate
Petroleum
Steatite

Note:
1. The town/village names on the map are major mining centres of the mineral/minerals indicated alongside the town/village name.
2. All other mineral symbols are located in the taluks/tahsils where the respective minerals are mined.

ECONOMIC DEVELOPMENT

ESTIMATED INDEX OF LEVELS OF ECONOMIC DEVELOPMENT — 1989

AVERAGE

State 135
All India 100

Over 200
150–200
100–150
75–100
50–75

SALEM — Other Districts
Madurai — Notified Backward Districts

Note:
The levels of economic development are based on the following indicators: Per capita value of output of major crops, Per capita bank credit to agriculture, Number of mining and factory workers per 100,000 population, Number of households of manufacturing workers per 100,000 population, Per capita bank credit for manufacturing sector, Per capita bank deposit, Per capita bank credit to services, Literacy and Urbanisation.

Scale 1:3 300 000   1 cm = '33 kms

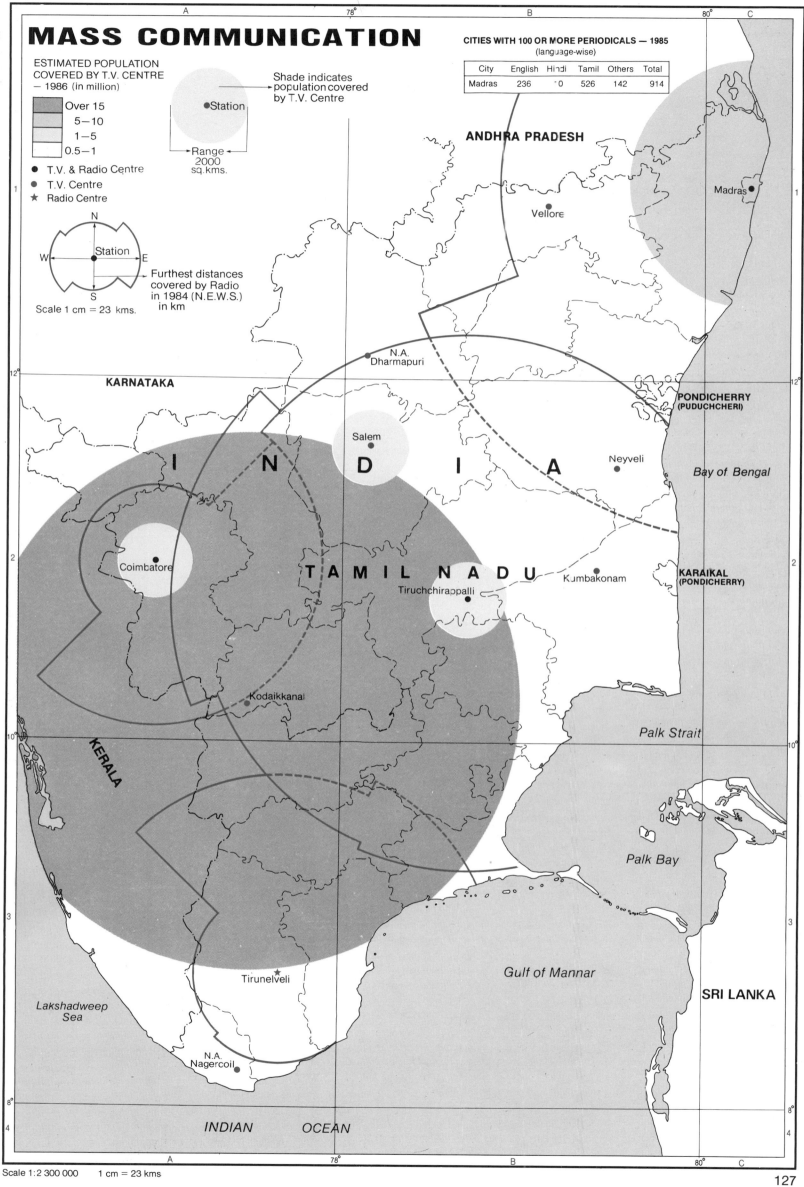

MASS COMMUNICATION

ESTIMATED POPULATION
COVERED BY T.V. CENTRE
— 1986 (in million)

Over 15
5 — 10
1 — 5
0.5 — 1

● T.V. & Radio Centre
● T.V. Centre
★ Radio Centre

Scale 1 cm = 23 kms.

●Station

Shade indicates
population covered
by T.V. Centre

← Range →
2000
sq.kms.

Station

Furthest distances
covered by Radio
in 1984 (N.E.W.S.)
in km

ANDHRA PRADESH

Madras

Vellore

KARNATAKA

I N D

N.A.
Dharmapuri

PONDICHERRY
(PUDUCHCHERI)

Salem

Neyveli

Bay of Bengal

I A

Coimbatore

T A M I L   N A D U

Tiruchchirappalli

Kumbakonam

KARAIKAL
(PONDICHERRY)

Kodaikkanal

KERALA

Palk Strait

Palk Bay

Tirunelveli

Gulf of Mannar

SRI LANKA

Lakshadweep
Sea

N.A.
Nagercoil

INDIAN   OCEAN

Scale 1:2 300 000   1 cm = 23 kms

TRIPURA

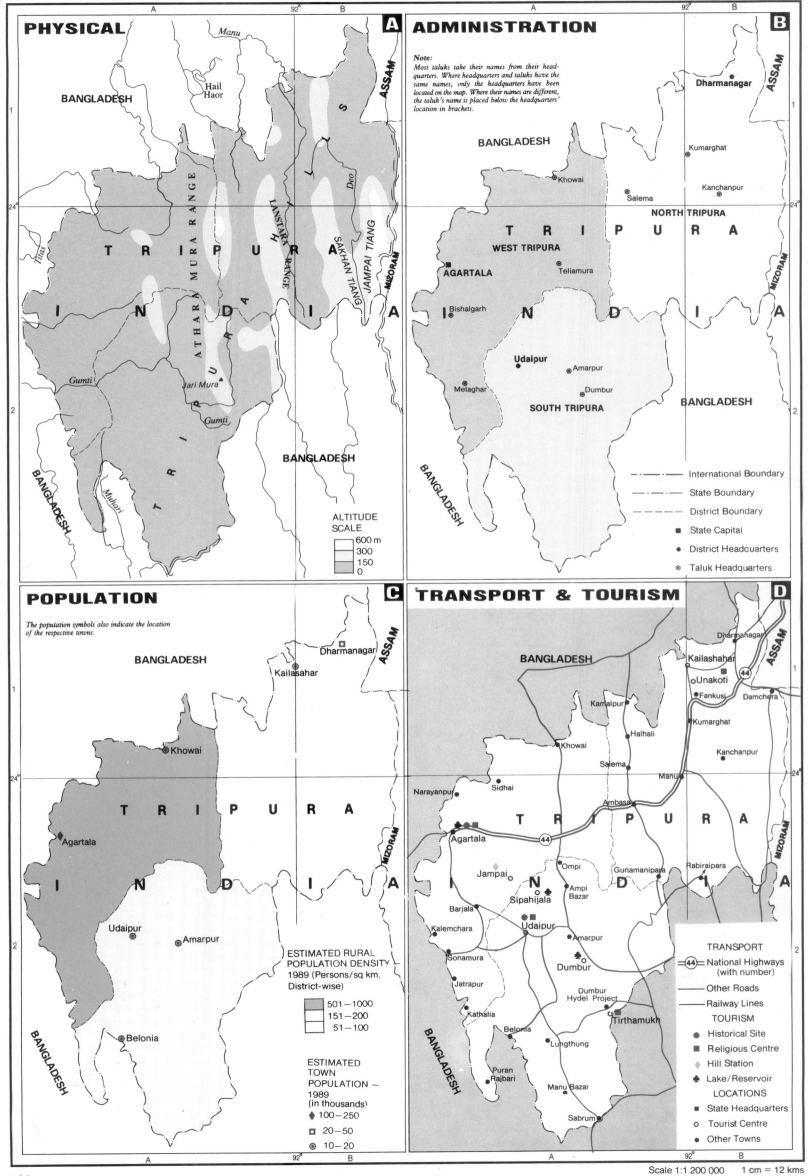

PHYSICAL A

Manu
BANGLADESH
Hail
Haor
ASSAM
Titas
ATHARA MURA RANGE
LANSTARA RANGE
SAKHAN TIANG
Deo
JAMPAI TIANG
MIZORAM
T R I P U R A
I N D I A
Gumti
Jari Mura
Gumti
BANGLADESH
Muhari
BANGLADESH

ALTITUDE SCALE
600 m
300
150
0

ADMINISTRATION B

Note:
Most taluks take their names from their headquarters. Where headquarters and taluks have the
same names, only the headquarters have been
located on the map. Where their names are different,
the taluk's name is placed below the headquarters'
location in brackets.

Dharmanagar
ASSAM
BANGLADESH
Kumarghat
Khowai
Kanchanpur
Salema
NORTH TRIPURA
T R I P U R A
WEST TRIPURA
AGARTALA
Teliamura
Bishalgarh
I N D I A
MIZORAM
Udaipur
Amarpur
Melaghar
Dumbur
SOUTH TRIPURA
BANGLADESH
BANGLADESH

International Boundary
State Boundary
District Boundary
■ State Capital
● District Headquarters
◉ Taluk Headquarters

POPULATION C

The population symbols also indicate the location
of the respective towns.

BANGLADESH
Dharmanagar
ASSAM
Kailasahar
Khowai
T R I P U R A
Agartala
I N D I A
MIZORAM
Udaipur
Amarpur
BANGLADESH
Belonia

ESTIMATED RURAL
POPULATION DENSITY
1989 (Persons/sq km,
District-wise)
501–1000
151–200
51–100

ESTIMATED
TOWN
POPULATION –
1989
(in thousands)
◆ 100–250
□ 20–50
◉ 10–20

TRANSPORT & TOURISM D

Dharmanagar
Kailashahar
BANGLADESH
Unakoti
ASSAM
44
Fankusi
Damchera
Kamalpur
Kumarghat
Halhali
Kanchanpur
Khowai
Salema
Manu
Narayanpur
Sidhai
Ambasa
T R I P U R A
Agartala
44
I N D I A
MIZORAM
Jampai
Ompi
Gunamanipara
Rabiraipara
Ampi
Sipahijala
Bazar
Barjala
Kalemchara
Udaipur
Amarpur
Sonamura
Dumbur
Jatrapur
Dumbur
Hydel Project
Kathalia
Tirthamukh
Belonia
Lungthung
Puran
Rajbari
Manu Bazar
BANGLADESH
Sabrum

TRANSPORT
(44) National Highways
(with number)
Other Roads
Railway Lines
TOURISM
● Historical Site
■ Religious Centre
◆ Hill Station
♣ Lake/Reservoir
LOCATIONS
■ State Headquarters
○ Tourist Centre
● Other Towns

Scale 1:1 200 000   1 cm = 12 kms

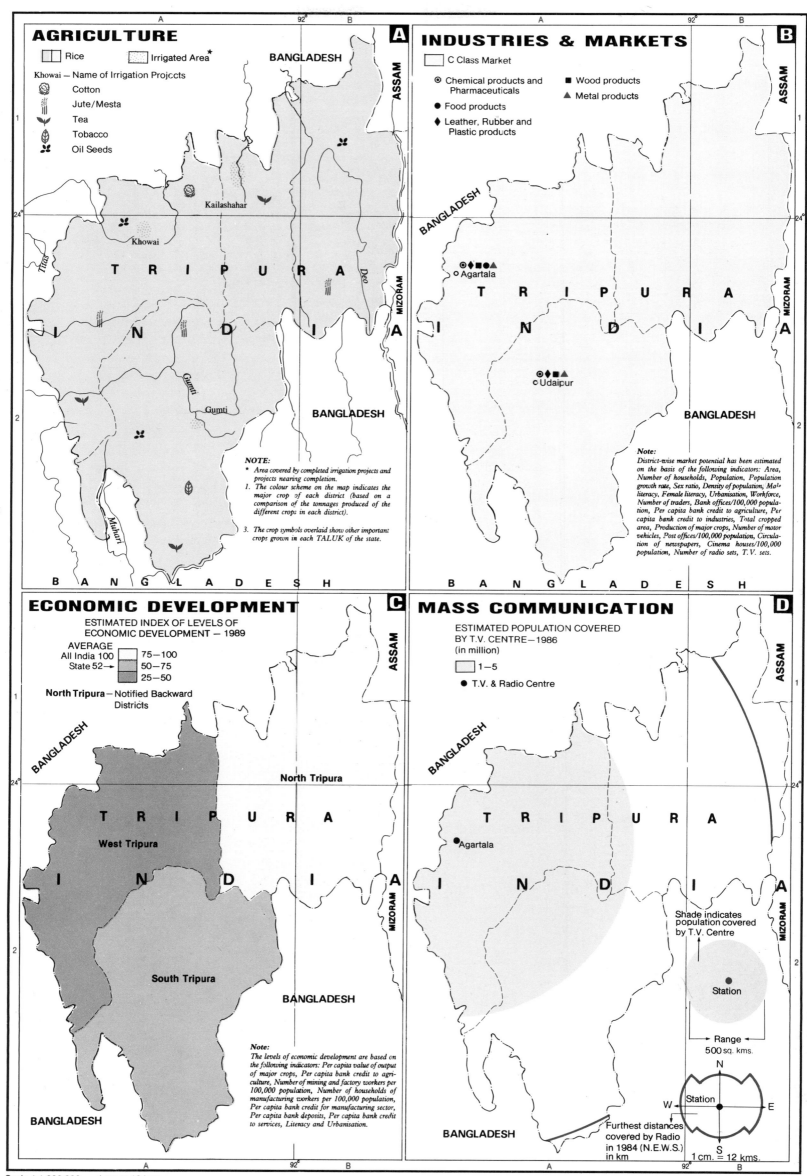

AGRICULTURE

| | Rice | | Irrigated Area |

Khowai — Name of Irrigation Projects

- Cotton
- Jute/Mesta
- Tea
- Tobacco
- Oil Seeds

BANGLADESH

ASSAM

Kailashahar

Khowai

Titas

T R I P U R A

I N D I A

Deo

MIZORAM

Gumti

Gumti

BANGLADESH

Muhari

NOTE:

* Area covered by completed irrigation projects and projects nearing completion.
1. The colour scheme on the map indicates the major crop of each district (based on a comparison of the tonnages produced of the different crops in each district).
3. The crop symbols overlaid show other important crops grown in each TALUK of the state.

B A N G L A D E S H

INDUSTRIES & MARKETS

| | C Class Market

- ⊙ Chemical products and Pharmaceuticals
- ● Food products
- ◆ Leather, Rubber and Plastic products
- ■ Wood products
- ▲ Metal products

ASSAM

BANGLADESH

⊙◆■▲
○ Agartala

T R I P U R A

I N D I A

MIZORAM

⊙◆■▲
○ Udaipur

BANGLADESH

Note:

District-wise market potential has been estimated on the basis of the following indicators: Area, Number of households, Population, Population growth rate, Sex ratio, Density of population, Male literacy, Female literacy, Urbanisation, Workforce, Number of traders, Bank offices/100,000 population, Per capita bank credit to agriculture, Per capita bank credit to industries, Total cropped area, Production of major crops, Number of motor vehicles, Post offices/100,000 population, Circulation of newspapers, Cinema houses/100,000 population, Number of radio sets, T.V. sets.

B A N G L A D E S H

ECONOMIC DEVELOPMENT

ESTIMATED INDEX OF LEVELS OF
ECONOMIC DEVELOPMENT — 1989

AVERAGE
All India 100
State 52 →

| | 75—100
| | 50—75
| | 25—50

North Tripura — Notified Backward Districts

ASSAM

BANGLADESH

North Tripura

T R I P U R A

West Tripura

I N D I A

South Tripura

MIZORAM

BANGLADESH

Note:

The levels of economic development are based on the following indicators: Per capita value of output of major crops, Per capita bank credit to agriculture, Number of mining and factory workers per 100,000 population, Number of households of manufacturing workers per 100,000 population, Per capita bank credit for manufacturing sector, Per capita bank deposits, Per capita bank credit to services, Literacy and Urbanisation.

BANGLADESH

MASS COMMUNICATION

ESTIMATED POPULATION COVERED
BY T.V. CENTRE—1986
(in million)

| | 1—5
- ● T.V. & Radio Centre

ASSAM

BANGLADESH

● Agartala

T R I P U R A

I N D I A

MIZORAM

Shade indicates population covered by T.V. Centre

Station

→ Range
500 sq. kms.

Furthest distances covered by Radio in 1984 (N.E.W.S.) in km

N
W — Station — E
S

1 cm. = 12 kms.

BANGLADESH

UTTAR PRADESH

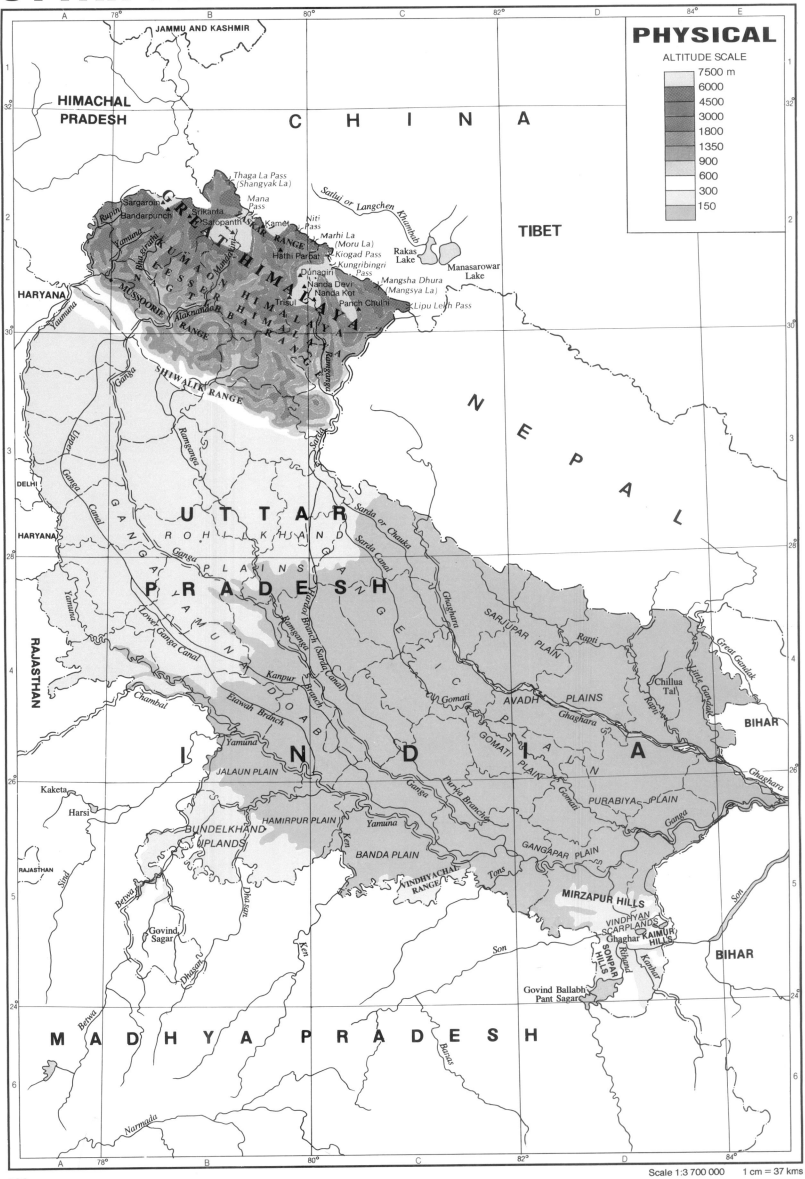

PHYSICAL

ALTITUDE SCALE

| 7500 m |
| 6000 |
| 4500 |
| 3000 |
| 1800 |
| 1350 |
| 900 |
| 600 |
| 300 |
| 150 |

JAMMU AND KASHMIR

HIMACHAL PRADESH

CHINA

TIBET

Thaga La Pass (Shangyak La)
Mana Pass
Sargaroin
Bandarpunch
Rupin
Srikanta
Satopanth
Kamet
Niti Pass
Marhi La (Moru La)
Kiogad Pass
Kungribingri Pass
Hathi Parbat
Dunagiri
Nanda Devi
Nanda Kot
Panch Chulhi
Mangsha Dhura (Mangsya La)
Lipu Lekh Pass
Rakas Lake
Manasarowar Lake
Satluj or Langchen Khambab
Trisul
GREAT HIMALAYA
KUMAON LESSER HIMALAYA RANGE
BHAGIRATHI
MUSSOORIE
SHIWALIK RANGE
Alaknanda
Ramganga
Yamuna
Bhagirathi
Mandakini

HARYANA
Yamuna
Ganga
Ganga Canal
Ramganga
DELHI
HARYANA

NEPAL

UTTAR
ROHILKHAND
PRADESH
PLAINS
Sarda or Chauka
Sarda Canal
Sarda
Ghaghara
SARJUPAR PLAIN
Rapti
Great Gandak
Chillua Tal
Little Gandak
Rapti

RAJASTHAN
Yamuna
Lower Ganga Canal
DOAB
Kanpur
GANGA
Hardoi Branch
Ramganga (Sarda Canal)
Etawah Branch
JALAUN PLAIN
Chambal
Yamuna
GANGETIC
Gomati
AVADH PLAINS
Ghaghara
GOMATI PLAIN
PLAIN
INDIA
BIHAR
Ghaghara

Kaketa
Harsi
HAMIRPUR PLAIN
Yamuna
Ken
BANDA PLAIN
Ganga
Purka Branch
Gomati
PURABIYA PLAIN
Ganga
RAJASTHAN
BUNDELKHAND UPLANDS
Sind
Betwa
Dhasan
GANGAPAR PLAIN
Tons
VINDHYACHAL RANGE
MIRZAPUR HILLS
VINDHYAN SCARPLANDS
Ghaghar
KAIMUR HILLS
Govind Sagar
Ken
Dhasan
Son
SONPAR HILLS
Rihand
Kanhar
BIHAR
Govind Ballabh Pant Sagar

MADHYA PRADESH
Betwa
Narmada
Banas
Son

Scale 1:3 700 000   1 cm = 37 kms

POPULATION

ESTIMATED RURAL
POPULATION
DENSITY — 1989
(Persons/sq km, District-wise)

	Above 1000
	501—1000
	301—500
	201—300
	151—200
	101—150
	51—100
	1—50
	Urban

Vrindavan — Town
Jaitpur — Village

ESTIMATED
TOWN/VILLAGE
POPULATION — 1989
(in thousands)

■	Above 1000
●	500—1000
▲	250—500
◆	100—250
★	50—100
◻	20—50
◉	10—20

Note:
The population symbols also indicate the location
of the respective towns/villages.

NAMES FOR LOCATIONS NUMBERED ON THE MAP

1. Jhalu
2. Bhokarheri
3. Hastinapur
4. Parichhatnagar
5. Tikri
6. Sisauli
7. Phalavda
8. Tajpur
9. Faridnagar
10. Thana Bhawan
11. Meerut
12. Sewalkhas

MAJOR CITIES — 1989

Towns with more than 250,000 population
(Estimated in thousands)

Kanpur	2,126
Lucknow	1,197
Varanasi	989
Agra	902
Allahabad	772
Meerut	738
Ghaziabad	592
Bareilly	558
Moradabad	424
Dehra Dun	397
Aligarh	389
Jhansi	376
Saharanpur	366
Gorakhpur	348
Shahjahanpur	275

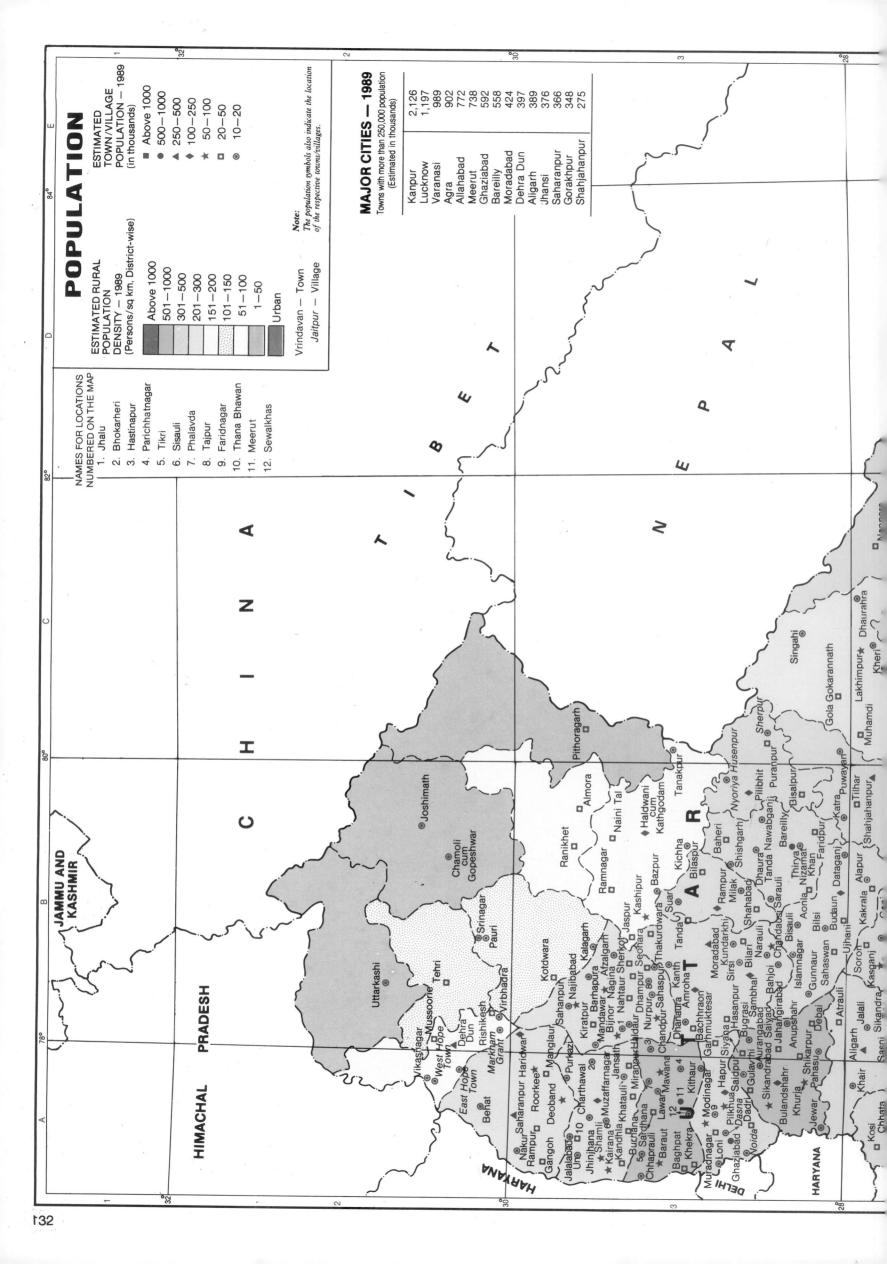

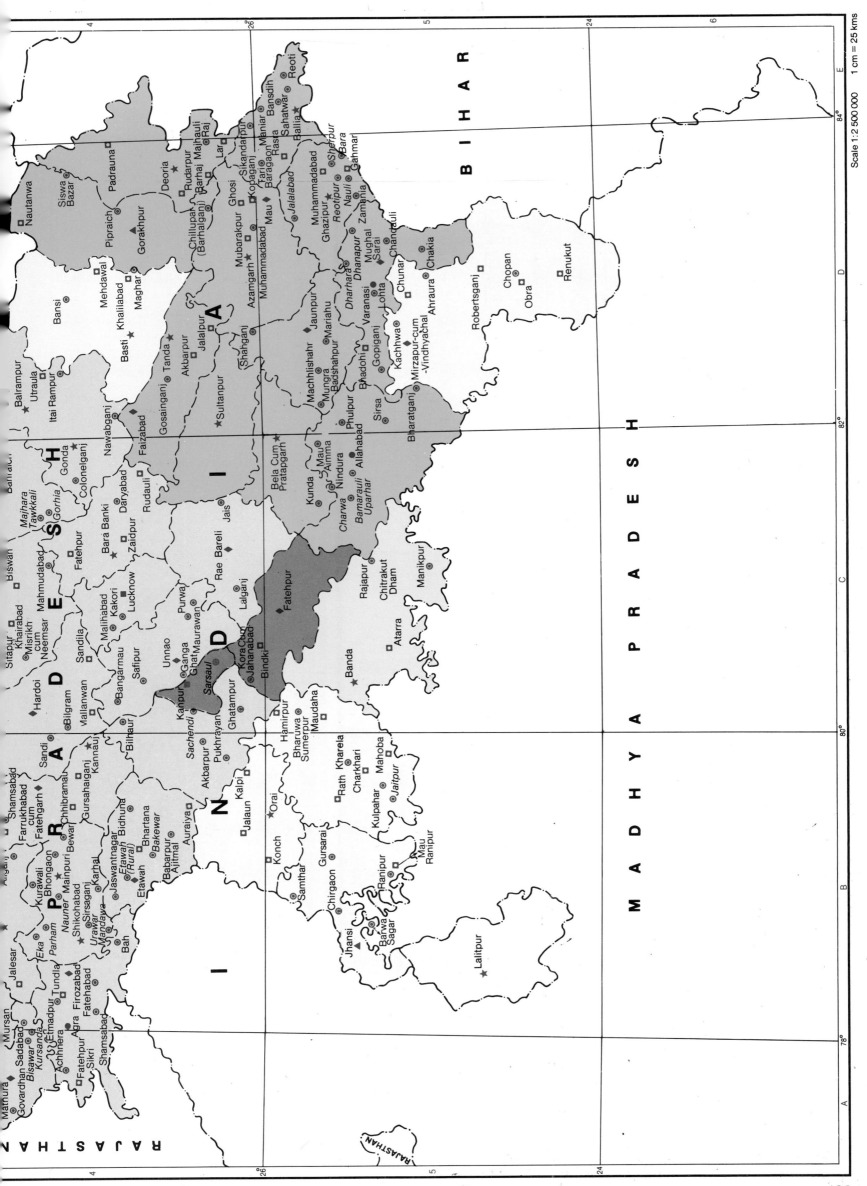

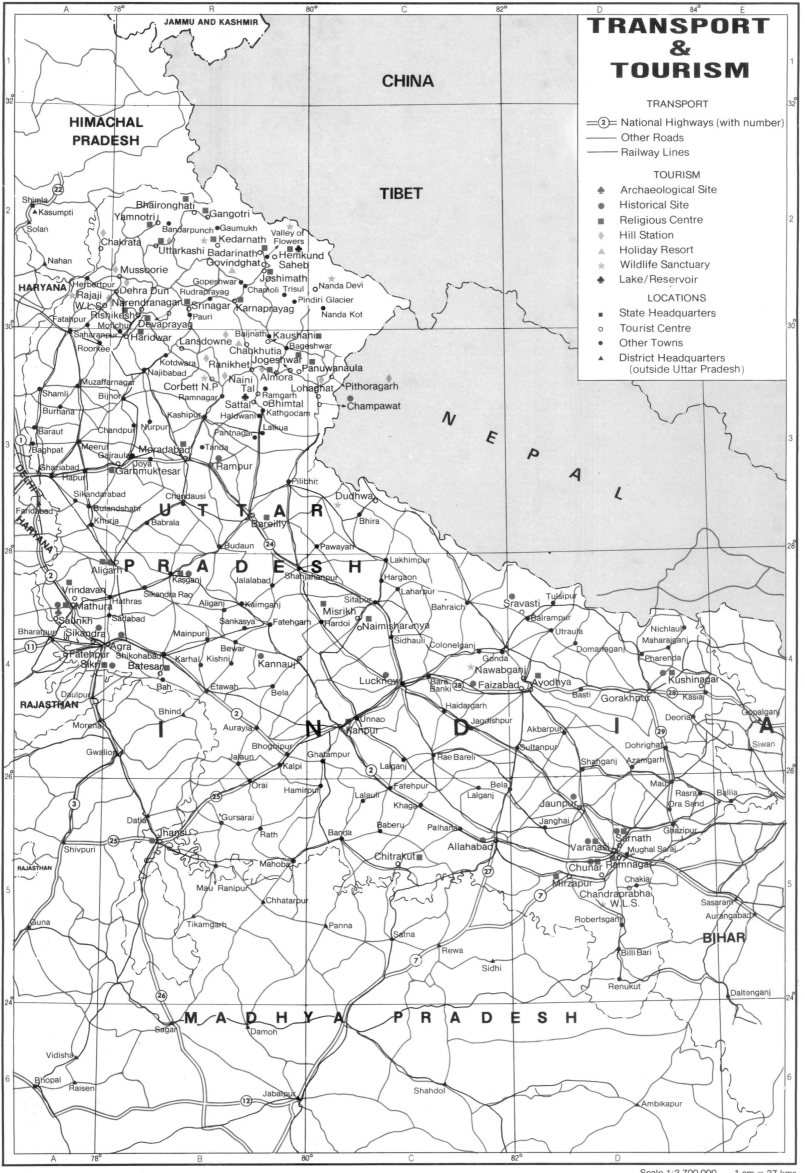

TRANSPORT & TOURISM

JAMMU AND KASHMIR

CHINA

TIBET

HIMACHAL PRADESH

TRANSPORT
═2═ National Highways (with number)
─── Other Roads
─── Railway Lines

TOURISM
♣ Archaeological Site
● Historical Site
■ Religious Centre
♦ Hill Station
▲ Holiday Resort
★ Wildlife Sanctuary
♣ Lake/Reservoir

LOCATIONS
■ State Headquarters
○ Tourist Centre
● Other Towns
▲ District Headquarters
  (outside Uttar Pradesh)

Shimla
Kasumpti
Solan
Nahan

HARYANA

Bhaironghati
Yamnotri
Bandarpunch
Gangotri
Gaumukh
Chakrata
Kedarnath
Valley of Flowers
Uttarkashi
Badarinath
Hemkund
Govindghat
Saheb
Mussoorie
Gopeshwar
Joshimath
Dehra Dun
Chamoli
Trisul
Nanda Devi
Herbertpur
Rajaji
Rudraprayag
Narendranagar
Srinagar
Karnaprayag
Pindiri Glacier
Nanda Kot
Fatehpur
Rishikesh
Pauri
Motichur
Devaprayag
Saharanpur
Haridwar
Lansdowne
Baijnath
Kaushani
Roorkee
Chaukhutia
Bageshwar
Kotdwara
Ranikhet
Jogeshwar
Najibabad
Naini
Almora
Panuwanaula
Muzaffarnagar
Corbett N.P
Tal
Lohaghat
Pithoragarh
Shamli
Bijnor
Ramnagar
Ramgarh
Burhana
Ramganar
Sattal
Bhimtal
Champawat
Baraut
Kashipur
Haldwani
Kathgodam
Baghpat
Chandpur
Nurpur
Pantnagar
Lalkua
Meerut
Gajraula
Moradabad
Tanda
Ghaziabad
Joya
Rampur
Hapur
Garhmuktesar
Faridabad
Sikandarabad
Chandausi
Pilibhit
Bulandshahr
Khurja
Babrala
Bareilly
Dudhwa
Bhira
Aligarh
Kasganj
Budaun
Pawayan
Lakhimpur
Vrindavan
Jalalabad
Shahjahanpur
Hargaon
Laharpur
Bahraich
Sravasti
Tulsipur
Mathura
Hathras
Sikandra Rao
Aliganj
Sitapur
Balrampur
Nichlaul
Saunkh
Sadabad
Aligarh
Kaimganj
Misrikh
Sidhauli
Gonda
Utraula
Domariganj
Maharajganj
Bharatpur
Sikandra
Mainpuri
Sankasya
Fatehgarh
Naimisharanya
Hardoi
Colonelganj
Nawabganj
Pharenda
Agra
Fatehpur
Shikohabad
Bewar
Karhal
Kishni
Kannauj
Lucknow
Bara Banki
Faizabad
Ayodhya
Basti
Gorakhpur
Kushinagar
Sikri
Batesar
Bah
Etawah
Bela
Haidargarh
Jagdishpur
Akbarpur
Kasia
Daulpur
Bhind
Auraiya
Unnao
Kanpur
Sultanpur
Shahganj
Deoria
Gopalganj
Morena
Bhognipur
Jalaun
Ghatampur
Rae Bareli
Azamgarh
Dohrighat
Siwan
Gwalior
Kalpi
Fatehpur
Bela
Jaunpur
Mau
Datia
Orai
Hamirpur
Lalganj
Lalganj
Janghai
Rasra
Ballia
Jhansi
Gursarai
Lalauli
Khaga
Palhana
Ghazipur
Ora Sand
Shivpuri
Rath
Banda
Baberu
Allahabad
Sarnath
Varanasi
Mughal Sarai
Mahoba
Chitrakut
Chunar
Ramnagar
Mau Ranipur
Mirzapur
Chakia
Chhatarpur
Panna
Chandraprabha
W.L.S.
Sasaram
Guna
Tikamgarh
Satna
Robertsganj
Aurangabad
BIHAR
Billi Bari
Bhopal
Raisen
Sagar
Damoh
Vidisha
Renukut
Daltenganj
Jabalpur
Shahdol
Ambikapur

RAJASTHAN

UTTAR PRADESH

INDIA

NEPAL

RAJASTHAN

MADHYA PRADESH

Scale 1:3 700 000   1 cm = 37 kms

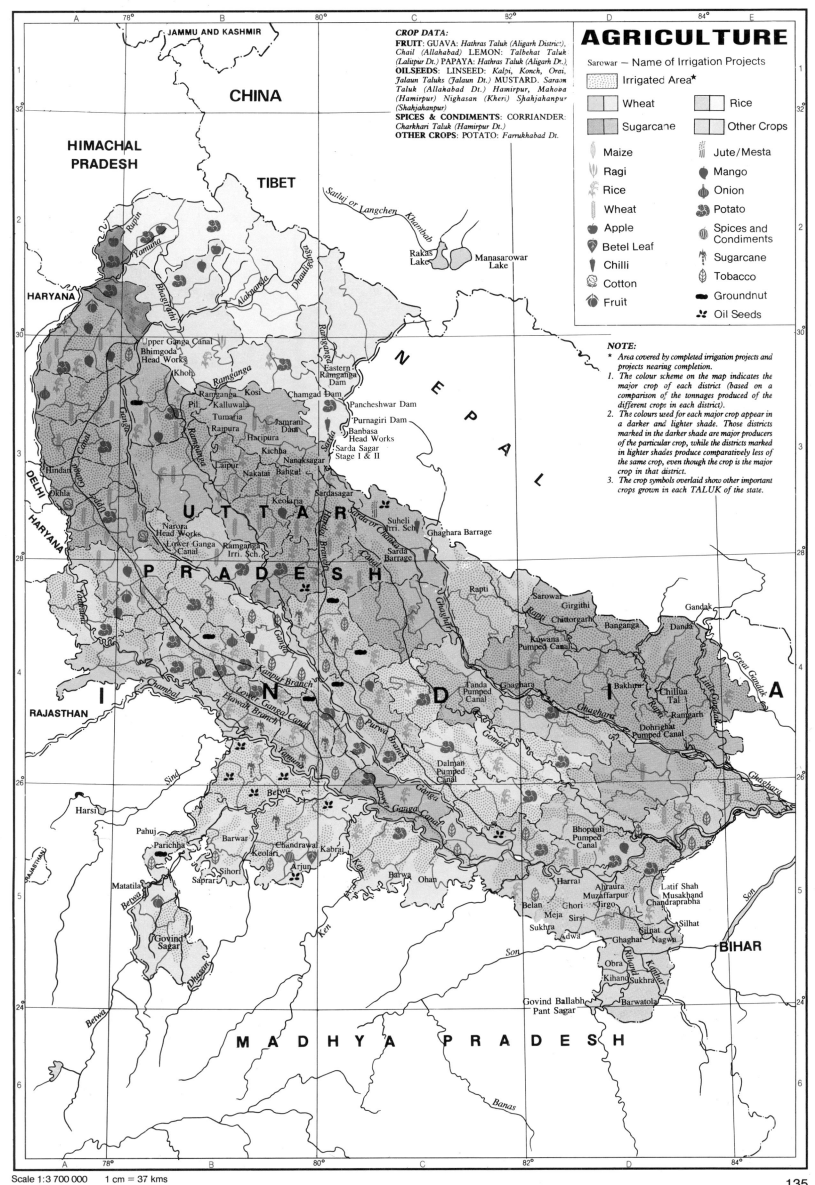

AGRICULTURE

Sarowar — Name of Irrigation Projects

CROP DATA:
FRUIT: GUAVA: *Hathras Taluk (Aligarh District)*, *Chail (Allahabad)* LEMON: *Talbehat Taluk (Lalitpur Dt.)* PAPAYA: *Hathras Taluk (Aligarh Dt.)*
OILSEEDS: LINSEED: *Kalpi, Konch, Orai, Jalaun Taluks (Jalaun Dt.)* MUSTARD: *Saraon Taluk (Allahabad Dt.)* Hamirpur, Mahoba (Hamirpur) Nighasan (Kheri) Shahjahanpur (Shahjahanpur)*
SPICES & CONDIMENTS: CORRIANDER: *Charkhari Taluk (Hamirpur Dt.)*
OTHER CROPS: POTATO: *Farrukhabad Dt.*

LEGEND:
- Irrigated Area*
- Wheat
- Rice
- Sugarcane
- Other Crops
- Maize
- Jute/Mesta
- Ragi
- Mango
- Rice
- Onion
- Wheat
- Potato
- Apple
- Spices and Condiments
- Betel Leaf
- Sugarcane
- Chilli
- Tobacco
- Cotton
- Groundnut
- Fruit
- Oil Seeds

NOTE:
* Area covered by completed irrigation projects and projects nearing completion.
1. The colour scheme on the map indicates the major crop of each district (based on a comparison of the tonnages produced of the different crops in each district).
2. The colours used for each major crop appear in a darker and lighter shade. Those districts marked in the darker shade are major producers of the particular crop, while the districts marked in lighter shades produce comparatively less of the same crop, even though the crop is the major crop in that district.
3. The crop symbols overlaid show other important crops grown in each TALUK of the state.

Scale 1:3 700 000   1 cm = 37 kms

135

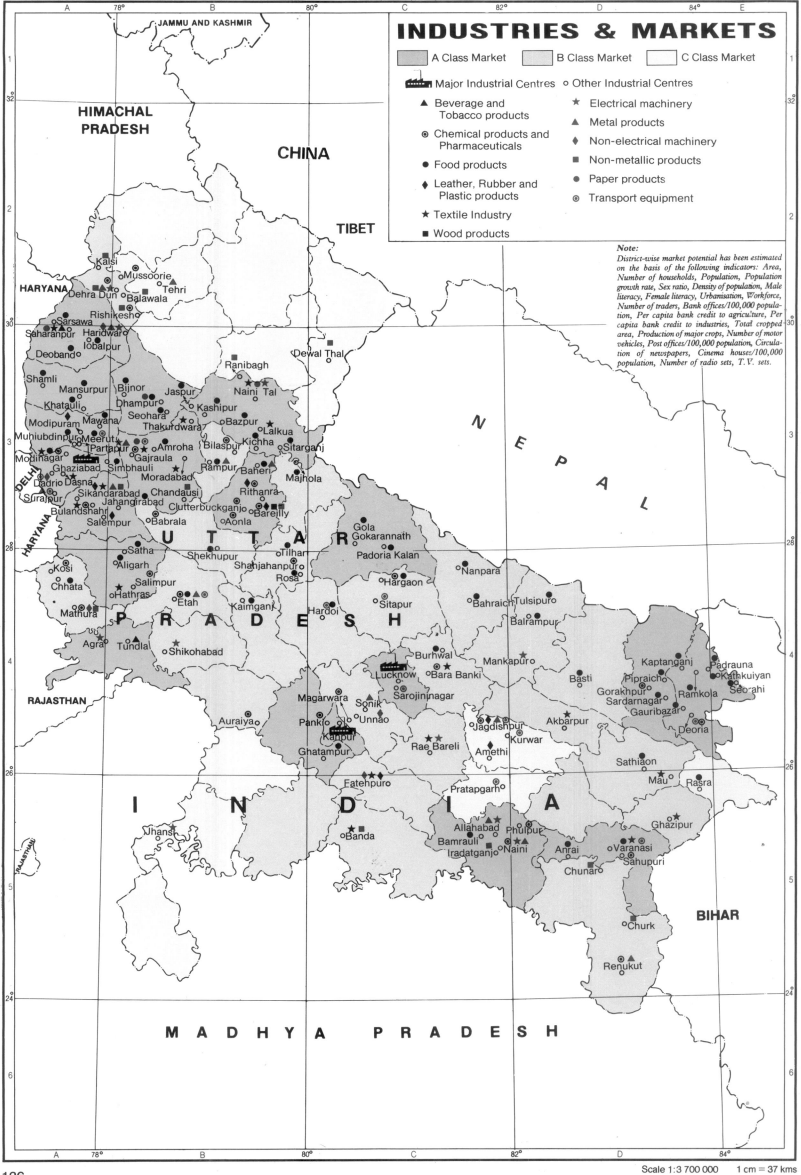

INDUSTRIES & MARKETS

A Class Market   B Class Market   C Class Market

Major Industrial Centres   ○ Other Industrial Centres

▲ Beverage and Tobacco products

◉ Chemical products and Pharmaceuticals

● Food products

◆ Leather, Rubber and Plastic products

★ Textile Industry

■ Wood products

★ Electrical machinery

▲ Metal products

◆ Non-electrical machinery

■ Non-metallic products

● Paper products

◉ Transport equipment

Note:
District-wise market potential has been estimated on the basis of the following indicators: Area, Number of households, Population, Population growth rate, Sex ratio, Density of population, Male literacy, Female literacy, Urbanisation, Workforce, Number of traders, Bank offices/100,000 population, Per capita bank credit to agriculture, Per capita bank credit to industries, Total cropped area, Production of major crops, Number of motor vehicles, Post offices/100,000 population, Circulation of newspapers, Cinema houses/100,000 population, Number of radio sets, T.V. sets.

JAMMU AND KASHMIR

HIMACHAL PRADESH

CHINA

TIBET

HARYANA

HIMACHAL PRADESH

RAJASTHAN

HARYANA

DELHI

NEPAL

UTTAR PRADESH

INDIA

BIHAR

MADHYA PRADESH

Kalsi
Mussoorie
Tehri
Dehra Dun
Balawala
Rishikesh
Sarsawa
Saharanpur
Haridwar
Iqbalpur
Deoband
Dewal Thal
Shamli
Mansurpur
Bijnor
Ranibagh
Khatauli
Dhampur
Jaspur
Naini Tal
Modipuram
Mawana
Seohara
Kashipur
Thakurdwara
Bazpur
Muhiubdinpur
Meerut
Amroha
Bilaspur
Kichha
Lalkua
Modinagar
Partapur
Gajraula
Rampur
Sitarganj
Ghaziabad
Simbhauli
Baheri
Majhola
Dadri
Dasna
Moradabad
Rithanra
Surajpur
Sikandarabad
Chandausi
Bulandshahr
Jahangirabad
Clutterbuckganj
Bareilly
Salempur
Babrala
Aonla
Gola
Gokarannath
Satha
Shekhupur
Tilhar
Padoria Kalan
Kosi
Aligarh
Salimpur
Shahjahanpur
Nanpara
Chhata
Hathras
Rosa
Sitapur
Bahraich
Tulsipur
Etah
Kaimganj
Hardoi
Balrampur
Mathura
Hargaon
Agra
Tundla
Shikohabad
Burhwal
Mankapur
Kaptanganj
Padrauna
Lucknow
Basti
Pipraich
Kathkuiyan
Bara Banki
Gorakhpur
Ramkola
Seorahi
Magarwara
Sarojininagar
Sardarnagar
Auraiya
Sonik
Unnao
Akbarpur
Gauribazar
Pankhi
Jagdishpur
Deoria
Kanpur
Kurwar
Ghatampur
Rae Bareli
Amethi
Sathiaon
Fatehpur
Mau
Rasra
Pratapgarh
Jhansi
Ghazipur
Allahabad
Phulpur
Banda
Bamrauli
Naini
Anrai
Varanasi
Iradatganj
Sahupuri
Chunar
Renukut
Churk

Scale 1:3 700 000   1 cm = 37 kms

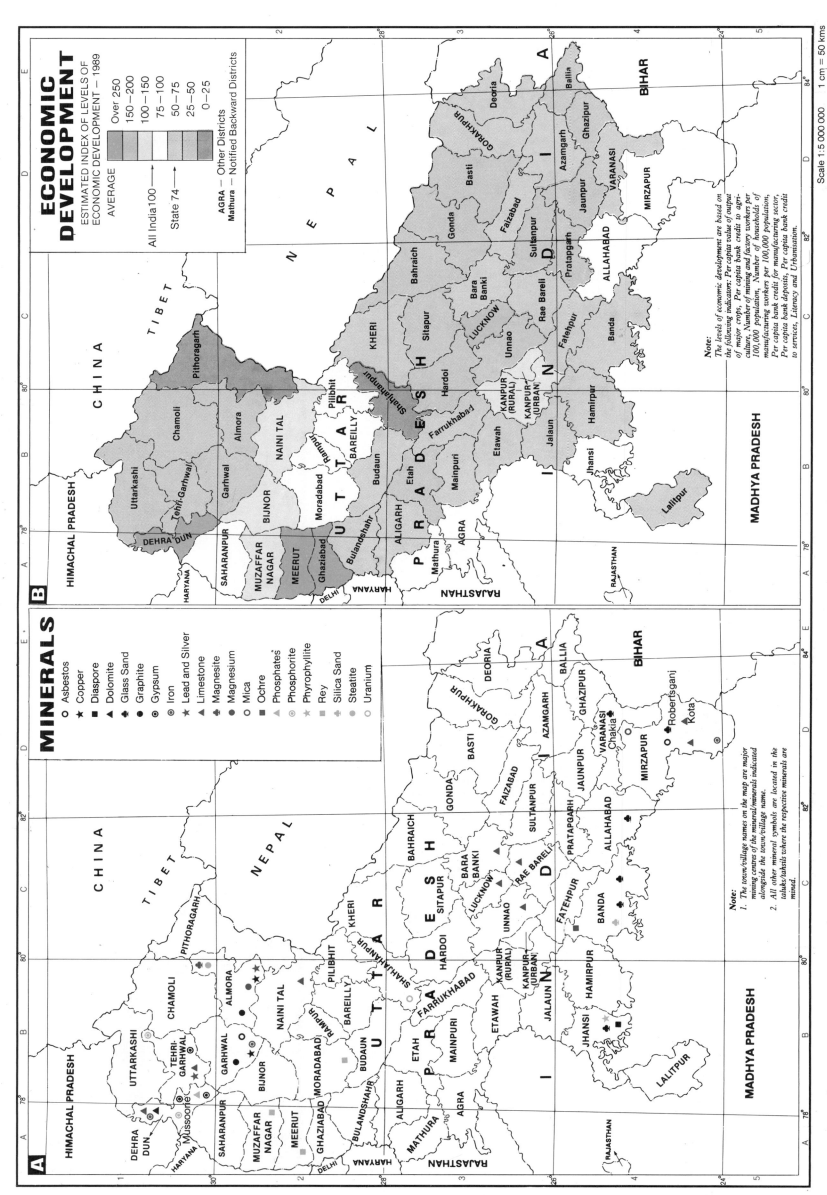

MINERALS

Minerals legend:
- ○ Asbestos
- ★ Copper
- ■ Diaspore
- ▲ Dolomite
- ♣ Glass Sand
- ● Graphite
- ◉ Gypsum
- ⊚ Iron
- ★ Lead and Silver
- ▲ Limestone
- ♣ Magnesite
- ● Magnesium
- ○ Mica
- ■ Ochre
- ▲ Phosphates
- ◉ Phosphorite
- ★ Phyrophyllite
- ■ Rey
- ♣ Silica Sand
- ● Steatite
- ○ Uranium

Note:
1. The town/village names on the map are major mining centres of the mineral/minerals indicated alongside the town/village name.
2. All other mineral symbols are located in the taluks/tahsils where the respective minerals are mined.

ECONOMIC DEVELOPMENT

ESTIMATED INDEX OF LEVELS OF ECONOMIC DEVELOPMENT – 1989

AVERAGE

Legend (Estimated Index):
- Over 250
- 150–200
- 100–150
- 75–100
- 50–75
- 25–50
- 0–25

All India 100 —→
State 74 —→

AGRA — Other Districts
Mathura — Notified Backward Districts

Note:
The levels of economic development are based on the following indicators: Per capita value of output of major crops, Per capita bank credit to agriculture, Number of mining and factory workers per 100,000 population, Number of households of manufacturing workers per 100,000 population, Per capita bank credit for manufacturing sector, Per capita bank deposits, Per capita bank credit to services, Literacy and Urbanisation.

Scale 1:5 000 000   1 cm = 50 kms

137

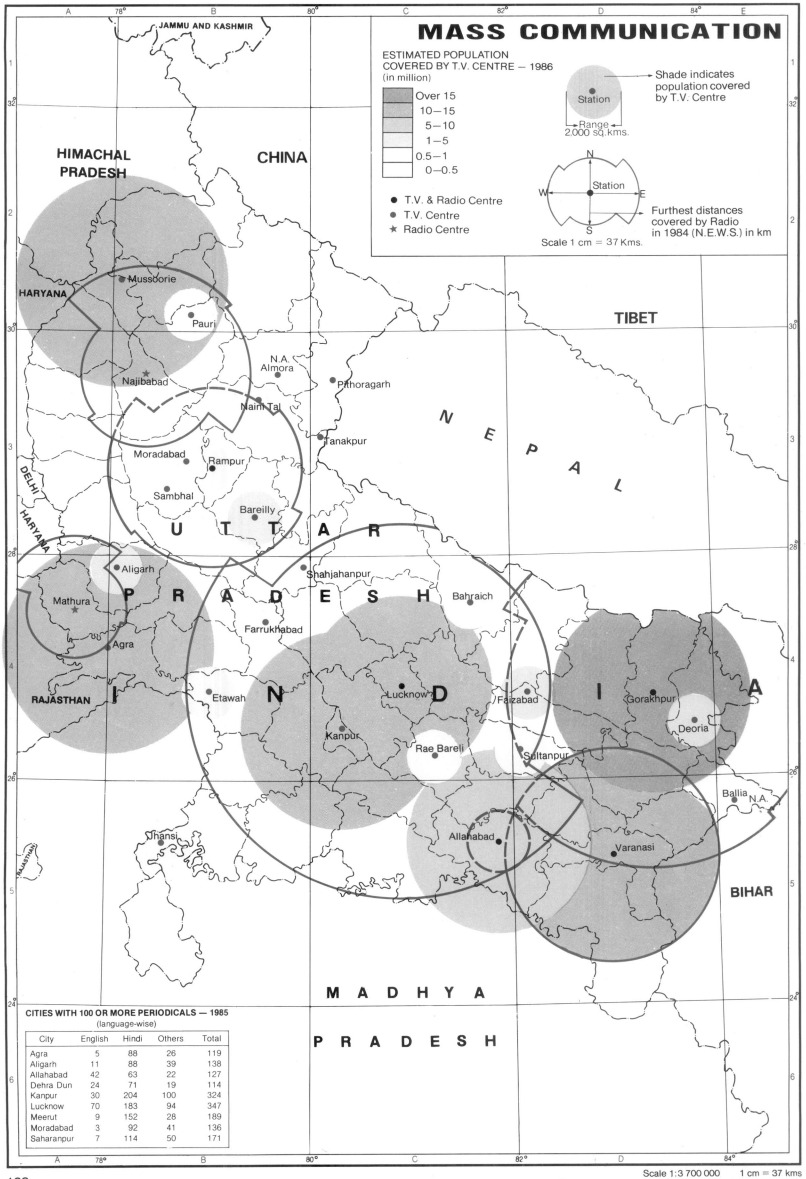

MASS COMMUNICATION

ESTIMATED POPULATION
COVERED BY T.V. CENTRE — 1986
(in million)

- Over 15
- 10—15
- 5—10
- 1—5
- 0.5—1
- 0—0.5

Shade indicates
population covered
by T.V. Centre

Station

Range
2,000 sq.kms.

- ● T.V. & Radio Centre
- ● T.V. Centre
- ★ Radio Centre

Station

Furthest distances
covered by Radio
in 1984 (N.E.W.S.) in km

Scale 1 cm = 37 Kms.

JAMMU AND KASHMIR

HIMACHAL
PRADESH

CHINA

TIBET

HARYANA

Mussoorie

Pauri

N.A.
Almora

Pithoragarh

N E P A L

Najibabad

Naini Tal

DELHI

Tanakpur

Moradabad

Rampur

Sambhal

Bareilly

HARYANA

U T T A R

Aligarh

Shahjahanpur

Bahraich

Mathura

P R A D E S H

Farrukhabad

RAJASTHAN

Agra

I

Etawah

N

Lucknow

D

Faizabad

I

Gorakhpur

A

Deoria

Kanpur

Rae Bareli

Sultanpur

Ballia N.A.

RAJASTHAN

Jhansi

Allahabad

Varanasi

BIHAR

M A D H Y A

P R A D E S H

CITIES WITH 100 OR MORE PERIODICALS — 1985
(language-wise)

City	English	Hindi	Others	Total
Agra	5	88	26	119
Aligarh	11	88	39	138
Allahabad	42	63	22	127
Dehra Dun	24	71	19	114
Kanpur	30	204	100	324
Lucknow	70	183	94	347
Meerut	9	152	28	189
Moradabad	3	92	41	136
Saharanpur	7	114	50	171

Scale 1:3 700 000   1 cm = 37 kms

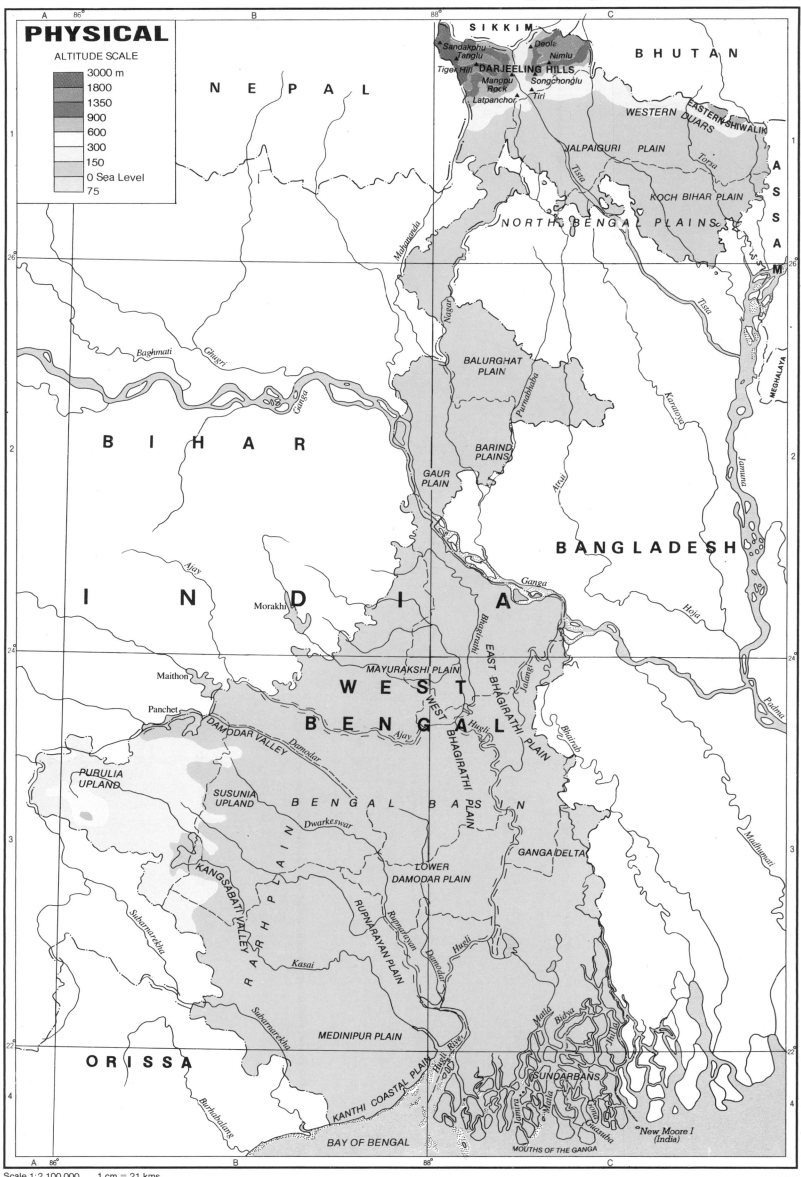

PHYSICAL

ALTITUDE SCALE

	3000 m
	1800
	1350
	900
	600
	300
	150
	0 Sea Level
	75

NEPAL

SIKKIM

BHUTAN

▲Sandakphu
Tanglu
Tiger Hill
DARJEELING HILLS
Mangpu Rock
Latpanchor ▲Tiri

▲Deola ▲Nimlu
▲Songchonglu

EASTERN SHIWALIK

WESTERN DUARS

JALPAIGURI PLAIN

Torsa

KOCH BIHAR PLAIN

NORTH BENGAL PLAINS

ASSAM

Mahananda

Tista

Tista

Jamuna

MEGHALAYA

Nagar

BALURGHAT PLAIN

Purnabhaba

BARIND PLAINS

GAUR PLAIN

Atrai

Karatoya

BANGLADESH

BIHAR

Baghmati

Ghugri

Ganga

Ganga

Hoja

Ajay

Morakhi

I N D I A

MAYURAKSHI PLAIN

Bhagirathi

Jalangi

EAST BHAGIRATHI PLAIN

Bhairab

Padma

Maithon

W E S T

Panchet

DAMODAR VALLEY

Damodar

Ajay

Hugli

WEST BHAGIRATHI PLAIN

B E N G A L

PURULIA UPLAND

SUSUNIA UPLAND

B E N G A L   B A S I N

Dwarkeswar

GANGA DELTA

LOWER DAMODAR PLAIN

R A R H   P L A I N

KANGSABATI VALLEY

Subarnarekha

Kasai

Rupnarayan

RUPNARAYAN PLAIN

Rupnarayan

Hugli

Damodar

Matla
Bidya

Subarnarekha

MEDINIPUR PLAIN

Hugli River

SUNDARBANS

ORISSA

Burhabalang

KANTHI COASTAL PLAIN

BAY OF BENGAL

MOUTHS OF THE GANGA

Matla
Mala
Gasuba

°New Moore I (India)

Madhumati

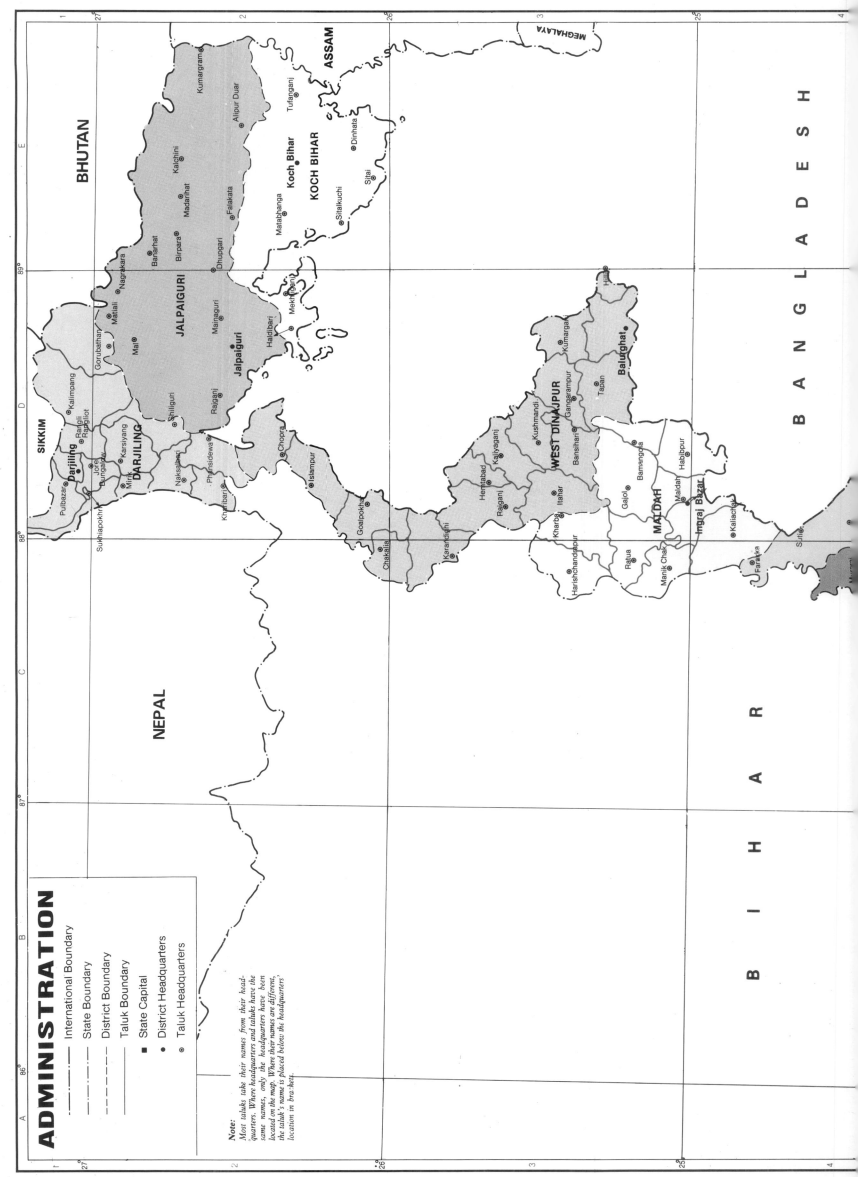

ADMINISTRATION

International Boundary
State Boundary
District Boundary
Taluk Boundary
■ State Capital
● District Headquarters
⊛ Taluk Headquarters

Note:
Most taluks take their names from their head-quarters. Where headquarters and taluks have the same names, only the headquarters have been located on the map. Where their names are different, the taluk's name is placed below the headquarters' location in brackets.

BHUTAN

SIKKIM

NEPAL

ASSAM

MEGHALAYA

BANGLADESH

B I H A R

DARJILING
Pulbazar
Sukhiapokhri
Darjiling
Jore Bungalow
Mirik
Karsiyang
Kalimpong
Rangli Rangliot

JALPAIGURI
Gorubathar
Matiali
Mal
Nagrakata
Banarhat
Madarihat
Kalchini
Kumargram
Alipur Duar
Falakata
Dhupgari
Birpara
Mainaguri
Jalpaiguri
Haldibari
Rajganj
Shiliguri

KOCH BIHAR
Mekhliganj
Matabhanga
Koch Bihar
Tufanganj
Dinhata
Sitai
Sitalkuchi

Phansidewa
Chopra
Islampur
Kharibari
Naksalbari
Goalpokhar
Chakalia
Karandighi

WEST DINAJPUR
Kushmandi
Kaliyaganj
Bansihari
Hemtabad
Raiganj
Itahar
Kharba
Kumarganj
Gangarampur
Tapan
Balurghat
Hili

MALDAH
Gajol
Bamangola
Maldah
Habibpur
Ratua
Manik Chak
Harishchandrapur
Old Maldah
Kaliachak
English Bazar
Farakka
Sitie

140

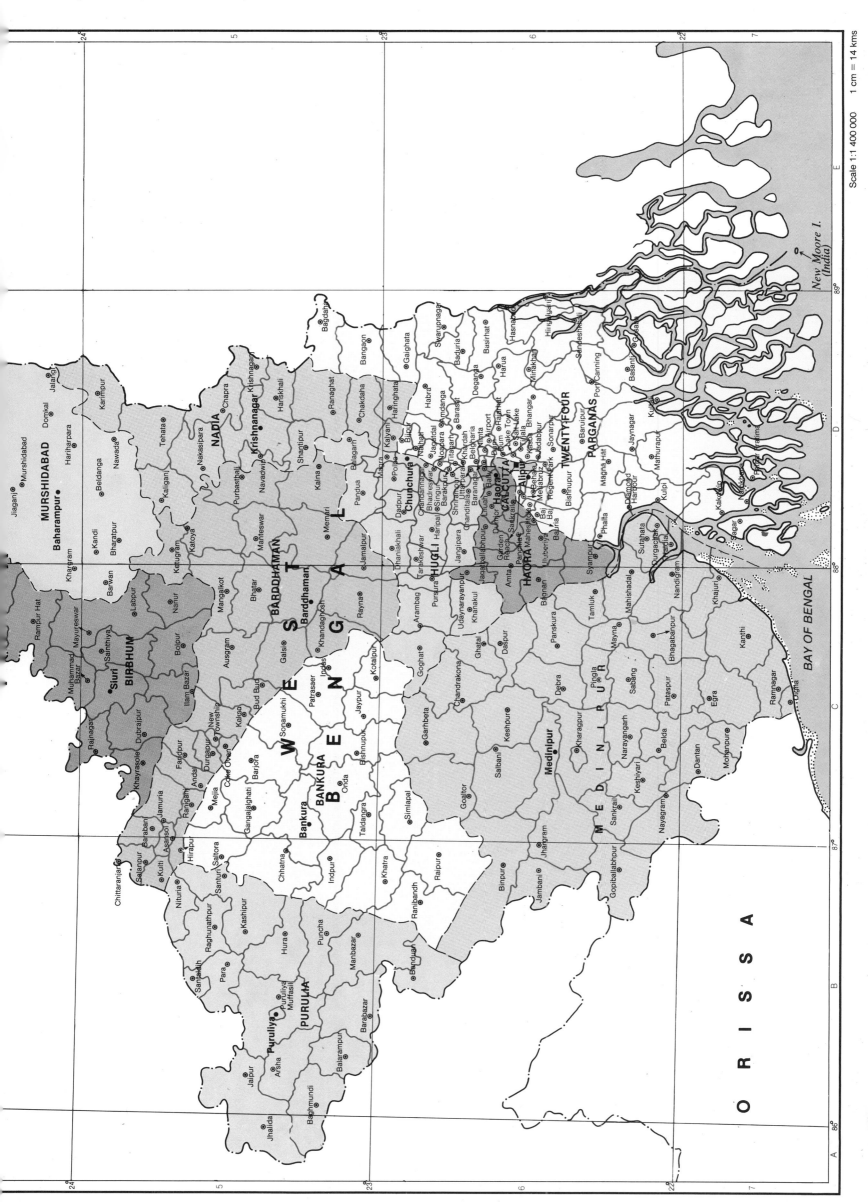

Scale 1:1 400 000   1 cm = 14 kms

141

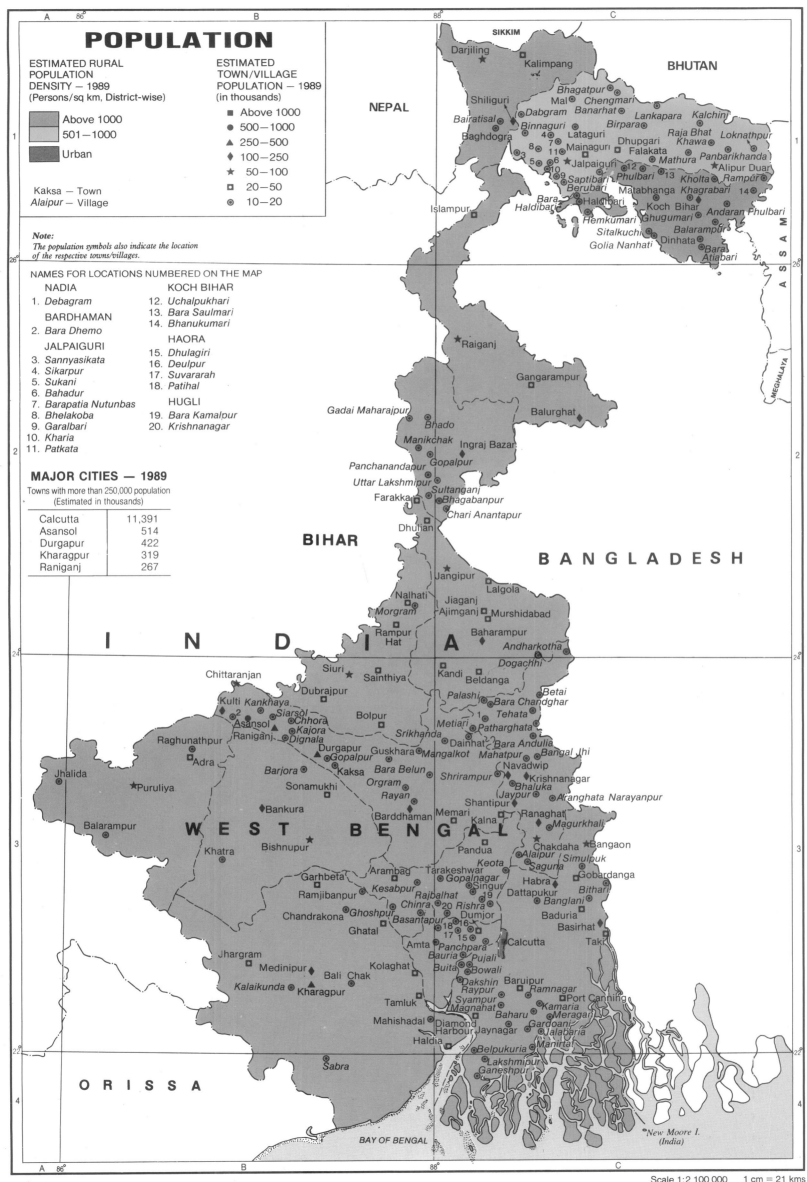

POPULATION

ESTIMATED RURAL POPULATION DENSITY — 1989
(Persons/sq km, District-wise)

- Above 1000
- 501—1000
- Urban

Kaksa — Town
Alaipur — Village

ESTIMATED TOWN/VILLAGE POPULATION — 1989
(in thousands)

- ■ Above 1000
- ● 500—1000
- ▲ 250—500
- ◆ 100—250
- ★ 50—100
- ◻ 20—50
- ◉ 10—20

Note:
The population symbols also indicate the location of the respective towns/villages.

NAMES FOR LOCATIONS NUMBERED ON THE MAP

NADIA
1. *Debagram*

BARDHAMAN
2. *Bara Dhemo*

JALPAIGURI
3. *Sannyasikata*
4. *Sikarpur*
5. *Sukani*
6. *Bahadur*
7. *Barapatia Nutunbas*
8. *Bhelakoba*
9. *Garalbari*
10. *Kharia*
11. *Patkata*

KOCH BIHAR
12. *Uchalpukhari*
13. *Bara Saulmari*
14. *Bhanukumari*

HAORA
15. *Dhulagiri*
16. *Deulpur*
17. *Suvararah*
18. *Patihal*

HUGLI
19. *Bara Kamalpur*
20. *Krishnanagar*

MAJOR CITIES — 1989

Towns with more than 250,000 population
(Estimated in thousands)

Calcutta	11,391
Asansol	514
Durgapur	422
Kharagpur	319
Raniganj	267

Scale 1:2 100 000   1 cm = 21 kms

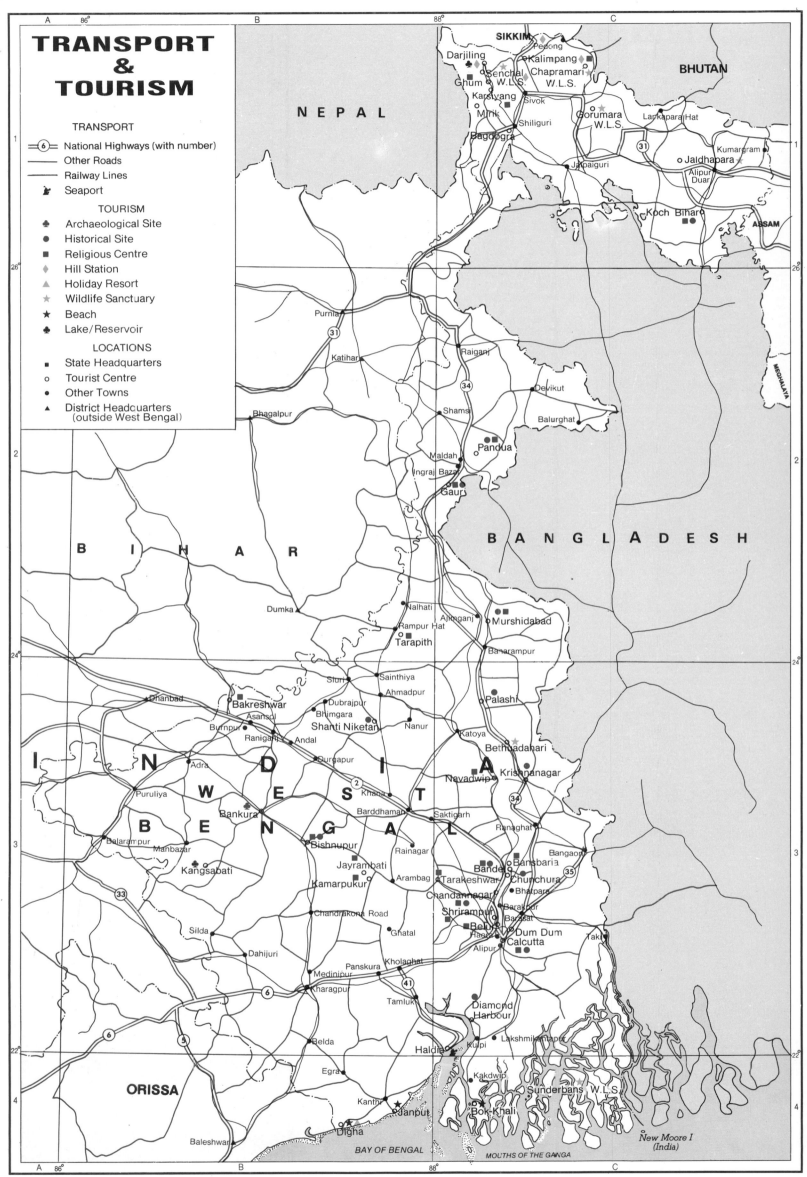

TRANSPORT & TOURISM

TRANSPORT
═(6)═ National Highways (with number)
— Other Roads
— Railway Lines
⚓ Seaport

TOURISM
♣ Archaeological Site
● Historical Site
■ Religious Centre
◆ Hill Station
▲ Holiday Resort
★ Wildlife Sanctuary
★ Beach
♣ Lake/Reservoir

LOCATIONS
■ State Headquarters
○ Tourist Centre
● Other Towns
▲ District Headquarters
  (outside West Bengal)

NEPAL

SIKKIM
Pedong
Darjiling
Kalimpang
Senchal Chapramari
Ghum W.L.S. W.L.S.
Karsiyang
Mirik Sivok
Shiliguri
Bagdogra
Gorumara
W.L.S.
Larkapara Hat
BHUTAN
Kumargram
Jalpaiguri
Jaldhapara
Alipur
Duar
Koch Bihar
ASSAM

Purnia
Katihar
Raiganj
Devikut
Shams
Balurghat
Bhagalpur
MEGHALAYA

Maldah
Ingraj Bazar
Gauri
Pandua

BANGLADESH

BIHAR

Dumka
Nalhati
Ajimganj
Murshidabad
Rampur Hat
Tarapith
Baharampur

Suri
Sainthiya
Ahmadpur
Palashi
Dhanbad
Bakreshwar
Dubrajpur
Bhimgara
Nanur
Asansol
Shanti Niketan
Katoya
Burnpur
Raniganj
Andal
Bethuadahari
Adra
Durgapur
Krishnanagar
Navadwip
INDIA WEST
Puruliya
Khana
2
Barddhaman
Saktigarh
Bankura
Ranaghat
BENGAL
Balarampur
Rainagar
Bangaon
Manbazar
Bishnupur
Bansbaria
35
Kangsabati
Jayrambati
Bandel
Chunchura
Kamarpukur
Tarakeshwar
Chandannagar
Bhatpara
Chandrakona Road
Shrirampur
Barakpur
Silda
Belur
Barasat
Ghatal
Dum Dum
Dahijuri
Haora
Calcutta
Takı
Panskura
Alipur
Medinipur
Kholaghat
Kharagpur
41
Tamluk
Diamond
Harbour
Belda
Kulpi
Lakshmikantapur
ORISSA
Egra
Haldia
Kakdwip
Kanthi
Sunderbans W.L.S.
Bok-Khali
Janput
Digha
New Moore I
(India)
Baleshwar
BAY OF BENGAL
MOUTHS OF THE GANGA

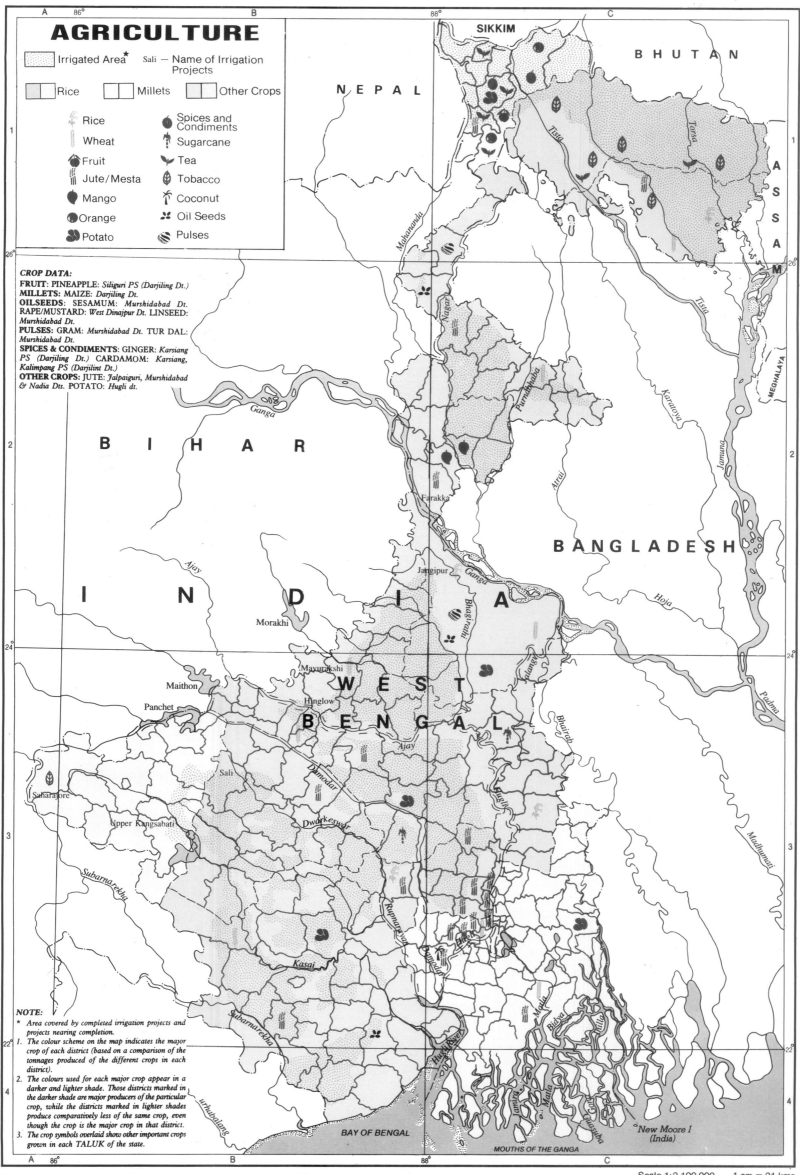

AGRICULTURE

- Irrigated Area *   Sali — Name of Irrigation Projects
- Rice   Millets   Other Crops

Rice
Wheat
Fruit
Jute/Mesta
Mango
Orange
Potato

Spices and Condiments
Sugarcane
Tea
Tobacco
Coconut
Oil Seeds
Pulses

CROP DATA:
FRUIT: PINEAPPLE: *Siliguri PS (Darjiling Dt.)*
MILLETS: MAIZE: *Darjiling Dt.*
OILSEEDS: SESAMUM: *Murshidabad Dt.*
RAPE/MUSTARD: *West Dinajpur Dt.* LINSEED: *Murshidabad Dt.*
PULSES: GRAM: *Murshidabad Dt.* TUR DAL: *Murshidabad Dt.*
SPICES & CONDIMENTS: GINGER: *Karsiang PS (Darjiling Dt.)* CARDAMOM: *Karsiang, Kalimpang PS (Darjilint Dt.)*
OTHER CROPS: JUTE: *Jalpaiguri, Murshidabad & Nadia Dts.* POTATO: *Hugli dt.*

SIKKIM

BHUTAN

NEPAL

A S S A M

MEGHALAYA

B I H A R

BANGLADESH

I N D I A

W E S T

B E N G A L

Morakhi
Maithon
Panchet
Hinglow
Mayurakshi
Saharajore
Upper Kangsabati
Sali
Farakka
Jangipur

Ganga
Ajay
Mahananda
Nagar
Purnabhaba
Tista
Karatoya
Jamuna
Atrai
Bhagirathi
Jalangi
Bhairab
Hugli
Damodar
Dwarkeswar
Ajay
Kasai
Rupnarayan
Subarnarekha
Urubalang
Hoja
Padma
Madhumati
Mohananda
Mohla
Bidya
Matla
Jamira
Gosaba
Saptamukhi
Hugli River

New Moore I (India)

BAY OF BENGAL

MOUTHS OF THE GANGA

NOTE:
* Area covered by completed irrigation projects and projects nearing completion.
1. The colour scheme on the map indicates the major crop of each district (based on a comparison of the tonnages produced of the different crops in each district).
2. The colours used for each major crop appear in a darker and lighter shade. Those districts marked in the darker shade are major producers of the particular crop, while the districts marked in lighter shades produce comparatively less of the same crop, even though the crop is the major crop in that district.
3. The crop symbols overlaid show other important crops grown in each TALUK of the state.

144

Scale 1:2 100 000   1 cm = 21 kms

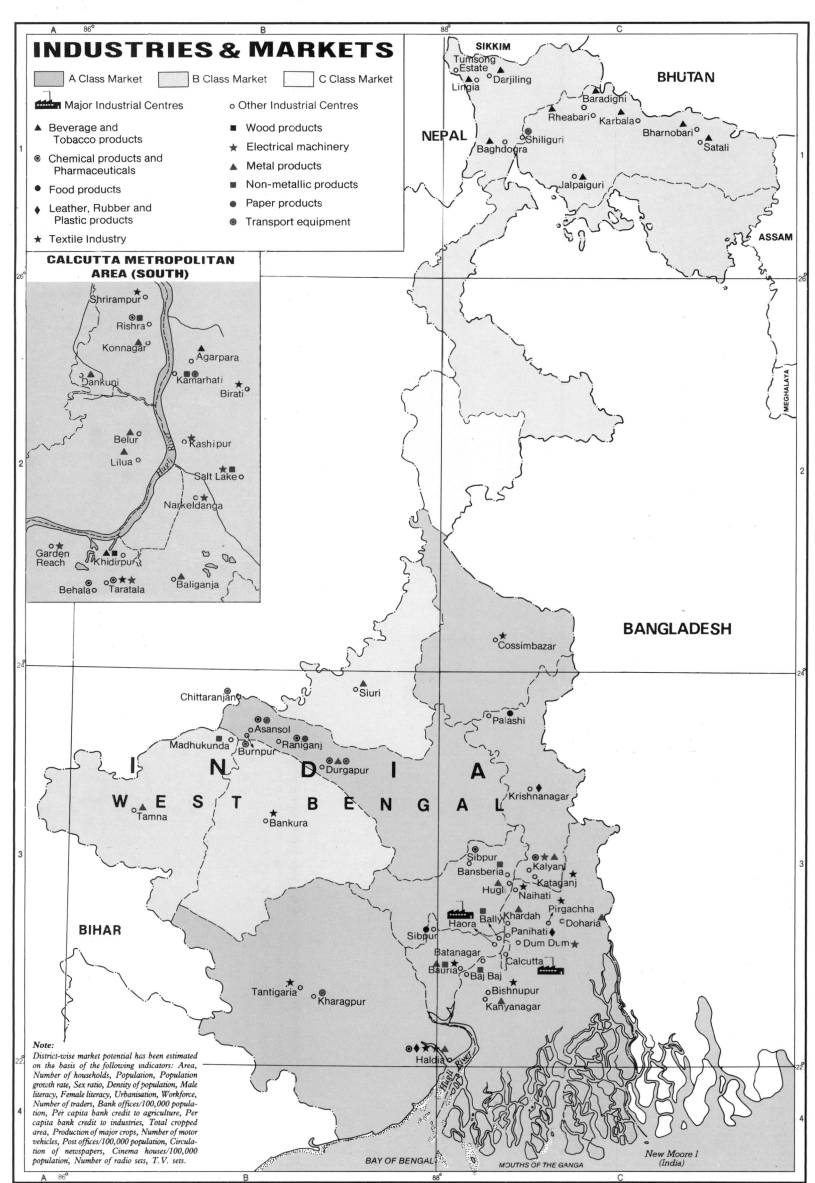

INDUSTRIES & MARKETS

A Class Market   B Class Market   C Class Market

Major Industrial Centres   o Other Industrial Centres

▲ Beverage and Tobacco products
◉ Chemical products and Pharmaceuticals
● Food products
♦ Leather, Rubber and Plastic products
★ Textile Industry

■ Wood products
★ Electrical machinery
▲ Metal products
■ Non-metallic products
● Paper products
◉ Transport equipment

CALCUTTA METROPOLITAN AREA (SOUTH)

Shrirampur
Rishra
Konnagar
Agarpara
Dankuni
Kamarhati
Birati
Belur
Kashipur
Lilua
Salt Lake
Narkeldanga
Garden Reach
Khidirpur
Behala   Taratala   Baliganja

SIKKIM
Tumsong Estate
Darjiling
Lingia
BHUTAN
Baradighi
Rheabari   Karbala
Bharnobari
NEPAL
Shiliguri   Satali
Baghdogra
ASSAM
Jalpaiguri

BANGLADESH

Cossimbazar

Siuri
Palashi

Chittaranjan
Asansol
Madhukunda   Raniganj
Burnpur
Durgapur

I N D I A

W E S T   B E N G A L
Krishnanagar

Tamna
Bankura

Sibpur
Bansberia   Kalyani
Hugli   Katagarj
Naihati
Pirgachha
Haora   Bally   Khardah   Doharia
BIHAR
Sibpur   Panihati
Batanagar   Dum Dum
Bauria   Calcutta
Baj Baj
Tantigaria   Bishnupur
Kharagpur   Kanyanagar

Haldia

MEGHALAYA

Hugli River

BAY OF BENGAL   MOUTHS OF THE GANGA   New Moore I (India)

Note:
District-wise market potential has been estimated on the basis of the following indicators: Area, Number of households, Population, Population growth rate, Sex ratio, Density of population, Male literacy, Female literacy, Urbanisation, Workforce, Number of traders, Bank offices/100,000 population, Per capita bank credit to agriculture, Per capita bank credit to industries, Total cropped area, Production of major crops, Number of motor vehicles, Post offices/100,000 population, Circulation of newspapers, Cinema houses/100,000 population, Number of radio sets, T.V. sets.

Scale 1:2 100 000   1 cm = 21 kms

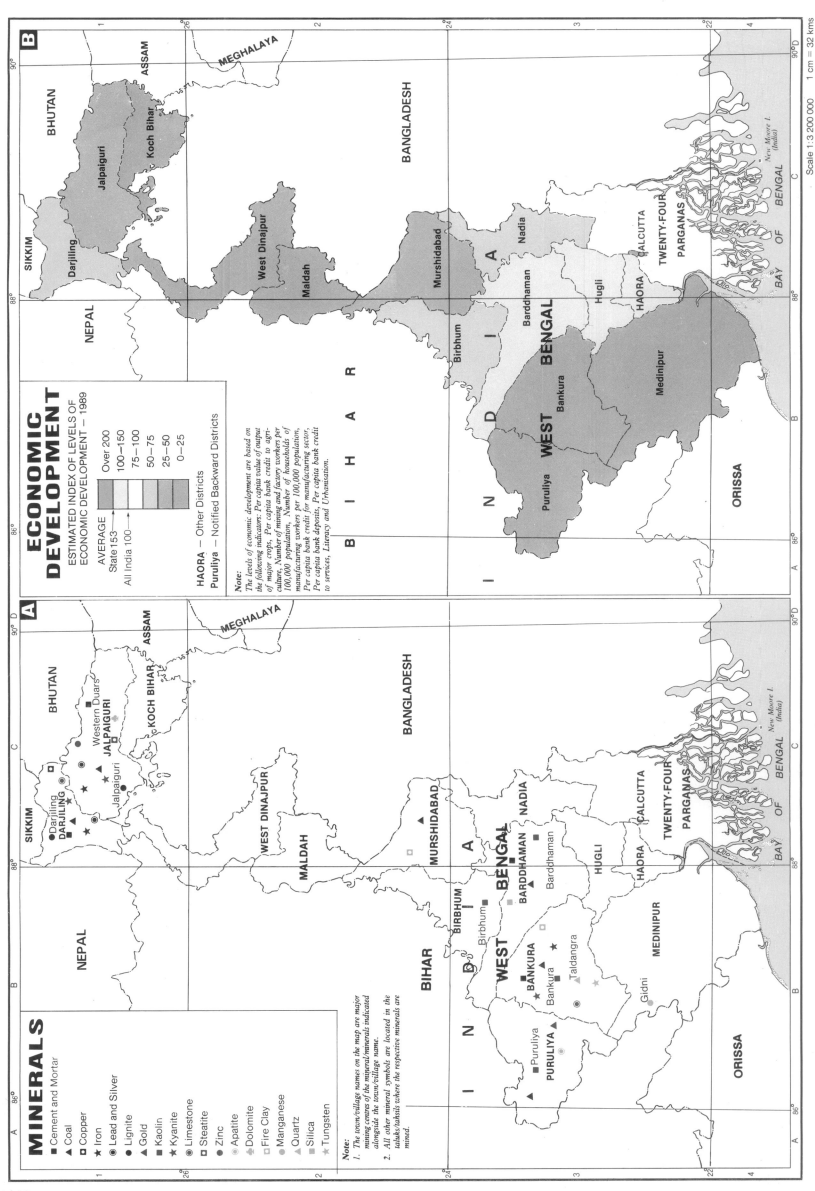

MINERALS

- ■ Cement and Mortar
- ▲ Coal
- □ Copper
- ★ Iron
- ◉ Lead and Silver
- ● Lignite
- ▲ Gold
- ■ Kaolin
- ★ Kyanite
- ◉ Limestone
- □ Steatite
- ■ Zinc
- ◉ Apatite
- ✦ Dolomite
- □ Fire Clay
- ● Manganese
- ▲ Quartz
- ■ Silica
- ★ Tungsten

Note:
1. *The town/village names on the map are major mining centres of the mineral/minerals indicated alongside the town/village name.*
2. *All other mineral symbols are located in the taluks/tahsils where the respective minerals are mined.*

A

BHUTAN

ASSAM

SIKKIM

NEPAL

MEGHALAYA

Western Duars
JALPAIGURI
Jalpaiguri
KOCH BIHAR

Darjiling
DARJILING

WEST DINAJPUR

MALDAH

BANGLADESH

BIHAR

MURSHIDABAD

BIRBHUM
Birbhum

NADIA

BARDDHAMAN
Barddhaman

I N D I A

WEST BENGAL

BANKURA
Bankura
Taldangra

PURULIYA
Puruliya

HUGLI

HAORA

CALCUTTA

TWENTY-FOUR PARGANAS

MEDINIPUR
Gidni

ORISSA

BAY OF BENGAL

New Moore I. (India)

Scale 1:3 200 000   1 cm = 32 kms

ECONOMIC DEVELOPMENT

ESTIMATED INDEX OF LEVELS OF ECONOMIC DEVELOPMENT – 1989

AVERAGE
State 153
All India 100

- Over 200
- 100–150
- 75–100
- 50–75
- 25–50
- 0–25

HAORA — Other Districts
Puruliya — Notified Backward Districts

Note:
The levels of economic development are based on the following indicators: Per capita value of output of major crops, Per capita bank credit to agri-culture, Number of mining and factory workers per 100,000 population, Number of households of manufacturing workers per 100,000 population, Per capita bank credit for manufacturing sector, Per capita bank deposits, Per capita bank credit to services, Literacy and Urbanisation.

B

BHUTAN

ASSAM

SIKKIM

NEPAL

MEGHALAYA

Darjiling

Jalpaiguri

Koch Bihar

West Dinajpur

Maldah

BANGLADESH

B I H A R

Murshidabad

Birbhum

Nadia

Barddhaman

I N D I A

WEST BENGAL

Bankura

Puruliya

Medinipur

Hugli

HAORA

CALCUTTA

TWENTY-FOUR PARGANAS

ORISSA

BAY OF BENGAL

New Moore I. (India)

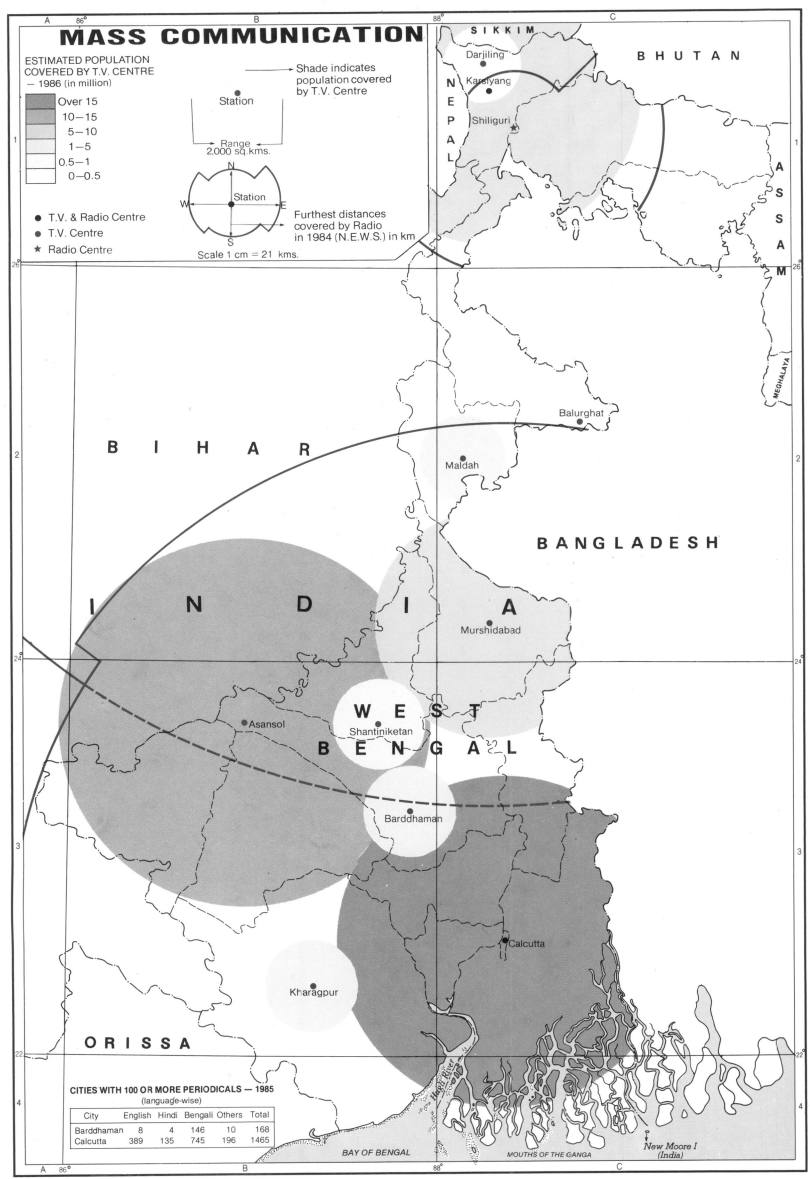

MASS COMMUNICATION

ESTIMATED POPULATION
COVERED BY T.V. CENTRE
— 1986 (in million)

Over 15
10—15
5—10
1—5
0.5—1
0—0.5

● T.V. & Radio Centre
● T.V. Centre
★ Radio Centre

Shade indicates
population covered
by T.V. Centre

Station

Range
2,000 sq.kms.

N
W — Station — E
S

Furthest distances
covered by Radio
in 1984 (N.E.W.S.) in km

Scale 1 cm = 21 kms.

SIKKIM

BHUTAN

Darjiling

Karsiyang

NEPAL

Shiliguri ★

ASSAM

MEGHALAYA

BIHAR

Balurghat

Maldah

BANGLADESH

I   N   D   I   A

Murshidabad

W E S T

Asansol

Shantiniketan

B E N G A L

Barddhaman

Calcutta

Kharagpur

ORISSA

Hugli River

Bay of Bengal

MOUTHS OF THE GANGA

New Moore I
(India)

CITIES WITH 100 OR MORE PERIODICALS — 1985

(language-wise)

City	English	Hindi	Bengali	Others	Total
Barddhaman	8	4	146	10	168
Calcutta	389	135	745	196	1465

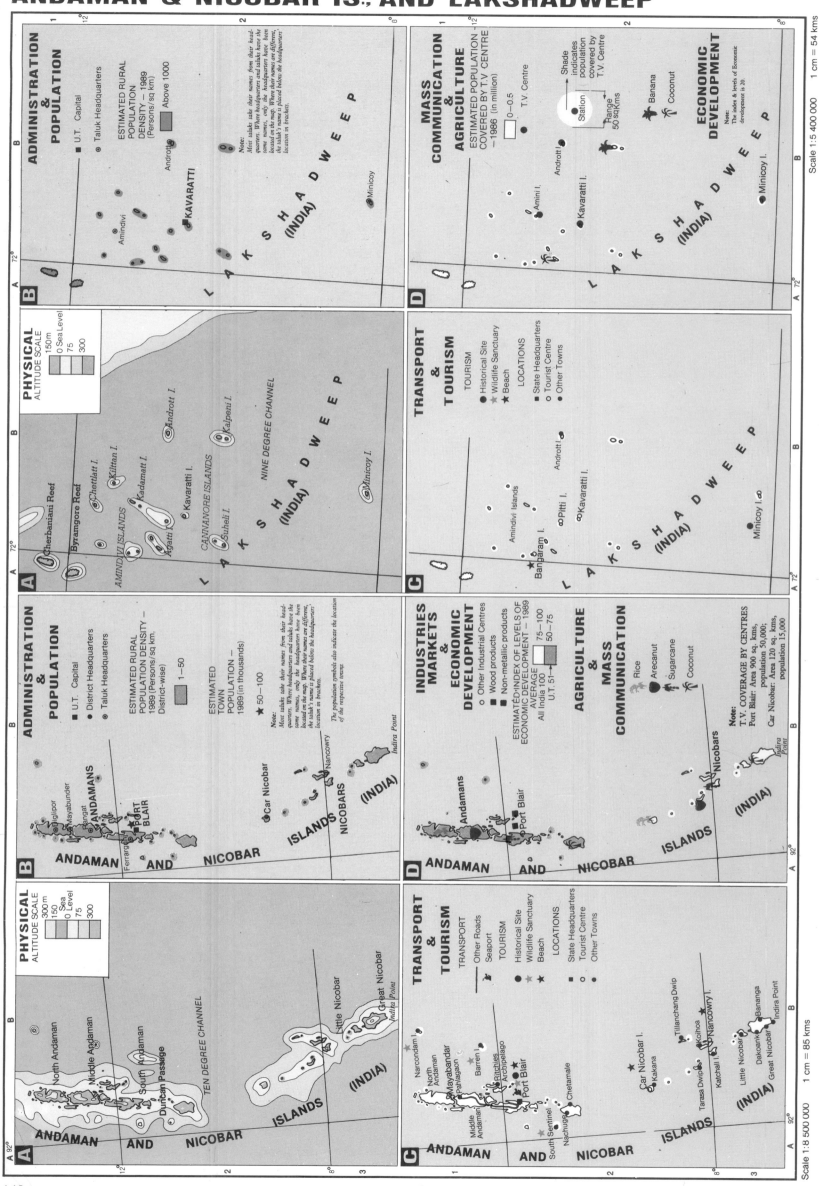

B — ADMINISTRATION & POPULATION

- ■ U.T. Capital
- ⊙ Taluk Headquarters

ESTIMATED RURAL POPULATION DENSITY – 1989 (Persons/sq km)
Above 1000

Note:
Most taluks take their names from their head-quarters. Where headquarters and taluks have the same names, only the headquarters have been located on the map. Where their names are different, the taluk's name is placed below the headquarters location in brackets.

Androtte
KAVARATTI
Amindivi
Minicoy

L A K S H A D W E E P (INDIA)

A — PHYSICAL
ALTITUDE SCALE
150m
0 Sea Level
75
300

Cherbaniani Reef
Byramgore Reef
Chettlatt I.
Kiltan I.
Kadamatt I.
AMINDIVI ISLANDS
Agatti I.
Andrott I.
Kavaratti I.
Kalpeni I.
CANNANORE ISLANDS
Suheli I.
NINE DEGREE CHANNEL
Minicoy I.

L A K S H A D W E E P (INDIA)

D — MASS COMMUNICATION & AGRICULTURE

ESTIMATED POPULATION COVERED BY T.V. CENTRE –1986 (in million)
0 – 0.5

● T.V. Centre

Station
Shade indicates population covered by T.V. Centre
Range 50 sq.Kms

Amini I.
Andrott I.
Kavaratti I.
Minicoy I.

ECONOMIC DEVELOPMENT
Banana
Coconut

Note:
The index & levels of Economic development is 20.

L A K S H A D W E E P (INDIA)

C — TRANSPORT & TOURISM

TOURISM
- ● Historical Site
- ★ Wildlife Sanctuary
- ★ Beach

LOCATIONS
- ■ State Headquarters
- ⊙ Tourist Centre
- ○ Other Towns

Amindivi Islands
Pitti I.
Kavaratti I.
Bangaram I.
Minicoy I.

L A K S H A D W E E P (INDIA)

B — ADMINISTRATION & POPULATION

- ■ U.T. Capital
- ● District Headquarters
- ⊙ Taluk Headquarters

ESTIMATED RURAL POPULATION DENSITY – 1989 (Persons/sq km, District-wise)
1 – 50

ESTIMATED TOWN POPULATION – 1989 (in thousands)
★ 50 – 100

Note:
Most taluks take their names from their head-quarters. Where headquarters and taluks have the same names, only the headquarters have been located on the map. Where their names are different, the taluk's name is placed below the headquarters location in brackets.
The population symbols also indicate the location of the respective towns.

Diglipor
Mayabunder
Rangat
ANDAMANS
Ferrarganj
PORT BLAIR
Car Nicobar
Nancowry
NICOBARS
Indira Point

ANDAMAN AND NICOBAR ISLANDS (INDIA)

A — PHYSICAL
ALTITUDE SCALE
300m
150
0 Sea Level
75
300

North Andaman
Middle Andaman
South Andaman
Duncan Passage
TEN DEGREE CHANNEL
Little Nicobar
Great Nicobar
Indira Point

ANDAMAN AND NICOBAR ISLANDS (INDIA)

D — INDUSTRIES MARKETS & ECONOMIC DEVELOPMENT

- ○ Other Industrial Centres
- ■ Wood products
- ■ Non-metallic products

ESTIMATED INDEX OF LEVELS OF ECONOMIC DEVELOPMENT – 1989
AVERAGE All India 100 — 75–100, 50–75
U.T. 51 —

AGRICULTURE & MASS COMMUNICATION
Rice
Arecanut
Sugarcane
Coconut

Note:
T.V. COVERAGE BY CENTRES
Port Blair: Area 900 sq. kms, population 50,000;
Car Nicobar: Area 120 sq. kms, population 15,000

Andamans
Port Blair
Nicobars
Indira Point

ANDAMAN AND NICOBAR ISLANDS (INDIA)

C — TRANSPORT & TOURISM

TRANSPORT
- —— Other Roads
- Seaport

TOURISM
- ● Historical Site
- ★ Wildlife Sanctuary
- ★ Beach

LOCATIONS
- ■ State Headquarters
- ⊙ Tourist Centre
- ○ Other Towns

Narcondam I.
North Andaman
Mayabundar
Pahlagaon
Barren I.
Middle Andaman
Ritchies Archipelago
Port Blair
South Sentinel
Nachugam
Chetamale
Car Nicobar I.
Kakana
Tillanchang Dwip
Kolhoa
Nancowry I.
Katchall I.
Tarasa Dwip
Little Nicobar
Dakoankt
Bananga
Great Nicobar
Indira Point

ANDAMAN AND NICOBAR ISLANDS (INDIA)

Scale 1:8 500 000   1 cm = 85 kms

Scale 1:5 400 000   1 cm = 54 kms

CHANDIGARH, DADRA & NAGAR HAVELI AND DAMAN & DIU

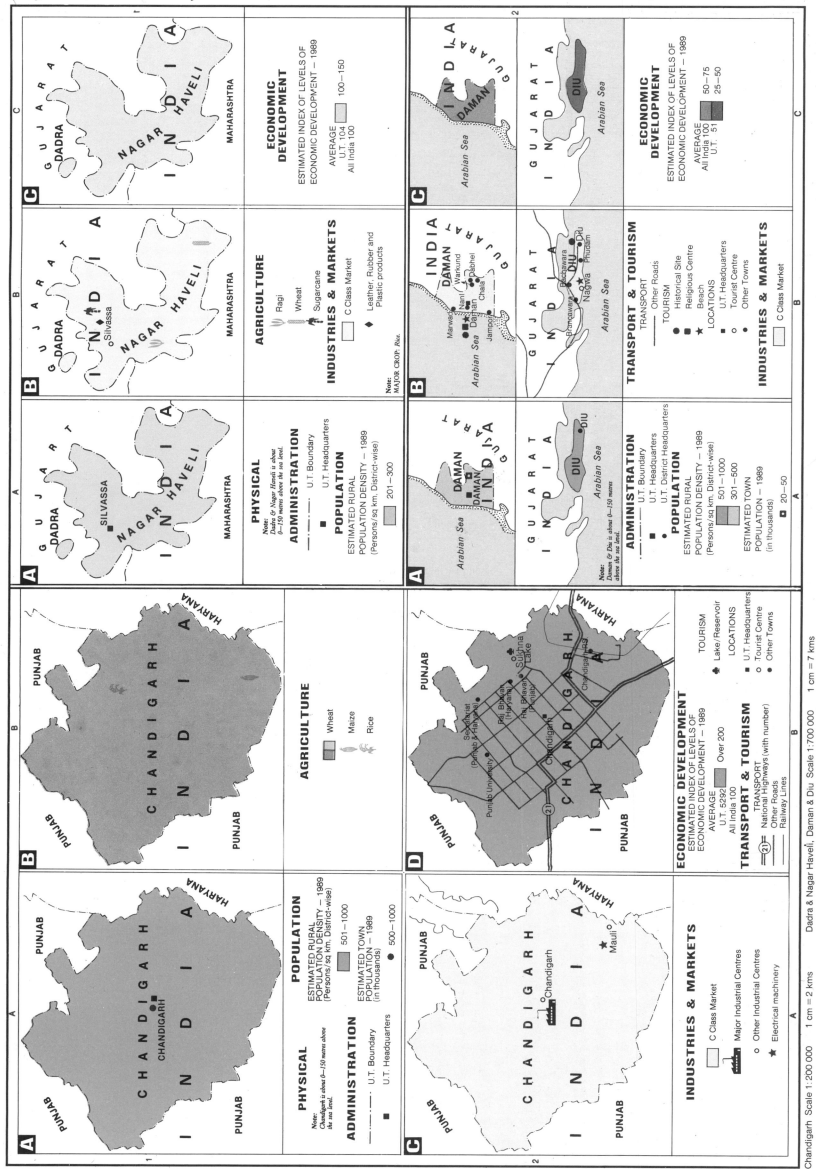

Dadra & Nagar Haveli — C (ECONOMIC DEVELOPMENT)

GUJARAT

DADRA

NAGAR HAVELI

MAHARASHTRA

INDIA

ECONOMIC DEVELOPMENT

ESTIMATED INDEX OF LEVELS OF
ECONOMIC DEVELOPMENT – 1989

AVERAGE
U.T. 104
All India 100

100–150

Dadra & Nagar Haveli — B (AGRICULTURE / INDUSTRIES & MARKETS)

GUJARAT

DADRA ● Silvassa

NAGAR HAVELI

MAHARASHTRA

INDIA

AGRICULTURE

Ragi

Wheat

Sugarcane

INDUSTRIES & MARKETS

C Class Market

◆ Leather, Rubber and
Plastic products

Note:
MAJOR CROP: Rice.

Dadra & Nagar Haveli — A (PHYSICAL / ADMINISTRATION / POPULATION)

GUJARAT

DADRA

■ SILVASSA

NAGAR HAVELI

MAHARASHTRA

INDIA

PHYSICAL

Note:
Dadra & Nagar Haveli is about
0–150 metres above the sea level.

ADMINISTRATION

---- U.T. Boundary

■ U.T. Headquarters

POPULATION

ESTIMATED RURAL
POPULATION DENSITY – 1989
(Persons/sq. km, District-wise)

201–300

Daman & Diu — C (ECONOMIC DEVELOPMENT)

INDIA

GUJARAT

DAMAN

Arabian Sea

DIU

GUJARAT

INDIA

Arabian Sea

ECONOMIC DEVELOPMENT

ESTIMATED INDEX OF LEVELS OF
ECONOMIC DEVELOPMENT – 1989

AVERAGE
All India 100
U.T. 51

50–75
25–50

Daman & Diu — B (TRANSPORT & TOURISM / INDUSTRIES & MARKETS)

INDIA

GUJARAT

DAMAN

Nani
Warkund
Dalvari Dabhel
Marwad Daman Chala
Jampore

DIU
DIU Diu
DIU Phudam
Bucharwada
Nagwa

Bramanwara

Arabian Sea

GUJARAT

INDIA

TRANSPORT & TOURISM

TRANSPORT
Other Roads

TOURISM
● Historical Site
■ Religious Centre
★ Beach

LOCATIONS
■ U.T. Headquarters
○ Tourist Centre
● Other Towns

INDUSTRIES & MARKETS

C Class Market

Daman & Diu — A (ADMINISTRATION / POPULATION)

Arabian Sea

GUJARAT

DAMAN
■ DAMAN

INDIA

DIU
● Diu

Arabian Sea

ADMINISTRATION

---- U.T. Boundary
■ U.T. Headquarters
■ U.T. District Headquarters

POPULATION

ESTIMATED RURAL
POPULATION DENSITY – 1989
(Persons/sq. km, District-wise)

501–1000
301–500

ESTIMATED TOWN
POPULATION – 1989
(in thousands)

● 20–50

Note:
Daman & Diu is about
0–150 metres above the sea level.

Chandigarh — B (PHYSICAL / AGRICULTURE)

HARYANA

PUNJAB

CHANDIGARH

INDIA

PUNJAB

AGRICULTURE

Wheat
Maize
Rice

Chandigarh — D (ECONOMIC DEVELOPMENT / TRANSPORT & TOURISM)

HARYANA

PUNJAB

Sukhna Lake
Secretariat
Raj Bhavan
(Haryana)
Raj Bhavan
(Punjab)
Chandigarh
Chandigarh

Punjab University
(Punjab & Haryana)

CHANDIGARH

INDIA

PUNJAB

TOURISM
♣ Lake/Reservoir

LOCATIONS
■ U.T. Headquarters
○ Tourist Centre
● Other Towns

ECONOMIC DEVELOPMENT

ESTIMATED INDEX OF LEVELS OF
ECONOMIC DEVELOPMENT – 1989

AVERAGE
U.T. 5292
All India 100

Over 200

TRANSPORT & TOURISM

TRANSPORT
⑳ National Highways (with number)
Other Roads
Railway Lines

Chandigarh — A (PHYSICAL / ADMINISTRATION / POPULATION)

HARYANA

PUNJAB

CHANDIGARH
● CHANDIGARH

INDIA

PUNJAB

PHYSICAL

Note:
Chandigarh is about 0–150 metres above
the sea level.

ADMINISTRATION

---- U.T. Boundary
■ U.T. Headquarters

POPULATION

ESTIMATED RURAL
POPULATION DENSITY – 1989
(Persons/sq. km, District-wise)

501–1000

ESTIMATED TOWN
POPULATION – 1989
(in thousands)

● 500–1000

Chandigarh — C (INDUSTRIES & MARKETS)

HARYANA

PUNJAB

○ Chandigarh

★ Mauli

CHANDIGARH

INDIA

PUNJAB

INDUSTRIES & MARKETS

C Class Market

■ Major Industrial Centres
○ Other Industrial Centres
★ Electrical machinery

DELHI

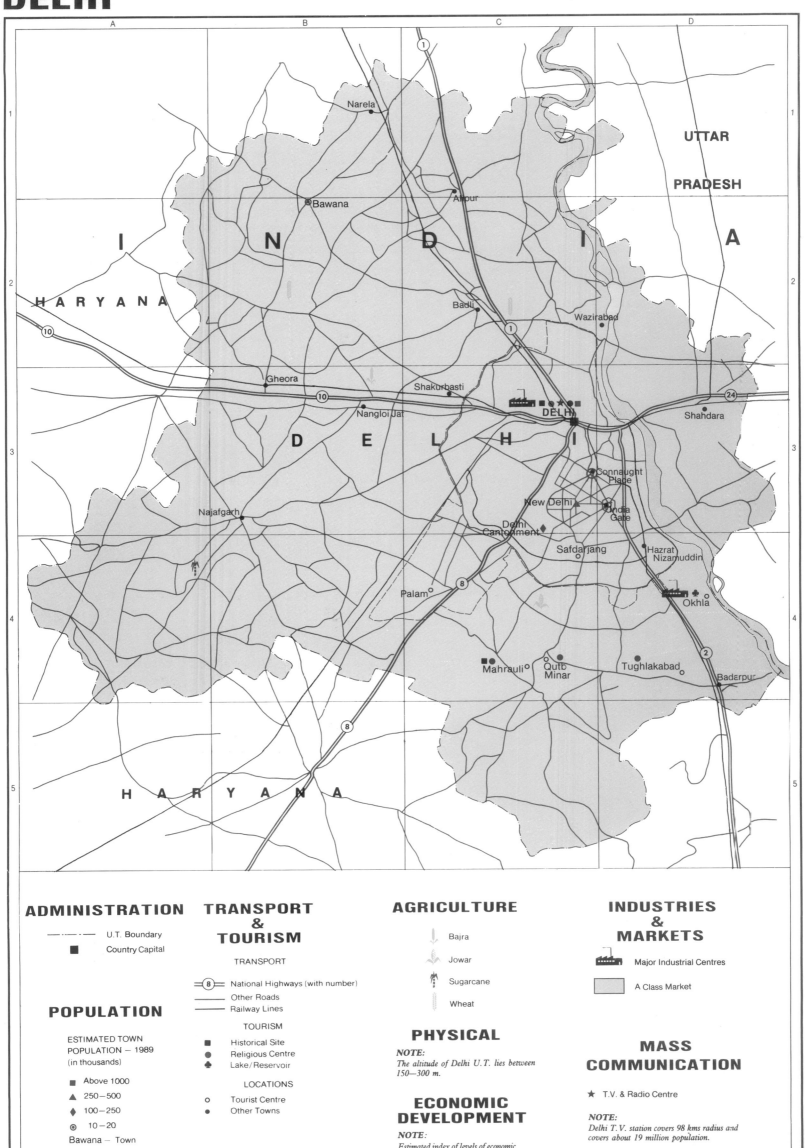

ADMINISTRATION

— · — · — U.T. Boundary

■ Country Capital

POPULATION

ESTIMATED TOWN
POPULATION – 1989
(in thousands)

■ Above 1000

▲ 250 – 500

◆ 100 – 250

◉ 10 – 20

Bawana – Town

TRANSPORT & TOURISM

TRANSPORT

⟨8⟩ National Highways (with number)

——— Other Roads

——— Railway Lines

TOURISM

■ Historical Site

● Religious Centre

♣ Lake/Reservoir

LOCATIONS

○ Tourist Centre

● Other Towns

AGRICULTURE

↓ Bajra

Jowar

Sugarcane

Wheat

PHYSICAL

NOTE:
*The altitude of Delhi U.T. lies between
150—300 m.*

ECONOMIC DEVELOPMENT

NOTE:
*Estimated index of levels of economic
development in 1989 was 772.*

INDUSTRIES & MARKETS

🏭 Major Industrial Centres

▨ A Class Market

MASS COMMUNICATION

★ T.V. & Radio Centre

NOTE:
*Delhi T.V. station covers 98 kms radius and
covers about 19 million population.*

Scale 1:420 000   1 Cm = 4.2 Km.

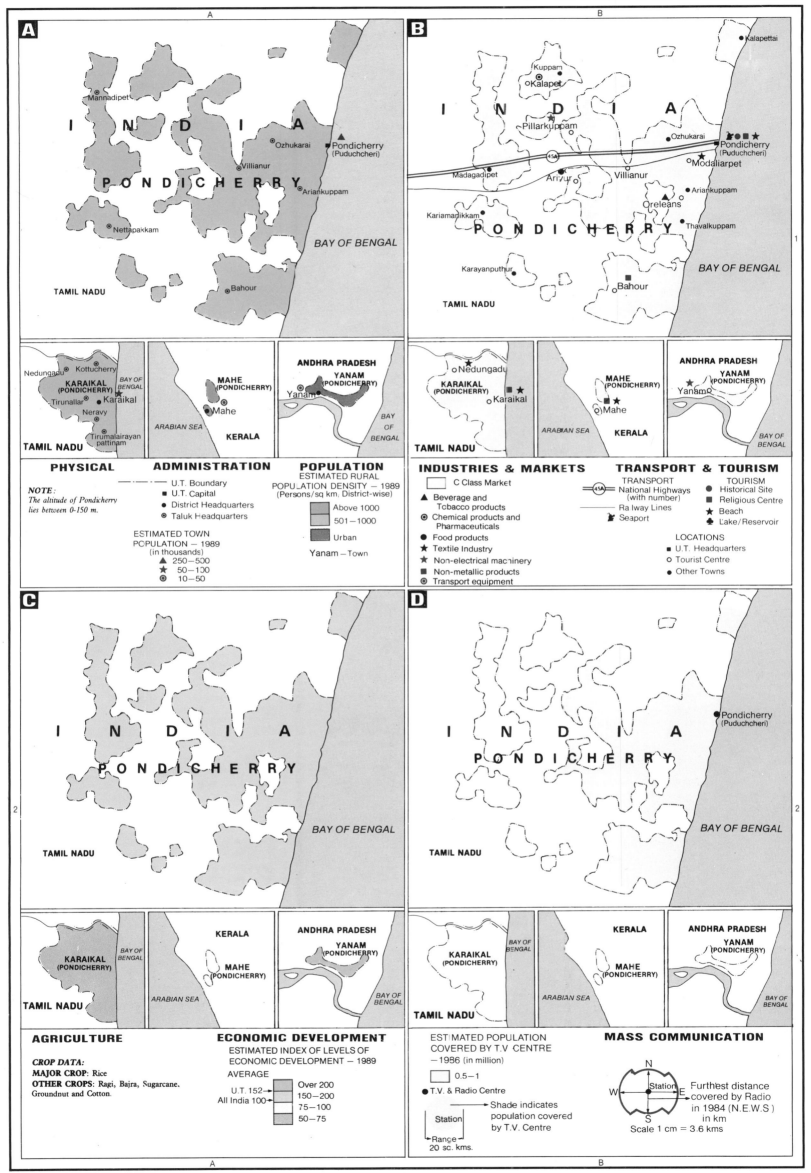

A

INDIA

Mannadipet

PONDICHERRY

Ozhukarai ▲ Pondicherry (Puduchcheri)

Villianur

Ariankuppam

Nettapakkam

TAMIL NADU

Bahour

BAY OF BENGAL

Nedungadu ⊙ Kottucherry

KARAIKAL (PONDICHERRY)

BAY OF BENGAL

Tirunallar ● Karaikal

Neravy

Tirumalairayan pattinam

TAMIL NADU

MAHE (PONDICHERRY)

● Mahe

ARABIAN SEA

KERALA

ANDHRA PRADESH

YANAM (PONDICHERRY)

Yanam

BAY OF BENGAL

PHYSICAL

NOTE :
The altitude of Pondicherry
lies between 0-150 m.

ADMINISTRATION

— — U.T. Boundary
■ U.T. Capital
● District Headquarters
⊙ Taluk Headquarters

ESTIMATED TOWN POPULATION — 1989 (in thousands)
▲ 250—500
★ 50—100
⊙ 10—50

POPULATION

ESTIMATED RURAL POPULATION DENSITY — 1989 (Persons/sq.km, District-wise)

Above 1000
501—1000
Urban

Yanam — Town

B

I N D I A

Kalapettai ●

Kuppam
⊙ Kalapet

Pillarkuppam ★

45A

Ozhukarai ⊙   ⚓●■★ Pondicherry (Puduchcheri)

Madagadipet ●   Arizur ⊙   Villianur   ★ Modaliarpet

Kariamanikkam ●   ▲ Oreleans   ● Ariankuppam

PONDICHERRY   Thavalkuppam ●

Karayanputhur ●   BAY OF BENGAL

■ Bahour

TAMIL NADU

⊙ Nedungadu   ■★

KARAIKAL (PONDICHERRY)   ⊙ Karaikal

TAMIL NADU

MAHE (PONDICHERRY)   ■★

⊙ Mahe

ARABIAN SEA   KERALA

ANDHRA PRADESH

YANAM (PONDICHERRY)

★ Yanam ⊙

BAY OF BENGAL

INDUSTRIES & MARKETS

☐ C Class Market
▲ Beverage and Tobacco products
⊙ Chemical products and Pharmaceuticals
● Food products
★ Textile Industry
★ Non-electrical machinery
■ Non-metallic products
⊙ Transport equipment

TRANSPORT & TOURISM

TRANSPORT
45A National Highways (with number)
— Railway Lines
⚓ Seaport

TOURISM
● Historical Site
■ Religious Centre
★ Beach
♣ Lake/Reservoir

LOCATIONS
■ U.T. Headquarters
○ Tourist Centre
● Other Towns

C

I N D I A

PONDICHERRY

TAMIL NADU

BAY OF BENGAL

KARAIKAL (PONDICHERRY)

BAY OF BENGAL

TAMIL NADU

KERALA

MAHE (PONDICHERRY)

ARABIAN SEA

ANDHRA PRADESH

YANAM (PONDICHERRY)

BAY OF BENGAL

AGRICULTURE

CROP DATA:
MAJOR CROP: Rice
OTHER CROPS: Ragi, Bajra, Sugarcane, Groundnut and Cotton.

ECONOMIC DEVELOPMENT

ESTIMATED INDEX OF LEVELS OF ECONOMIC DEVELOPMENT — 1989

AVERAGE

U.T. 152 →
All India 100 →

Over 200
150—200
75—100
50—75

D

I N D I A

● Pondicherry (Puduchcheri)

PONDICHERRY

TAMIL NADU

BAY OF BENGAL

KARAIKAL (PONDICHERRY)

BAY OF BENGAL

TAMIL NADU

KERALA

MAHE (PONDICHERRY)

ARABIAN SEA

ANDHRA PRADESH

YANAM (PONDICHERRY)

BAY OF BENGAL

ESTIMATED POPULATION COVERED BY T.V CENTRE

—1986 (in million)

☐ 0.5—1
● T.V. & Radio Centre

— Shade indicates population covered by T.V. Centre

Station
Range
20 sc. kms.

MASS COMMUNICATION

N
W ⊕ Station E
S

Furthest distance covered by Radio in 1984 (N.E.W.S) in km

Scale 1 cm = 3.6 kms

RAINFALL AND TEMPERATURE REPORT (1931-60 AVERAGE)

Station	Temperature (in °C) Daily Max.	Daily Min.	Total Rainfall (in Cm) In wettest month	In driest month
ANDHRA PRADESH				
Anantapur(17B3)	33.3	21.9	77.8	23.3
Cuddapah(17C3)	34.5	24.0	127.4	29.9
Gannavaram(17D2)	33.5	23.5	148.8	67.6
Hanamkonda(19C1)	33.1	22.3	161.6	38.4
Hyderabad(17C2)	31.7	20.0	143.1	45.5
Kakinada(17E2)	31.7	23.8	182.6	40.8
Khammam(17D2)	34.2	22.9	134.2	53.1
Kurnool(17B3)	34.0	21.1	109.6	28.1
Machilipatnam(17D2)	31.7	24.0	176.5	54.2
Nellore(17C3)	34.0	24.4	163.8	29.6
Nizamabad(17C1)	33.1	20.9	172.3	49.5
Ongole(17D3)	32.8	24.3	135.8	58.9
Vishakhapatnam(17E2)	31.0	23.5	144.2	47.3
ASSAM				
Dhubri(26A2)	28.3	20.4	366.9	170.4
Dibrugarh(26D2)	27.7	18.7	330.0	216.5
Guwahati(26B2)	29.5	19.7	212.1	87.4
Guwahati(26B2)*	29.3	19.4	212.1	156.8
Lumding(26C3)	29.9	18.3	172.3	94.8
Nagaon(26C2)	32.6	18.1	186.8	34.5
Sibsagar(26D2)	28.3	19.2	320.9	174.4
Silchar(26C3)	29.9	19.9	559.8	193.9
Tezpur(26C2)	29.1	19.6	288.3	133.4
BIHAR				
Bhagalpur(32D4)	31.6	21.0	159.5	83.8
Chaibasa(32C7)	32.3	20.6	190.7	74.0
Daltenganj(32B5)	31.8	19.0	185.7	62.2
Darbhanga(32C3)	30.8	19.7	194.2	80.0
Dehri(36B3)	32.4	21.0	171.9	56.8
Dhanbad(32D6)	31.1	20.4	178.2	86.9
Dumka(32E5)	31.6	20.0	199.3	8.4
Gaya(32BC5)	32.3	20.1	189.3	58.8
Hazaribag(32C6)	29.4	18.2	214.6	87.4
Jamshedpur(36C4)	32.4	20.7	186.3	74.6
Jamui(36C3)	32.2	20.6	135.0	67.2
Motihari(32B3)	30.6	18.7	202.1	25.5
Patna(32C4)	31.6	20.8	195.9	64.2
Purnia(32E4)	30.8	19.0	233.8	69.1
Ranchi(32C6)	29.5	18.0	210.6	97.9
GUJARAT				
Ahmadabad(44C2)	34.2	20.5	199.8	12.0
Bharuch(44C3)	34.9	21.1	166.4	42.6
Bhavnagar(44C3)	33.4	21.0	143.8	14.7
Bhuj(44A2)	33.0	20.1	177.0	2.2
Dwarka(44A2)	29.2	22.9	108.0	2.6
Jamnagar(44B2)*	32.1	20.0	112.0	7.8
Rajkot(44B2)	33.9	19.6	205.0	15.4
Surat(44C3)	33.6	21.9	228.4	36.3
Veraval(44B3)	30.6	21.4	132.7	6.9
HARYANA				
Ambala(50C2)	31.8	17.5	204.8	34.8
Hisar(50B3)	32.8	17.4	104.8	15.8
Karnal(50C3)	30.9	17.7	115.1	38.4
HIMACHAL PRADESH				
Chamba(54B2)	33.2	21.5	178.5	117.8
Dharmsala(54B2)	23.6	15.5	472.1	221.9
Shimla(54C3)	17.1	10.1	278.7	99.3
JAMMU & KASHMIR				
Dras(60B2)	8.6	− 4.7	127.3	35.7
Jammu(59B3)	30.0	18.7	196.5	64.6
Kargil(60C2)	14.5	3.3	63.7	8.5
Leh(59C2)	12.4	− 1.4	23.1	2.5
Srinagar(59B2)	19.5	7.3	129.2	39.9
KARNATAKA				
Balehonnur(66B4)	26.9	17.5	366.6	197.6
Bangalore(64C4)	28.8	18.4	134.9	54.4
Bangalore(64C4)*	29.6	18.6	108.7	54.3
Belgaum(64B3)	30.1	18.0	284.8	73.0
Bellary(64C3)	32.9	22.2	94.9	20.8
Bidar(64C2)	31.1	20.7	183.8	43.8
Bijapur(64B2)	32.5	20.3	99.1	24.3
Chitradurga(64C3)	30.8	20.1	106.0	29.5
Gadag(66B3)	31.7	20.1	104.1	34.5
Gulbarga(64C2)	33.4	21.0	143.2	35.8
Hassan(64C4)	28.5	17.9	129.4	51.0

Station	Temperature (in °C) Daily Max.	Daily Min.	Total Rainfall (in Cm) In wettest month	In driest month
Honavar(64B3)	30.9	22.9	573.0	229.3
Madikeri(66B4)	24.1	16.5	576.9	214.7
Mangalore(66B4)	30.5	23.7	470.3	227.0
Mysore(64C4)	29.6	19.2	129.6	45.1
Raichur(64C2)	33.3	22.4	127.1	26.9
Shimoga(64B4)	30.6	19.2	119.0	67.4
KERALA				
Alleppey(71B3)	30.7	23.9	383.9	246.2
Calicut(71A2)	30.9	23.7	462.4	188.7
Fort Cochin(72B3)	29.8	24.4	421.5	204.7
Palghat(71B2)	32.3	23.4	290.7	150.8
Trivandrum(71B3)	30.7	23.5	303.6	102.9
MADHYA PRADESH				
Ambikapur(79G3)	30.2	17.4	233.9	108.2
Betul(77D4)	30.7	17.9	166.2	70.9
Bhopal(77D3)	31.5	18.5	178.0	50.9
Chhindwara(77E3)	30.6	18.2	181.0	48.7
Guna(77D2)	31.8	17.6	181.3	67.5
Gwalior(77E1)	32.5	18.8	151.4	22.4
Hoshangabad(77D3)	32.4	20.0	204.6	58.8
Indore(77C3)	31.3	17.5	174.3	40.0
Jabalpur(77E3)	32.1	18.3	240.7	77.4
Jagdalpur(79F5)	31.2	18.9	239.5	88.2
Kanker(79F4)	31.6	19.9	197.6	87.9
Khandwa(79D4)	33.6	19.6	151.8	22.6
Mandla(77F3)	31.7	17.1	164.9	115.3
Nimachi(79C2)	31.5	18.6	175.1	30.5
Pachmarhi(79E3)	26.9	15.9	336.6	100.3
Raigarh(77G4)	33.5	21.5	204.7	98.7
Raipur(77F4)	32.6	21.2	218.1	68.9
Ratlam(77C3)	31.9	19.2	151.2	56.3
Sagar(77E3)	30.9	19.5	200.6	59.2
Satna(77F2)	32.1	18.7	168.3	67.5
Seoni(77E3)	30.8	18.5	212.7	62.1
Sheopur(79D2)	32.9	18.5	132.0	67.2
Umaria(79F3)	31.4	18.0	212.4	71.2
MAHARASHTRA				
Ahmadnagar(85C2)	32.2	18.5	111.9	28.7
Akola(85D1)	34.0	20.4	168.3	30.7
Alibag(87B2)	30.0	22.6	339.7	72.1
Aurangabad(85C2)	32.4	19.7	117.2	26.7
Bombay(85B2)	31.0	23.6	348.2	84.9
Bombay(85B2)*	31.6	21.9	378.5	181.4
Buldana(85D1)	30.9	20.5	127.3	53.0
Devgarh(85B3)	30.5	23.9	336.8	190.9
Gondia(85F1)	32.7	20.9	166.8	98.6
Jalgaon(85C1)	34.6	19.9	135.2	20.6
Kolhapur(85C3)	31.2	19.2	140.1	80.6
Mahabaleshwar(85B3)	24.1	16.1	1,022.1	354.5
Malegaon(87C1)	33.5	18.6	97.6	20.5
Nagpur(85E1)	33.5	20.3	193.1	36.5
Parbhani(85D2)	33.7	19.9	149.6	27.3
Pune(85C3)	32.0	18.2	124.2	26.9
Ratnagiri(85B3)	30.8	23.1	409.3	140.7
Sironcha(85E2)	33.9	22.3	196.1	24.7
Solapur(85C3)	33.7	20.5	124.0	32.5
Vengurla(85B4)	31.2	22.4	395.6	258.3
Yavatmal(85E1)	32.8	21.0	174.6	72.1
MEGHALAYA				
Cherrapunji(94C2)	20.6	14.3	1,570.7	717.8
Tura(94B2)	28.4	19.4	413.3	281.7
MIZORAM				
Aizawl(96A2)	23.9	16.3	284.6	158.4
ORISSA				
Anugul(101C2)	32.5	21.5	183.1	75.6
Baleshwar(101D2)	31.5	22.0	295.7	107.7
Chandbali(101D2)	33.5	22.1	233.5	86.7
Cuttack(101C2)	32.9	22.5	236.4	88.5
Gopalpur(103C3)	33.3	23.0	204.1	65.5
Koraput(101B3)	28.1	17.8	215.8	139.5
Puri(101C3)	30.0	24.1	198.4	60.0
Sambalpur(101B2)	32.9	20.8	230.8	93.5
Titlagarh(101B2)	33.0	21.3	202.5	108.7
PUNJAB				
Amritsar(107B2)	30.5	15.9	101.4	25.8
Firozpur(107B3)	31.3	16.7	82.1	36.1
Ludhiana(107C3)	31.9	17.3	140.2	23.7

Station	Temperature (in °C) Daily Max.	Daily Min.	Total Rainfall (in Cm) In wettest month	In driest month
Pathankot(107C1)	29.8	16.6	206.6	98.7
Patiala(107D3)	31.4	17.6	120.4	33.6
RAJASTHAN				
Abu(114C3)	24.9	16.5	399.1	29.0
Ajmer(112D2)	31.2	18.3	122.7	14.9
Barmer(112B3)	34.0	22.2	89.5	12.9
Bikaner(112B2)	33.6	18.2	77.1	2.9
Ganganagar(112C1)	32.9	17.0	64.1	12.3
Jaipur(112D2)	31.7	18.4	140.4	12.0
Jaisalmer(112B2)	34.0	19.0	45.4	10.4
Jhalawar(112E3)	32.6	18.9	169.8	44.5
Jodhpur(112C2)	33.6	19.8	117.7	2.4
Kota(112D3)	33.3	20.8	159.3	17.2
Sikar(112D2)	31.9	17.3	63.2	26.9
Udaipur(112C3)	31.1	17.5	122.3	25.2
TAMILNADU				
Coimbatore(121A2)	31.1	21.7	105.9	32.4
Coimbatore(121A2)*	31.7	21.2	88.0	24.4
Coonoor(121A2)	21.2	12.6	223.9	113.0
Cuddalore(121B2)	32.1	24.2	244.4	53.7
Kallakkurichchi(121B2)	33.9	23.6	136.0	56.7
Kodaikanal(121A2)	17.9	10.6	235.5	118.4
Madras(121C1)*	32.9	24.3	213.5	52.2
Madurai(121B3)	33.9	24.0	143.7	41.1
Nagapattinam(121B2)	31.9	25.2	219.6	60.4
Palayankottai(123A3)	33.7	24.9	131.1	32.7
Pamban(122B3)	30.6	25.7	177.2	41.1
Salem(121B2)	33.5	22.6	153.2	52.7
Tiruchchirappalli(121B2)	33.7	24.0	132.4	51.2
Udagamandalam(121A2)	19.5	9.1	192.7	90.3
Vellore(121B1)	33.1	22.8	170.2	46.8
UTTAR PRADESH				
Agra(131B4)	32.3	19.0	120.2	27.7
Aligarh(131B4)	31.9	18.7	134.3	20.5
Allahabad(131C5)	32.4	19.8	193.6	51.6
Azamgarh(131D4)	32.3	19.6	160.5	63.7
Bahraich(131C4)	31.5	19.1	244.3	46.9
Banda(131C5)	32.9	19.9	134.3	79.4
Bareilly(131B3)	31.5	18.9	197.4	49.2
Dehra Dun(131B2)	27.8	15.8	311.9	115.2
Fatehpur(131C5)	32.4	19.5	136.1	53.7
Gonda(131C4)	31.5	18.8	222.2	40.6
Gorakhpur(131D4)	31.4	20.0	245.6	65.0
Hardoi(131C4)	31.9	19.0	184.6	54.3
Jhansi(131B5)	32.8	19.7	149.6	31.5
Kanpur(131C4)*	32.1	19.3	156.7	42.1
Kheri(131C3)	31.5	19.0	151.7	74.8
Lucknow(131C4)	32.3	19.4	186.7	42.4
Lucknow(131C4)*	31.9	18.5	153.0	54.6
Mainpuri(131B4)	32.8	18.7	147.7	30.0
Meerut(131A3)	31.2	18.3	152.0	30.2
Mussoorie(134B2)	18.0	10.4	328.5	147.6
Orai(131B5)	32.6	19.1	172.5	73.8
Roorkee(134A3)	30.2	17.0	230.0	51.2
Varanasi(131D5)	32.2	19.8	210.9	44.6
WEST BENGAL				
Asansol(140B5)	32.0	20.8	213.5	96.4
Barddhaman(140C5)	31.7	21.4	225.8	88.6
Baghdogra(140C5)	29.7	18.3	442.9	238.0
Baharampur(142B2)	31.9	20.8	213.7	77.5
Calcutta(140D6)	31.8	22.1	250.1	90.9
Calcutta(140D6)*	31.4	21.4	262.6	86.7
Darjiling(140D1)	16.4	10.2	402.4	227.1
Jalpaiguri(140D2)	28.8	19.3	429.2	172.0
Kalimpong(142B1)	21.5	14.5	310.2	177.8
Koch Bihar(140E2)	29.7	19.4	653.4	211.2
Krishna Nagar(140D5)	32.4	20.6	229.6	88.5
Maldah(140D3)	30.6	20.2	230.6	85.9
Medinipur(140C6)	32.2	21.5	248.0	76.1
Puruliya(140B5)	31.9	21.0	157.0	92.8
ANDAMAN & NICOBAR				
Port Blair(148B2)	29.6	3.5	402.7	213.2
DELHI				
New Delhi(150C3)	31.7	18.8	153.3	26.3
GOA				
Marmagao(41A1)	29.5	23.7	350.0	184.4
LAKSHADWEEP				
Amini(148B2)	31.1	25.4	255.0	91.0
Minicoy(148B2)	29.9	24.6	224.8	105.0

Note: 1. * Airport   2. Within brackets by the side of the each town names are given the page and grid reference for the location of the town.

Tables

This section consists of five Tables and an Index. The Population table gives the projected estimates for the year 1990, showing the male, female and the rural-urban divide. These figures are likely to approximate the numbers the 1991 Census will reveal. The table on Crop Production gives by state and district the area sown and the actual production figures of 14 major crops of India for the year 1982-1983. The Rainfall and Temperature Report gives the average figures for a 30 year period from 1931-1960; this data is from over 200 meterological stations in India. The table on Mineral Production gives the production figures and the value of 59 major minerals found in India for the year 1982-1983. The table also shows the states and districts where these mineral deposits are found.

The table on Industrial Locations is grouped under 12 major manufacturing categories. It shows the states and districts in which these industries are located. The Index has over 12,000 entries with a cross-grid reference number to enable the user to locate any place marked on the map. All locations marked on the maps in the second section of this atlas are indexed.

POPULATION
(An estimate in 1990)

(in thousands) (in thousands)

STATE/DISTRICT	TOTAL	MALE	FEMALE	RURAL	URBAN
Andhra Pradesh	**65,275**	**33,680**	**31,595**	**46,550**	**18,725**
Adilabad	2,000	1,010	990	1,525	475
Anantapur	2,975	1,525	1,450	2,275	700
Chittoor	3,175	1,625	1,550	2,550	625
Cuddapah	2,300	1,300	1,000	1,700	600
East Godavari	4,300	2,200	2,100	3,200	1,100
Guntur	4,000	2,030	1,970	2,350	1,650
Hyderabad	2,950	1,550	1,400	—	2,950
Karimnagar	2,900	1,500	1,400	2,300	600
Khammam	2,150	1,100	1,050	1,750	400
Krishna	3,750	1,975	1,775	2,300	1,450
Kurnool	2,825	1,450	1,375	1,975	850
Mahbubnagar	2,975	1,500	1,475	2,500	475
Medak	2,150	1,100	1,050	1,800	350
Nalgonda	2,750	1,400	1,350	2,200	550
Nellore	2,425	1,325	1,100	1,700	725
Nizamabad	2,050	1,125	925	1,500	550
Prakasam	2,750	1,400	1,350	2,200	550
Rangareddi	2,075	1,050	1,025	1,450	625
Srikakulam	2,275	1,115	1,160	2,025	250
Vishakhapatnam	3,750	1,900	1,850	2,500	1,250
Vizianagaram	2,625	1,325	1,300	2,100	525
Warangal	2,750	1,400	1,350	2,150	600
West Godavari	3,375	1,775	1,600	2,500	875
Arunachal Pradesh	**806**	**433**	**373**	**731**	**75**
Dibang Valley	40	25	15	40	—
East Kameng	51	25	26	51	—
East Siang	81	44	37	68	13
Lohit	94	53	41	85	9
Lower Subansiri	161	86	75	129	32
Tirap	162	87	75	162	—
Upper Subansiri	56	29	27	52	4
West Kameng	75	39	36	71	4
West Siang	86	45	41	73	13
Assam	**24,123**	**12,721**	**11,402**	**21,994**	**2,129**
Cachar	2,826	1,470	1,356	2,602	224
Darrang	2,864	1,517	1,347	2,693	171
Goalpara	3,670	1,904	1,766	3,385	285
Kamrup	4,709	2,491	2,218	4,156	553
Karbi Anglong (Mikir Hills)	625	333	292	608	17
Lakhimpur	3,501	1,873	1,628	3,084	417
Nagaon	2,772	1,458	1,314	2,575	197
North Cachar Hills	125	68	57	116	9
Sibsagar	3,031	1,607	1,424	2,775	256
Bihar	**86,290**	**45,170**	**41,120**	**73,665**	**12,625**
Aurangabad	1,750	900	850	1,550	200
Begusarai	1,600	850	750	1,400	200
Bhagalpur	3,175	1,675	1,500	2,755	420
Bhojpur	2,500	1,300	1,200	2,100	400
Darbhanga	2,200	1,150	1,050	2,000	200
Dhanbad	2,875	1,575	1,300	1,275	1,600
Gaya	3,430	1,720	1,710	3,070	360
Giridih	1,875	975	900	1,575	300
Gopalganj	1,850	950	900	1,675	175
Hazaribag	3,050	1,630	1,420	2,525	525
Katihar	1,700	880	820	1,550	150
Madhubani	3,085	1,690	1,395	2,900	185
Munger	4,550	2,450	2,100	3,900	650
Muzaffarpur	2,850	1,470	1,380	2,600	250
Nalanda	2,450	1,290	1,160	1,830	620
Nawada	1,310	660	650	1,150	160
Palamu	2,350	1,200	1,150	2,150	200
Pashchim Champaran	2,850	1,500	1,350	2,600	250
Patna	3,325	1,750	1,575	2,200	1,125
Purba Champaran	3,775	1,950	1,825	3,600	175
Purnia	4,450	2,290	2,160	3,950	500
Ranchi	3,500	1,790	1,710	2,700	800
Rohtas	3,575	1,890	1,685	3,100	475
Saharsa	3,575	1,850	1,725	3,325	250
Samastipur	2,575	1,450	1,125	2,400	175
Santhal Pargana	4,225	2,150	2,075	3,800	425
Saran	2,375	1,275	1,100	2,195	180
Singhbhum	3,275	1,775	1,500	2,070	1,205
Sitamarhi	2,375	1,200	1,175	2,200	175
Siwan	1,995	995	1,000	1,840	155
Vaishali	1,820	940	880	1,680	140
Goa	**1,225**	**675**	**550**	**775**	**450**
Gujarat	**41,770**	**21,595**	**20,175**	**27,530**	**14,240**
Ahmadabad	4,910	2,560	2,350	1,210	3,700
Amreli	1,310	660	650	1,040	270
Banas Kantha	2,100	1,080	1,020	1,900	200
Bharuch	1,470	870	600	1,180	290

STATE/DISTRICT	TOTAL	MALE	FEMALE	RURAL	URBAN
Bhavnagar	2,390	1,210	1,180	1,570	820
Gandhinagar	390	200	190	270	120
Jamnagar	1,680	870	810	1,010	670
Junagadh	2,550	1,300	1,250	1,750	800
Kachchh	1,250	630	620	910	340
Kheda	3,575	1,840	1,735	2,850	725
Mahesana	2,995	1,520	1,475	2,360	635
Panch Mahals	2,800	1,450	1,350	2,500	300
Rajkot	2,580	1,350	1,230	1,450	1,130
Sabar Kantha	1,825	925	900	1,620	205
Surat	3,290	1,700	1,590	1,600	1,690
Surendranagar	1,220	630	590	750	470
The Dangs	135	75	60	135	—
Vadodara	3,175	1,650	1,525	1,825	1,350
Valsad	2,125	1,075	1,050	1,600	525
Haryana	**15,780**	**8,455**	**7,325**	**11,800**	**3,980**
Ambala	1,730	920	810	1,140	590
Bhiwani	1,110	590	520	800	310
Faridabad	1,070	600	470	610	460
Gurgaon	910	480	430	700	210
Hisar	1,850	990	860	1,460	390
Jind	1,350	730	620	1,100	250
Karnal	1,560	840	720	1,150	410
Kurukshetra	1,570	840	730	1,160	410
Mahendragarh	1,150	590	560	970	180
Rohtak	1,590	840	750	1,230	360
Sirsa	895	500	395	690	205
Sonipat	995	535	460	790	205
Himachal Pradesh	**5,120**	**2,584**	**2,536**	**4,554**	**566**
Bilaspur	300	150	150	280	20
Chamba	370	190	180	335	35
Hamirpur	370	175	195	340	30
Kangra	1,180	600	580	1,090	90
Kinnaur	69	36	33	69	—
Kullu	285	148	137	259	26
Lahul & Spiti	36	20	16	36	—
Mandi	780	395	385	700	80
Shimla	600	280	320	480	120
Sirmaur	370	200	170	330	40
Solan	370	190	180	320	50
Una	390	200	190	315	75
Jammu & Kashmir	**7,430**	**3,825**	**3,605**	**5,743**	**1,687**
Anantang	800	425	375	695	105
Badgam	475	260	215	400	75
Baramula	840	445	395	710	130
Doda	510	270	240	465	45
Jammu	1,175	500	675	790	385
Kargil	80	45	35	73	7
Kathua	470	250	220	410	60
Kupwara	400	215	185	381	19
Ladakh	85	45	40	73	12
Pulwama	496	262	234	442	54
Punch	280	148	132	257	23
Rajauri	396	207	189	368	28
Srinagar	853	455	398	175	678
Udhampur	570	298	272	504	66
Karnataka	**44,820**	**22,875**	**21,945**	**30,150**	**14,670**
Bangalore	6,810	3,545	3,265	1,790	5,020
Belgaum	3,530	1,800	1,730	2,670	860
Bellary	1,880	950	930	1,280	600
Bidar	1,160	590	570	930	230
Bijapur	2,810	1,450	1,360	2,100	710
Chikmagalur	1,090	550	540	880	210
Chitradurga	2,170	1,110	1,060	1,600	570
Dakshin Kannad	2,810	1,370	1,440	2,010	800
Dharwad	3,100	1,600	1,500	2,000	1,100
Gulbarga	2,410	1,220	1,190	1,760	650
Hassan	1,620	820	800	1,360	260
Kodagu	530	280	250	450	80
Kolar	2,300	1,170	1,130	1,740	560
Mandya	1,680	850	830	1,390	290
Mysore	3,120	1,590	1,530	2,210	910
Raichur	2,160	1,090	1,070	1,660	500
Shimoga	2,020	1,030	990	1,470	550
Tumkur	2,320	1,190	1,130	1,960	360
Uttar Kannad	1,300	670	630	890	410
Kerala	**29,245**	**14,320**	**14,925**	**23,170**	**6,075**
Alleppey	2,550	1,240	1,310	2,170	380
Cannanore	3,375	1,650	1,725	2,550	825
Ernakulam	2,660	1,320	1,340	1,600	1,060
Idukki	1,180	600	580	840	340
Kottayam	1,840	920	920	1,650	190
Kozhikode	2,670	1,320	1,350	1,920	750
Malappuram	2,970	1,440	1,530	2,730	240
Palghat	2,390	1,160	1,230	2,120	270

Population (Contd.)

(in thousands)

STATE / DISTRICT	TOTAL	MALE	FEMALE	RURAL	URBAN
Quilon	3,190	1,560	1,630	2,540	650
Trichur	2,730	1,290	1,440	2,100	630
Trivandrum	2,980	1,460	1,520	2,240	740
Wayanad	710	360	350	710	—
Madhya Pradesh	**63,090**	**32,550**	**30,540**	**47,795**	**15,295**
Balaghat	1,310	650	660	1,170	140
Bastar	2,170	1,090	1,080	1,890	280
Betul	1,120	570	550	870	250
Bhind	1,160	640	520	960	200
Bhopal	1,300	690	610	280	1,020
Bilaspur	3,450	1,740	1,710	2,760	690
Chhatarpur	1,070	570	500	850	220
Chhindwara	1,480	750	730	1,120	360
Damoh	880	480	400	740	140
Datia	370	200	170	280	90
Dewas	1,010	530	480	790	220
Dhar	1,280	650	630	1,080	200
Durg	2,330	1,180	1,150	1,540	790
East Nimar	1,450	740	710	1,010	440
Guna	1,230	650	580	1,040	190
Gwalior	1,370	740	630	580	790
Hoshangabad	1,210	630	580	870	340
Indore	1,840	970	870	700	1,140
Jabalpur	2,740	1,430	1,310	1,410	1,330
Jhabua	920	465	455	830	90
Mandla	1,200	605	595	1,090	110
Mandsaur	1,590	820	770	1,260	330
Morena	1,650	900	750	1,370	280
Narsimhapur	790	410	380	680	110
Panna	650	340	310	590	60
Raigarh	1,590	790	800	1,420	170
Raipur	3,520	1,750	1,770	2,740	780
Raisen	880	460	420	720	160
Rajgarh	960	490	470	805	155
Raj Nandgaon	1,340	660	680	1,150	190
Ratlam	940	490	450	640	300
Rewa	1,440	730	710	1,080	360
Sagar	1,590	840	750	1,100	490
Satna	1,390	730	660	1,000	390
Sehore	810	430	380	690	120
Seoni	950	480	470	860	90
Shahdol	1,680	870	810	1,290	390
Shajapur	1,010	520	490	830	180
Shivpuri	1,060	570	490	900	160
Sidhi	1,210	610	600	1,170	40
Surguja	1,940	990	950	1,740	200
Tikamgarh	920	480	440	650	270
Ujjain	1,390	720	670	840	550
Vidisha	910	490	420	730	180
West Nimar	1,990	1,010	980	1,680	310
Maharashtra	**75,540**	**38,880**	**36,660**	**46,085**	**29,455**
Ahmadnagar	3,130	1,590	1,540	2,670	460
Akola	2,150	1,090	1,060	1,590	560
Amravati	2,180	1,120	1,060	1,510	670
Aurangabad	2,890	1,480	1,410	2,030	860
Bhandara	2,080	1,045	1,035	1,770	310
Beed	1,680	850	830	1,360	320
Buldana	1,710	890	820	1,300	410
Chandrapur	2,480	1,270	1,210	2,100	380
Dhule	2,440	1,230	1,210	1,915	525
Greater Bombay	10,760	5,910	4,850	—	10,760
Jalgaon	3,110	1,590	1,520	2,280	830
Kolhapur	2,960	1,490	1,470	2,140	820
Nagpur	3,280	1,710	1,570	1,360	1,920
Nanded	2,110	1,070	1,040	1,660	450
Nashik	3,630	1,870	1,760	2,400	1,230
Osmanabad	2,550	1,290	1,260	2,090	460
Parbhani	2,150	1,100	1,050	1,590	560
Pune	5,200	2,680	2,520	2,420	2,780
Raigarh	1,700	840	860	1,420	280
Ratnagiri	2,220	995	1,225	2,040	180
Sangli	2,110	1,070	1,040	1,600	510
Satara	2,340	1,130	1,210	2,020	320
Solapur	2,950	1,510	1,440	2,030	920
Thane	4,610	2,460	2,150	2,290	2,320
Wardha	1,070	550	520	790	280
Yavatmal	2,050	1,050	1,000	1,710	340
Manipur	**1,740**	**895**	**845**	**1,110**	**630**
Manipur Central	1,110	560	550	650	460
Manipur East	105	55	50	90	15
Manipur North	200	110	90	140	60
Manipur South	170	90	80	115	55
Manipur West	85	40	45	60	25
Tengnoupal	70	40	30	55	15
Meghalaya	**1,682**	**861**	**821**	**1,379**	**303**
East Garo Hills	172	89	83	167	5
East Khasi Hills	644	331	313	416	228
Jaintia Hills	197	100	97	181	16

(in thousands)

STATE / DISTRICT	TOTAL	MALE	FEMALE	RURAL	URBAN
West Garo Hills	466	237	229	416	50
West Khasi Hills	203	104	99	199	4
Mizoram	**691**	**360**	**331**	**520**	**171**
Aizawl	477	247	230	340	137
Chhimtuipui	93	49	44	83	10
Lunglei	121	64	57	97	24
Nagaland	**1,082**	**581**	**501**	**915**	**167**
Kohima	350	195	155	256	94
Mokokchung	146	77	69	121	25
Mon	110	59	51	100	10
Phek	98	52	46	98	—
Tuensang	213	113	100	196	17
Wokha	80	42	38	69	11
Zunheboto	85	43	42	75	10
Orissa	**30,730**	**15,575**	**15,155**	**25,520**	**5,210**
Balangir	1,660	840	820	1,340	320
Baleshwar	2,670	1,350	1,320	2,290	380
Cuttack	5,410	2,760	2,650	4,700	710
Dhenkanal	1,860	960	900	1,650	210
Ganjam	3,020	1,490	1,530	2,340	680
Kalahandi	1,510	750	760	1,390	120
Kendujhar	1,270	640	630	1,010	260
Koraput	2,920	1,455	1,465	2,500	420
Mayurbhanj	1,720	870	850	1,590	130
Phulabani	810	410	400	740	70
Puri	3,500	1,800	1,700	2,750	750
Sambalpur	2,720	1,380	1,340	2,210	510
Sundargarh	1,660	870	790	1,010	650
Punjab	**19,440**	**10,280**	**9,160**	**13,265**	**6,175**
Amritsar	2,530	1,340	1,190	1,610	920
Bathinda	1,590	850	740	1,210	380
Faridkot	1,720	920	800	1,245	475
Gurdaspur	1,790	930	860	1,380	410
Firozpur	980	510	470	740	240
Hoshiarpur	1,430	740	690	1,190	240
Jalandhar	2,000	1,050	950	1,200	800
Kapurthala	670	350	320	430	240
Ludhiana	2,230	1,190	1,040	1,180	1,050
Patiala	1,940	1,030	910	1,310	630
Rupnagar	890	480	410	520	370
Sangrur	1,670	890	780	1,250	420
Rajasthan	**43,240**	**22,415**	**20,825**	**32,445**	**10,795**
Ajmer	1,740	900	840	910	830
Alwar	2,160	1,130	1,030	1,870	90
Banswara	1,140	580	560	1,060	80
Barmer	1,520	890	630	1,350	170
Bharatpur	2,290	1,250	1,040	1,830	460
Bhilwara	1,560	790	770	1,290	270
Bikaner	1,180	630	550	730	450
Bundi	730	380	350	590	140
Chittaurgarh	1,540	780	760	1,270	270
Churu	1,510	770	740	1,070	440
Dungarpur	850	410	440	780	70
Ganganagar	2,570	1,280	1,290	1,820	750
Jaipur	4,400	2,300	2,100	2,600	1,800
Jaisalmer	340	190	150	290	50
Jalor	1,160	590	570	1,010	150
Jhalawar	950	490	460	820	130
Jhunjhunun	1,510	760	750	1,150	360
Jodhpur	2,270	1,180	1,090	1,425	845
Kota	2,010	1,065	945	1,150	860
Nagaur	2,010	1,020	990	1,650	360
Pali	1,600	830	770	1,150	450
Sawai Madhopur	1,900	1,010	890	1,610	290
Sikar	1,730	880	850	1,320	410
Sirohi	670	340	330	520	150
Tonk	950	490	460	760	190
Udaipur	2,950	1,480	1,470	2,420	530
Sikkim	**398**	**217**	**181**	**334**	**64**
East	174	97	77	120	54
North	33	18	15	32	1
South	96	52	44	89	7
West	95	50	45	93	2
Tamil Nadu	**55,020**	**27,680**	**27,340**	**35,130**	**19,890**
Chengalpattu	4,330	2,210	2,120	2,490	1,840
Coimbatore	3,520	1,810	1,710	1,600	1,920
Dharmapuri	2,310	1,180	1,130	2,010	300
Kanniyakumari	1,610	820	790	1,330	280
Madras	4,130	2,110	2,020	—	4,130
Madurai	5,090	2,580	2,510	3,140	1,950
Nilgiri	770	390	380	400	370
North Arcot	5,040	2,550	2,490	3,780	1,260
Periyar	2,320	1,190	1,130	1,740	580
Pudukkottai	1,360	700	660	1,150	210
Ramanathapuram	3,780	1,880	1,900	2,650	1,130
Salem	3,860	1,990	1,870	2,570	1,290

Population (Contd.)

(in thousands)

STATE / DISTRICT	TOTAL	MALE	FEMALE	RURAL	URBAN
Tamil Nadu					
South Arcot	4,750	2,300	2,450	3,900	850
Thanjavur	4,250	2,150	2,100	3,180	1,070
Tiruchchirappalli	3,990	1,820	2,170	2,790	1,200
Tirunelveli	3,910	2,000	1,910	2,400	1,510
Tripura	**3,680**	**1,900**	**1,780**	**3,000**	**680**
North Tripura	600	320	280	540	60
South Tripura	2,080	1,060	1,020	1,610	470
West Tripura	1,000	520	480	850	150
Uttar Pradesh	**132,090**	**70,255**	**61,835**	**102,240**	**29,850**
Agra	3,390	1,850	1,540	2,060	1,330
Aligarh	3,030	1,640	1,390	2,210	820
Allahabad	4,690	2,480	2,210	3,660	1,030
Almora	770	370	400	700	70
Azamgarh	4,230	2,070	2,160	3,600	630
Bahraich	2,720	1,460	1,260	2,500	220
Ballia	2,290	1,150	1,140	2,000	290
Banda	1,890	1,020	870	1,620	270
Bara Banki	2,340	1,250	1,090	1,970	370
Bareilly	2,780	1,510	1,270	1,820	960
Basti	4,150	2,130	2,020	3,800	350
Bijnor	2,100	1,200	900	1,600	500
Budaun	2,290	1,270	1,020	1,700	590
Bulandshahr	2,620	1,390	1,230	2,000	620
Chamoli	440	225	215	380	60
Dehra Dun	960	520	440	470	490
Deoria	4,000	2,100	1,900	3,400	600
Etah	1,890	1,170	720	1,510	380
Etawah	2,030	1,100	930	1,550	480
Faizabad	2,840	1,560	1,380	2,400	440
Farrukhabad	2,340	1,260	1,080	1,840	500
Fatehpur	1,870	990	880	1,610	260
Garhwal	720	350	370	610	110
Ghaziabad	2,390	1,310	1,080	1,510	880
Ghazipur	2,370	1,190	1,180	2,010	360
Gonda	3,360	1,760	1,600	3,050	310
Gorakhpur	4,560	2,340	2,220	3,870	690
Hamirpur	1,390	760	630	1,020	370
Hardoi	2,690	1,470	1,220	2,300	390
Jalaun	1,160	640	520	820	340
Jaunpur	3,070	1,530	1,540	2,850	220
Jhansi	1,420	760	660	850	570
Kanpur	4,490	2,430	2,060	2,280	2,210
Kheri	2,440	1,310	1,130	2,100	340
Lalitpur	730	390	340	610	120
Lucknow	2,410	1,300	1,110	1,110	1,300
Mainpuri	1,900	1,000	900	1,500	400
Mathura	1,820	1,010	810	1,340	480
Meerut	3,330	1,810	1,520	2,220	1,110
Mirzapur	2,570	1,350	1,220	2,210	360

(in thousands)

STATE / DISTRICT	TOTAL	MALE	FEMALE	RURAL	URBAN
Moradabad	4,030	2,105	1,825	2,750	1,280
Muzaffarnagar	2,330	1,300	1,030	1,800	530
Naini Tal	1,540	820	720	1,040	500
Pilibhit	1,280	670	610	1,050	230
Pithoragarh	710	360	350	560	150
Pratapgarh	2,120	1,090	1,030	1,820	300
Rae Bareli	2,270	1,170	1,000	1,900	370
Rampur	1,470	790	680	990	480
Saharanpur	3,320	1,800	1,520	1,980	1,340
Shahjahanpur	2,020	1,105	915	1,560	460
Sitapur	2,500	1,400	1,100	2,200	300
Sultanpur	2,440	1,240	1,200	2,210	230
Tehri Garhwal	590	300	290	540	50
Unnao	2,160	1,140	1,020	1,700	460
Uttar Kashi	240	130	110	210	30
Varanasi	4,590	2,410	2,180	3,270	1,320
West Bengal	**65,230**	**33,940**	**31,290**	**46,280**	**18,950**
Barddhaman	5,740	3,010	2,730	3,610	2,130
Bankura	2,690	1,370	1,320	2,440	250
Birbhum	2,390	1,230	1,160	2,110	280
Calcutta	3,800	2,200	1,600	—	3,800
Darjiling	1,280	680	600	840	440
Haora	3,510	1,840	1,670	1,840	1,670
Hugli	4,240	2,200	2,040	2,940	1,300
Jalpaiguri	2,690	1,390	1,300	2,200	490
Koch Bihar	2,130	1,090	1,040	1,980	150
Maldah	2,460	1,260	1,200	2,300	160
Medinipur	7,960	4,060	3,900	7,150	810
Murshidabad	4,460	2,280	2,180	4,010	450
Nadia	3,750	1,930	1,820	2,850	900
Puruliya	2,090	1,070	1,020	1,810	280
24-Parganas	13,070	6,810	6,260	7,610	5,460
West Dinajpur	2,970	1,520	1,450	2,590	380
A. & N. Is.	**290**	**154**	**136**	**200**	**90**
Andamans	240	130	110	150	90
Nicobars	50	24	26	50	—
Chandigarh	**725**	**410**	**315**	**30**	**695**
Dadra & N. H.	**140**	**72**	**68**	**120**	**20**
Daman & Diu	**100**	**49**	**51**	**65**	**35**
Daman	60	30	30	35	25
Diu	40	19	21	30	10
Delhi	**8,860**	**4,890**	**3,970**	**480**	**8,380**
Lakshadweep	**50**	**26**	**24**	**20**	**30**
Pondicherry	**740**	**376**	**664**	**325**	**415**
Karaikal	140	71	69	80	60
Mahe	35	17	18	25	10
Pondicherry	550	280	270	220	330
Yanam	15	8	7	—	15

CROP PRODUCTION
(1982-83)

STATE/DISTRICT	AREA IN '000 ha	PRODUCTION IN '000 tns.
BAJRA		
Andhra Pradesh		
Anantapur	50	15
Chittoor	20	11
Cuddapah	23	18
East Godavari	14	10
Guntur	17	4
Khammam	5	5
Kurnool	34	17
Mahbubnagar	45	16
Nalgonda	113	43
Nellore	12	6
Prakasam	50	29
Rangareddi	7	3
Srikakulam	7	10
Vishakhapatnam	58	50
Vizianagaram	12	9
Warangal	20	15
Gujarat		
Ahmadabad	63	55
Amreli	87	87
Banas Kantha	259	95
Bhavnagar	168	226
Gandhinagar	19	18
Jamnagar	76	66
Junagadh	47	60
Kachchh	102	46
Kheda	150	194
Mahesana	150	168
Panch Mahals	26	18
Rajkot	85	62
Sabar Kantha	48	30
Surendranagar	91	31
Haryana		
Ambala	4	3
Bhiwani	231	145
Faridabad	37	31
Gurgaon	61	44
Hisar	108	99
Jind	73	62
Mahendragarh	140	31
Rohtak	76	47
Sirsa	14	7
Sonipat	19	18
Jammu & Kashmir		
Jammu	18	6
Karnataka		
Belgaum	76	17
Bellary	36	24
Bidar	22	13
Bijapur	193	56
Chitradurga	15	10
Gulbarga	92	46
Raichur	73	28
Madhya Pradesh		
Bhind	37	26
Morena	63	52
Maharashtra		
Amravati	6	2
Aurangabad	152	72
Beed	126	43
Buldana	10	4
Dhule	128	60
Jalgaon	108	52
Nashik	312	87
Pune	82	31
Sangli	77	10
Satara	84	12
Solapur	34	5
Punjab		
Bathinda	10	10
Faridkot	10	14
Sangrur	15	17
Rajasthan		
Ajmer	56	16
Alwar	178	109
Barmer	944	18
Bharatpur	172	130
Bikaner	213	20

STATE/DISTRICT	AREA IN '000 ha.	PRODUCTION IN '000 tns.
BAJRA (Contd.)		
Churu	476	167
Ganganagar	78	25
Jaipur	268	170
Jaisalmer	144	N.A.
Jalor	321	30
Jhunjhunun	254	138
Jodhpur	543	37
Nagaur	578	279
Pali	119	36
Sawai Madhopur	109	78
Sikar	248	95
Sirohi	21	6
Tonk	36	10
Tamil Nadu		
Chengalpattu	5	7
Dharmapuri	7	8
Madurai	24	33
North Arcot	14	11
Periyar	19	15
Ramanathapuram	23	13
Salem	22	12
South Arcot	55	54
Tiruchchirapalli	58	42
Tirunelveli	49	40
Uttar Pradesh		
Agra	131	107
Aligarh	102	115
Allahabad	54	32
Bareilly	11	5
Budaun	81	78
Bulandshahr	30	25
Etah	86	80
Etawah	69	43
Farrukhabad	17	9
Jalaun	16	5
Kanpur	21	13
Lucknow	8	5
Manipur	50	39
Mathura	75	45
Moradabad	42	26
Pratapgarh	19	21
Shahjahanpur	14	4
INDIA	10,260	4,852
COCONUT		
Andhra Pradesh		
Chittoor	1	5
East Godavari	26	105
Srikakulam	7	30
Karnataka		
Chikmagalur	20	48
Chitradurga	23	130
Dakshin Kannad	16	71
Hassan	35	194
Mandya	10	52
Tumkur	45	309
Uttar Kannad	5	25
Kerala		
Alleppey	61	234
Cannanore	73	219
Ernakulam	62	286
Idukki	17	30
Kottayam	49	134
Malappuram	60	194
Palghat	23	58
Quilon	81	267
Trichur	57	347
Trivandrum	74	266
Wynad	4	8
Orissa		
Baleshwar	3	11
Cuttack	5	25
Tamil Nadu		
Chengalpattu	4	30
Coimbatore	24	282
Dharmapuri	7	77
Kanniyakumari	21	135
Ramanathapuram	9	54

STATE/DISTRICT	AREA IN '000 ha.	PRODUCTION IN '000 tns.
COCONUT (Contd.)		
Thanjavur	23	241
Tirunelveli	10	64
INDIA	855	3,931
COFFEE		
Karnataka		
Chikmagalur	N.A.	35
Hassan	N.A.	14
Kodagu	N.A.	45
Kerala		
Wynad	N.A.	22
Tamil Nadu		
Nilgiri	N.A.	5
INDIA	N.A.	107
COTTON		
Andhra Pradesh		
Adilabad	133	52
Anantapur	14	23
Cuddapah	4	5
Guntur	121	316
Kurnool	88	33
Mahbubnagar	5	2
Prakasam	59	135
Assam		
Karbi Anglong	1	1
Gujarat		
Ahmadabad	194	97
Amreli	23	25
Banas Kantha	32	26
Bharuch	87	75
Bhavnagar	61	73
Jamnagar	21	19
Junagadh	32	39
Kachchh	65	36
Kheda	48	43
Mahesana	81	131
Panch Mahals	36	28
Rajkot	152	215
Sabar Kantha	121	155
Surat	31	43
Surendranagar	324	256
Vadodara	188	283
Haryana		
Ambala	5	8
Bhiwani	23	40
Hisar	188	440
Jind	30	54
Rohtak	8	13
Sirsa	124	255
Sonipat	4	6
Karnataka		
Belgaum	52	34
Bellary	76	75
Bidar	7	6
Bijapur	190	43
Chitradurga	23	20
Dharwad	230	191
Gulbarga	147	65
Hassan	4	12
Mysore	7	13
Raichur	34	182
Shimoga	8	10
Madhya Pradesh		
Chhindwara	11	17
Dewas	39	35
Dhar	58	34
East Nimar	145	64
Hoshangabad	40	24
Jhabua	21	9
Ratlam	32	24
Shajapur	32	10
Ujjain	17	9
West Nimar	170	81
Maharashtra		
Ahmadnagar	4	6
Akola	349	137

Crop Production (Contd.)

COTTON (Contd.)

STATE/DISTRICT	AREA IN '000 ha	PRODUCTION IN '000 tns.
Maharashtra (Contd.)		
Amravati	361	176
Aurangabad	56	36
Beed	29	22
Buldana	263	101
Chandrapur	49	27
Dhule	59	46
Jalgaon	188	140
Kolhapur	1	1
Nagpur	87	60
Nanded	216	178
Parbhani	232	156
Sangli	2	2
Satara	8	10
Solapur	11	17
Wardha	176	155
Yavatmal	410	249
Punjab		
Amritsar	15	18
Bathinda	263	470
Faridkot	188	382
Firozpur	90	157
Ludhiana	18	14
Patiala	19	15
Sangrur	116	149
Rajasthan		
Ajmer	18	21
Banswara	17	14
Bhilwara	23	20
Bikaner	1	1
Chittaurgarh	5	4
Ganganagar	292	447
Jhalawar	10	6
Pali	15	28
Sirohi	4	5
Tamil Nadu		
Coimbatore	18	42
Dharmapuri	4	6
Madurai	45	94
Periyar	7	11
Ramanathapuram	44	41
Salem	9	17
South Arcot	8	15
Tirunelveli	58	55
Uttar Pradesh		
Agra	1	N.A.
Aligarh	10	8
Bijnor	1	N.A.
INDIA	7,446	7,444

GROUNDNUT

STATE/DISTRICT	AREA IN '000 ha	PRODUCTION IN '000 tns.
Andhra Pradesh		
Anantapur	397	231
Chittoor	208	141
East Godavari	11	13
Guntur	27	33
Karimnagar	26	27
Khammam	26	26
Krishna	28	30
Kurnool	189	136
Medak	4	3
Nalgonda	50	47
Nellore	12	23
Nizamabad	9	12
Prakasam	25	25
Rangareddi	7	6
Srikakulam	6	37
Vizianagaram	59	46
Warangal	63	67
West Godavari	7	10
Gujarat		
Ahmadabad	3	2
Amreli	275	158
Bharuch	8	4
Bhavnagar	193	136
Jamnagar	383	186
Junagadh	385	266
Kachchh	78	46
Kheda	11	6
Panch Mahals	28	9

GROUNDNUT (Contd.)

STATE/DISTRICT	AREA IN '000 ha.	PRODUCTION IN '000 tns.
Rajkot	393	111
Surat	27	17
Surendranagar	15	10
The Dangs	1	1
Vadodara	15	8
Karnataka		
Bangalore	12	9
Belgaum	110	66
Bellary	39	42
Bidar	10	6
Bijapur	112	48
Chikmagalur	6	5
Chitradurga	58	50
Dharwad	125	73
Gulbarga	87	47
Kolar	47	30
Mysore	31	46
Raichur	117	75
Shimoga	11	11
Tumkur	79	43
Kerala		
Palghat	9	8
Madhya Pradesh		
Chhindwara	17	20
Dhar	36	8
East Nimar	16	9
Jhabua	22	6
Mandsaur	33	15
Raigarh	11	6
Shajapur	23	9
Shivpuri	12	8
West Nimar	43	26
Maharashtra		
Ahmadnagar	13	10
Akola	14	11
Amravati	24	21
Aurangabad	13	7
Bhandara	1	1
Buldana	22	7
Dhule	92	34
Jalgaon	65	33
Kolhapur	53	63
Nagpur	14	13
Nanded	6	4
Nashik	55	18
Pune	31	23
Sangli	32	26
Satara	61	71
Solapur	17	12
Wardha	9	12
Yavatmal	18	7
Orissa		
Balangir	12	15
Baleshwar	9	15
Cuttack	82	153
Dhenkanal	32	27
Ganjam	33	36
Kendujhar	4	2
Puri	21	26
Sambalpur	36	28
Punjab		
Ludhiana	22	17
Patiala	13	12
Sangrur	20	18
Rajasthan		
Ajmer	6	1
Banswara	2	1
Bhilwara	16	5
Chittaurgarh	45	31
Jaipur	23	16
Sawai Madhopur	27	19
Tonk	14	8
Tamil Nadu		
Chengalpattu	95	96
Coimbatore	61	80
Dharmapuri	54	32
Kanniyakumari	1	1
Madurai	77	80
North Arcot	212	155
Periyar	51	59

GROUNDNUT (Contd.)

STATE/DISTRICT	AREA IN '000 ha.	PRODUCTION IN '000 tns.
Pudukkottai	45	49
Ramanathapuram	35	24
Salem	114	123
South Arcot	165	147
Thanjavur	45	56
Tiruchchirapalli	56	42
Tirunelveli	19	22
Uttar Pradesh		
Bahraich	10	6
Bara Banki	4	3
Bareilly	25	14
Basti	2	1
Budaun	62	50
Hardoi	48	22
Saharanpur	14	4
INDIA	6,187	4,395

JOWAR

STATE/DISTRICT	AREA IN '000 ha.	PRODUCTION IN '000 tns.
Andhra Pradesh		
Adilabad	216	113
Anantapur	38	28
Chittoor	19	17
Cuddapah	103	111
East Godavari	5	3
Guntur	18	7
Karimnagar	65	49
Khammam	144	122
Krishna	32	32
Kurnool	248	222
Mahbubnagar	337	156
Medak	136	105
Nalgonda	125	47
Nellore	35	10
Nizamabad	52	38
Prakasam	112	66
Rangareddi	103	60
Warangal	137	84
West Godavari	5	3
Bihar		
Bhojpur	2	2
Gujarat		
Ahmadabad	59	7
Amreli	55	20
Banas Kantha	107	30
Bharuch	100	92
Bhavnagar	80	14
Jamnagar	70	12
Junagadh	25	27
Kachchh	77	8
Kheda	3	2
Mahesana	106	52
Rajkot	32	16
Sabar Kantha	11	6
Surat	88	102
Surendranagar	86	10
The Dangs	Neg.	Neg.
Vadodara	59	67
Valsad	13	7
Haryana		
Bhiwani	5	1
Faridabad	13	2
Gurgaon	14	3
Jind	7	2
Rohtak	45	12
Sonipat	19	5
Karnataka		
Belgaum	203	138
Bellary	99	110
Bidar	111	136
Bijapur	477	236
Chikmagalur	23	19
Chitradurga	80	123
Dharwad	246	287
Gulbarga	365	175
Hassan	11	19
Mysore	88	63
Raichur	333	222
Shimoga	23	44
Tumkur	6	7

Crop Production (Contd.)

STATE/DISTRICT	AREA IN '000 ha	PRODUCTION IN '000 tns.
JOWAR (Contd.)		
Madhya Pradesh		
Betul	87	68
Bhind	21	16
Chhatarpur	28	15
Chhindwara	89	79
Damoh	20	12
Datia	18	10
Dewas	122	148
Dhar	73	33
East Nimar	112	96
Guna	161	52
Gwalior	28	28
Hoshangabad	39	27
Indore	52	31
Jabalpur	18	10
Jhabua	35	15
Mandsaur	150	90
Morena	24	13
Panna	7	4
Ratlam	60	40
Rewa	17	9
Sagar	31	18
Satna	11	5
Sehore	60	52
Seoni	19	10
Shahdol	6	3
Shajapur	152	132
Shivpuri	61	22
Tikamgarh	38	13
Ujjain	180	194
Vidisha	56	23
West Nimar	173	116
Maharashtra		
Ahmadnagar	630	237
Akola	274	203
Amravati	194	261
Aurangabad	323	162
Bhandara	31	16
Bid	424	219
Buldana	245	305
Chandrapur	206	95
Dhule	171	102
Jalgaon	255	291
Kolhapur	45	60
Nagpur	213	184
Nanded	289	394
Nashik	99	56
Osmanabad	529	440
Parbhani	349	272
Pune	522	195
Sangli	197	147
Satara	248	197
Solapur	803	151
Wardha	127	162
Yavatmal	259	358
Orissa		
Kalahandi	7	6
Koraput	15	13
Rajasthan		
Ajmer	128	17
Alwar	12	7
Banswara	7	3
Bharatpur	15	7
Bundi	37	9
Chittaurgarh	40	20
Jaipur	36	6
Jhalawar	108	41
Kota	145	115
Nagaur	47	4
Sawai Madhopur	77	56
Tonk	142	31
Udaipur	18	7
Tamil Nadu		
Coimbatore	114	80
Dharmapuri	43	45
Madurai	97	84
North Arcot	30	32
Periyar	71	24
Ramanathapuram	12	9
Salem	81	67

STATE/DISTRICT	AREA IN '000 ha.	PRODUCTION IN '000 tns.
JOWAR (Contd.)		
South Arcot	17	15
Tiruchchirapalli	91	56
Tirunelveli	23	25
Uttar Pradesh		
Allahabad	20	7
Banda	61	33
Farrukhabad	11	5
Fatehpur	30	11
Hamirpur	69	24
Hardoi	19	15
Jalaun	24	14
Kanpur	34	17
Lalitpur	45	2
Moradabad	10	7
Rai Bareli	19	15
Rampur	13	13
Unnao	17	15
INDIA	**15,238**	**10,189**
JUTE		
Assam		
Cachar	1	5
Darrang	17	137
Goalpara	37	320
Kamrup	26	217
Karbi Anglong	2	21
Nowgong	29	215
Sibsagar	3	23
Bihar		
Katihar	18	109
Purnia	84	517
Saharsa	9	59
Orissa		
Baleshwar	8	85
Cuttack	32	222
Uttar Pradesh		
Bahraich	1	10
West Bengal		
Barddhaman	12	148
Darjiling	3	19
Haora	3	36
Hugli	25	315
Jalpaiguri	44	328
Koch Bihar	59	388
Maldah	21	131
Medinipur	10	131
Murshidabad	68	671
Nadia	89	823
24 Parganas	44	468
West Dinajpur	59	313
INDIA	**704**	**5,711**
MAIZE		
Andhra Pradesh		
Adilabad	22	22
Karimnagar	94	235
Khammam	6	9
Medak	69	118
Nizamabad	64	173
Rangareddi	9	20
Vishakhapatnam	6	17
Warangal	48	93
Assam		
Dibrugarh	5	3
Karbi Anglong	8	6
Lakhimpur	3	3
Bihar		
Aurangabad	1	1
Begusarai	68	71
Bhagalpur	64	58
Bhojpur	5	4
Darbhanga	12	22
Dhanbad	5	4
Gaya	12	9
Giridih	13	10
Gopalganj	25	31
Hazaribag	25	18
Katihar	23	27
Madhubani	5	6

STATE/DISTRICT	AREA IN '000 ha.	PRODUCTION IN '000 tns.
MAIZE (Contd.)		
Munger	35	35
Muzaffarpur	27	22
Nalanda	10	16
Nawada	3	2
Palamu	34	35
Pashchini Chanparn	9	18
Patna	11	12
Purba Champaran	27	50
Purnia	38	52
Ranchi	13	12
Saharsa	30	33
Samastipur	54	94
Saran	37	47
Singhbhum	8	9
Sitamarhi	15	36
Siwan	30	35
Vaishali	27	22
Gujarat		
Banas Kantha	8	7
Kheda	22	23
Panch Mahals	159	189
Sabar Kantha	48	30
The Dangs	Neg.	Neg.
Vadodara	20	16
Haryana		
Ambala	30	29
Karnal	9	9
Kurukshetra	9	10
Himachal Pradesh		
Bilaspur	24	24
Chamba	26	42
Hamirpur	30	34
Kangra	50	75
Kullu	13	28
Mandi	42	57
Simla	22	28
Sirmaur	26	46
Solan	24	28
Una	29	37
Jammu & Kashmir		
Anantang	22*	24 *
Badgam	14*	9 *
Baramula	28*	34 *
Doda	42*	56 *
Jammu	10	9
Kathua	15	30
Kupwara	19	19
Pulwama	11	9
Punch	22	59
Rajauri	38	72
Srinagar	4	4
Udhampur	48	83
Karnataka		
Bangalore	27	182
Belgaum	53	117
Chitradurga	9	27
Hassan	8	14
Kolar	15	39
Mysore	9	24
Madhya Pradesh		
Bastar	28	37
Bilaspur	24	24
Chhindwara	31	75
Guna	34	13
Indore	10	7
Jhabua	75	69
Mandla	28	37
Mandsaur	113	109
Ratlam	43	40
Shajapur	20	20
Shivpuri	29	21
Sidhi	21	13
Surguja	42	46
Ujjain	20	18
Vidisha	9	10
West Nimar	39	42
Maharashtra		
Chandrapur	4	4
Raigarh	Neg.	Neg.

* 1981-82.

159

Crop Production (Contd.)

STATE/DISTRICT	AREA IN '000 ha	PRODUCTION IN '000 tns.
MAIZE (Contd.)		
Orissa		
Balangir	49	46
Dhenkanal	13	12
Kalahandi	20	18
Koraput	48	65
Phulabani	21	29
Punjab		
Amritsar	17	33
Faridkot	6	11
Gurudaspur	17	23
Jalandhar	56	120
Kapurthala	43	105
Patiala	16	17
Rupnagar	33	37
Sangrur	27	55
Rajasthan		
Alwar	9	13
Banswara	88	59
Bhilwara	135	75
Bundi	29	10
Chittaurgarh	118	150
Dungarpur	57	49
Jaipur	18	12
Jhalawar	52	17
Kota	28	6
Pali	37	25
Sirohi	28	24
Tonk	20	18
Udaipur	215	172
Tamil Nadu		
Coimbatore	12	22
Uttar Pradesh		
Almora	3	3
Azamgarh	17	16
Bahraich	111	47
Bara Banki	8	3
Bijnor	3	2
Budaun	36	27
Bulandshahr	107	154
Dehra Dun	14	19
Deoria	14	13
Etah	41	24
Etawah	25	11
Farrukhabad	69	43
Ghaziabad	37	66
Gonda	79	42
Hardoi	40	11
Jaunpur	56	71
Kheri	23	12
Lalitpur	16	8
Mainpuri	39	22
Meerut	29	42
Muzaffarnagar	15	14
Naini Tal	15	18
Saharanpur	27	27
Sitapur	21	4
Unnao	32	9
West Bengal		
Bankura	2	2
Darjiling	27	40
INDIA	4,663	5,549

MESTA

STATE/DISTRICT	AREA IN '000 ha	PRODUCTION IN '000 tns.
Andhra Pradesh		
Srikakulam	20	189
Vizianagaram	58	302
Assam		
Goalpara	5	23
Kamrup	3	12
Nowgong	2	8
Bihar		
Katihar	3	17
Purnia	13	52
Madhya Pradesh		
Raigarh	3	3
Maharashtra		
Nanded	7	13
Osmanabad	31	54

STATE/DISTRICT	AREA IN '000 ha.	PRODUCTION IN '000 tns.
MESTA (Contd.)		
Parbhani	9	18
West Bengal		
Bankura	2	8
Koch Bihar	5	24
INDIA	161	723

RICE

STATE/DISTRICT	AREA IN '000 ha.	PRODUCTION IN '000 tns.
Andhra Pradesh		
Adilabad	71*	107*
Anantapur	59	116
Chittoor	90	156
Cuddapah	45	97
East Godavari	368	902
Guntur	304	711
Hyderabad	Neg.	Neg.
Karimnagar	148	346
Krishna	399	853
Kurnool	98	182
Mahbubnagar	119	173
Medak	107	204
Nalgonda	201	459
Nellore	190	303
Nizamabad	156	335
Prakasam	105	243
Rangareddi	44	71
Srikakulam	204	357
Vishakhapatnam	71	62
Vizianagaram	112	106
Warangal	130	234
West Godavari	445	1,339
Assam		
Cachar	211	265
Darrang	266	279
Dibrugarh	140	171
Goalpara	381	343
Kamrup	470	471
Karbi Anglong	101	148
Lakhimpur	150	193
North Cachar Hills	13	18
Nowgong	278	267
Sibsagar	293	426
Bihar		
Aurangabad	116	131
Begusarai	15	9
Bhagalpur	149	110
Bhojpur	174	211
Darbhanga	80	43
Dhanbad	45	23
Gaya	188	119
Giridih	75	43
Gopalganj	90	83
Hazaribag	122	53
Katihar	122	72
Madhubani	142	66
Munger	118	65
Muzaffarpur	134	65
Nalanda	49	56
Nawada	46	32
Palamu	75	29
Paschim Champaran	202	172
Patna	89	110
Purba Champaran	203	151
Purnia	324	196
Ranchi	418	226
Rohtas	268	291
Saharsa	106	70
Samastipur	71	39
Saran	85	68
Singhbhum	327	158
Sitamarhi	113	33
Siwan	92	77
Vaishali	60	31
Gujarat		
Ahmadabad	34	45
Bharuch	16	5
Bhavnagar	1	3
Kheda	96	150
Mahesana	7	6
Panch Mahals	94	11

STATE/DISTRICT	AREA IN '000 ha.	PRODUCTION IN '000 tns.
RICE (Contd.)		
Sabar Kantha	16	8
Surat	52	73
The Dangs	5	3
Vadodara	51	16
Valsad	92	156
Haryana		
Ambala	61	119
Faridabad	3	5
Hisar	22	73
Jind	33	66
Karnal	157	437
Kurukshetra	173	481
Rohtak	3	4
Sirsa	20	58
Sonipat	17	32
Himachal Pradesh		
Bilaspur	4	2
Chamba	3	3
Hamirpur	5	2
Kangra	31	31
Kullu	3	4
Mandi	24	19
Shimla	5	3
Sirmaur	4	3
Solan	4	2
Una	2	1
Jammu & Kashmir		
Anantnag	41	120
Badgam	28	76
Baramula	35	81
Doda	4	6
Jammu	53	35
Kathua	28	29
Kupwara	18	35
Pulwama	30	87
Punch	3	9
Rajauri	6	11
Srinagar	13	40
Udhampur	12	8
Karnataka		
Bangalore	45*	61*
Belgaum	56	69
Bellary	32	71
Bidar	9	13
Chikmagalur	57	101
Chitradurga	32	69
Dakshin Kannad	148	253
Dharwad	82	108
Hassan	52	85
Kodagu	61	128
Kolar	26	46
Mandya	55	168
Mysore	73	192
Raichur	52	104
Shimoga	187	381
Tumkur	22	55
Uttar Kannad	89	163
Kerala		
Alleppey	85	165
Cannanore	60	80
Ernakulam	99	146
Idukki	8	16
Kottayam	34	69
Malappuram	77	97
Quilon	50	82
Trichur	115	155
Trivandrum	30	45
Wynad	29	54
Madhya Pradesh		
Balaghat	225	229
Bastar	528	372
Betul	37	37
Bhind	8	10
Bhopal	2	1
Bilaspur	4	2
Chhatarpur	16	12
Chhindwara	26	22
Damoh	50	31
Dhar	12	5

* 1981-82

Crop Production (Contd.)

STATE/DISTRICT	AREA IN '000 ha	PRODUCTION IN '000 tns.

RICE (Contd.)

Madhya Pradesh (Contd.)

STATE/DISTRICT	AREA IN '000 ha	PRODUCTION IN '000 tns.
Durg	356	314
East Nimar	33	29
Gwalior	22	28
Jabalpur	115	55
Jhabua	30	7
Mandla	136	91
Narsimhapur	16	18
Panna	51	20
Raigarh	371	270
Rajnandgaon	272	199
Rewa	116	60
Sagar	15	9
Satna	68	25
Seoni	92	80
Shahdol	202	126
Sidhi	78	25
Surguja	313	230
Tikamgarh	27	14
West Nimar	18	7
Maharashtra		
Ahmadnagar	7	10
Akola	8	5
Amravati	13	7
Aurangabad	5	3
Bhandara	291	254
Buldana	3	2
Chandrapur	261	318
Dhule	24	15
Greater Bombay	1	2
Jalgaon	9	13
Kolhapur	97	183
Nagpur	37	41
Nanded	35	25
Nashik	39	41
Osmanabad	39	25
Parbhani	23	15
Pune	51	72
Raigarh	152	283
Ratnagiri	81	137
Sangli	16	28
Satara	36	54
Thane	151	269
Wardha	9	11
Yavatmal	12	8
Orissa		
Balangir	291	226
Baleshwar	194	261
Cuttack	475	293
Dhenkanal	219	134
Ganjam	303	310
Kalahandi	290	205
Kendujhar	209	112
Koraput	363	347
Mayurbhanj	328	214
Phulabani	90	75
Puri	392	261
Sambalpur	496	438
Sundargarh	206	117
Punjab		
Amritsar	193	420
Bathinda	13	47
Faridkot	100	335
Firozpur	175	539
Gurdaspur	135	329
Hoshiarpur	43	101
Jalandhar	93	326
Kapurthala	69	201
Ludhiana	133	524
Patiala	216	751
Rupnagar	20	68
Sangrur	129	506
Rajasthan		
Bharatpur	2	3
Bundi	8	14
Chittaurgarh	2	1
Dungarpur	28	9
Ganganagar	16	42
Kota	5	7
Udaipur	11	4

RICE (Contd.)

Tamil Nadu

STATE/DISTRICT	AREA IN '000 ha	PRODUCTION IN '000 tns.
Chengalpattu	188	268
Coimbatore	31	92
Dharmapuri	22	37
Kanniyakumari	46	99
Madurai	101	197
Nilgiri	3	4
North Arcot	79	146
Periyar	60	171
Pudukkottai	69	81
Ramanathapuram	240	270
Salem	43	90
South Arcot	132	226
Thanjavur	566	1,189
Tiruchchirapalli	86	170
Tirunelveli	109	260
Uttar Pradesh		
Agra	1	1
Aligarh	17	22
Allahabad	165	187
Almora	36	39
Azamgarh	223	157
Bahraich	218	123
Ballia	103	88
Banda	82	200
Bara Banki	151	204
Bareilly	122	148
Basti	365	274
Bijnor	66	103
Budaun	36	37
Bulandshahr	6	5
Chamoli	24	29
Dehra Dun	16	15
Deoria	243	347
Etah	19	17
Etawah	65	90
Faizabad	163	207
Farrukhabad	29	30
Fatehpur	88	103
Garhwal	28	27
Ghaziabad	7	8
Ghazipur	111	107
Gonda	242	180
Gorakhpur	300	307
Hardoi	62	48
Jaunpur	95	107
Kanpur	70	98
Kheri	166	212
Lalitpur	11	4
Lucknow	44	48
Mainpuri	54	49
Meerut	11	17
Mirzapur	152	119
Moradabad	90	142
Muzaffarnagar	34	60
Naini Tal	98	214
Pilibhit	114	206
Pithoragarh	36	39
Pratapgarh	90	115
Rai Bareli	113	147
Rampur	73	138
Saharanpur	84	138
Shahjahanpur	114	158
Sitapur	123	96
Tehri Garhwal	17	19
Unnao	83	89
Uttarkashi	11	15
Varanasi	146	200
West Bengal		
Bankura	319	290
Barddhaman	437	717
Birbhum	308	323
Darjiling	41	37
Haora	88	113
Hugli	214	315
Jalpaiguri	252	182
Koch Bihar	285	242
Maldah	216	261
Medinipur	901	752
Murshidabad	287	266

RICE (Contd.)

STATE/DISTRICT	AREA IN '000 ha	PRODUCTION IN '000 tns.
Puruliya	209	135
24-Parganas	658	632
West Dinajpur	440	406
INDIA	**34,144**	**42,835**

SUGARCANE

Andhra Pradesh

STATE/DISTRICT	AREA IN '000 ha	PRODUCTION IN '000 tns.
Adilabad	1	77
Anantapur	3	219
Chittoor	21	1,398
Cuddapah	3	182
East Godavari	16	1,190
Krishna	17	1,565
Medak	14	1,009
Nizamabad	24	1,988
Srikakulam	6	400
Vishakhapatnam	25	1,337
Vzianagaram	7	307
West Godavari	22	1,960
Assam		
Cachar	6	244
Darrang	4	207
Dibrugarh	5	183
Kamrup	4	123
Karbi Anglong	5	244
Lakhimpur	3	110
Nowgong	10	421
S bsagar	12	497
Bihar		
Bhagalpur	4	45
Gaya	6	171
Gopalganj	12	644
Hazaribag	2	23
Madhubani	3	107
Muzaffarpur	4	228
Paschim Champaran	42	1,356
Patna	2	56
Purba Champaran	14	496
Purnia	2	44
Rohtas	3	93
Samastipur	4	197
Saran	3	124
S tamarhi	5	150
S wan	9	292
Gujarat		
Amreli	8*	346*
Banas Kantha	1*	5*
Bhavnagar	2*	98*
Junagadh	8*	435*
Rajkot	14*	863*
Surat	35*	2,256*
Valsad	8*	480*
Haryana		
Ambala	37	1,660
Faridabad	7	200
Hisar	7	170
Jind	14	440
Karnal	19	850
Kurukshetra	16	680
Rohtak	25	800
Sonipat	18	600
Jammu & Kashmir		
Udhampur	Neg.	Neg.
Karnataka		
Belgaum	57	4,093
Bijapur	15	1,404
Chikmagalur	2	138
Chitradurga	6	654
Mandya	28	2,667
Mysore	9	915
Shimoga	8	668
Kerala		
Alleppey	2	129
Idukki	2	145
Palghat	3	166
Madhya Pradesh		
Dewas	8	148
Gwalior	3	106

* 1981-82

161

Crop Production (Contd.)

STATE/DISTRICT	AREA IN '000 ha	PRODUCTION IN '000 tns.
SUGARCANE (Contd.)		
Madhya Pradesh (Contd.)		
Morena	5	142
Sehore	4	116
Maharashtra		
Ahmadnagar	56	6,085
Aurangabad	18	1,592
Bid	9	841
Buldana	1	86
Dhule	10	926
Jalgaon	9	868
Kolhapur	57	4,940
Nagpur	Neg.	13
Nanded	9	796
Nashik	20	2,080
Osmanabad	20	1,780
Parbhani	5	483
Pune	26	2,800
Ratnagiri	Neg.	40 *
Sangli	28	2,668
Satara	23	1,821
Solapur	24	2,780
Thane	Neg.	10 *
Orissa		
Balangir	5	315
Cuttack	5	241
Ganjam	10	640
Koraput	7	345
Puri	6	505
Sambalpur	5	300
Punjab		
Amritsar	9	580
Bathinda	2	100
Gurdaspur	23	1,190
Hoshiarpur	73	107
Jalandhar	16	1,150
Patiala	10	610
Rupnagar	10	630
Sangrur	10	620
Rajasthan		
Bharatpur	1	49
Bhilwara	1	63
Bundi	10	369
Chittaurgarh	4	116
Ganganagar	5	243
Udaipur	7	265
Tamil Nadu		
Chengalpattu	5	442
Coimbatore	16	1,634
Dharmapuri	11	783
Madurai	9	865
North Arcot	29	1,719
Periyar	20	2,086
South Arcot	30	2,951
Thanjavur	11	915
Tiruchchirapalli	17	1,808
Uttar Pradesh		
Agra	3	98
Aligarh	19	736
Allahabad	6	254
Azamgarh	45	1,480
Baharach	8	342
Ballia	16	525
Bara Banki	19	679
Bareilly	55	2,298
Basti	35	1,420
Bijnor	150	7,583
Budaun	24	983
Bulandshahr	65	3,075
Dehra Dun	5	251
Deoria	79	3,533
Etah	10	399
Faizabad	23	960
Ghaziabad	50	2,842
Ghazipur	18	578
Gonda	38	1,546
Gorakhpur	29	1,398
Jaunpur	20	834
Kheri	111	4,318
Mathura	20	632
Meerut	165	8,329

STATE/DISTRICT	AREA IN '000 ha.	PRODUCTION IN '000 tns.
SUGARCANE (Contd.)		
Moradabad	143	7,213
Muzaffarnagar	181	9,813
Naini Tal	41	1,728
Pilibhit	40	1,725
Rampur	31	1,344
Saharanpur	143	7,140
Shahjahanpur	38	1,768
Sitapur	53	2,064
Varanasi	15	485
West Bengal		
Barddhaman	1	65
Birbhum	3	173
Murshidabad	8	466
Nadia	9	465
INDIA	3,126	156,415

* 1981-82

TEA

STATE/DISTRICT	AREA IN '000 ha.	PRODUCTION IN '000 tns.
Assam		
Cachar	32	31
Darrang	34	62
Dibrugarh	59	110
Goalpara	3	4
Kamrup	3	5
Lakhimpur	4	7
Nowgong	7	9
Sibsagar	58	79
Tamil Nadu		
Coimbatore	10	24
Nilgiri	7	45
West Bengal		
Darjiling	18	12
INDIA	237	391

TOBACCO

STATE/DISTRICT	AREA IN '000 ha.	PRODUCTION IN '000 tns.
Andhra Pradesh		
East Godavari	16	26
Khammam	14	11
Krishna	24	28
Kurnool	30	25
Nellore	20	13
Prakasam	83	67
West Godavari	26	53
Gujarat		
Ahmadabad	1	2
Kheda	79	128
Vadodara	25	56
Karnataka		
Mysore	9	4
Tamil Nadu		
Periyar	3	9
West Bengal		
Koch Bihar	12	12
INDIA	342	434

WHEAT

STATE/DISTRICT	AREA IN '000 ha.	PRODUCTION IN '000 tns.
Andhra Pradesh		
Medak	4	2
Nizamabad	2	1
Rangareddi	3	2
Assam		
Darrang	20	22
Dibrugarh	8	9
Goalpara	23	27
Kamrup	26	33
Karbi Anglong	2	2
Nowgong	20	17
Sibsagar	4	5
Bihar		
Aurangabad	64	81
Begusarai	64	108
Bhagalpur	49	64
Bhojpur	129	174
Darbhanga	34	49
Gaya	94	95
Giridih	4	4
Gopalganj	69	96
Hazaribag	11	11

STATE/DISTRICT	AREA IN '000 ha.	PRODUCTION IN '000 tns.
WHEAT (Contd.)		
Katihar	34	54
Madhubani	34	43
Munger	62	78
Muzaffarpur	53	79
Nalanda	70	126
Nawada	45	57
Palamu	18	13
Paschim Champaran	61	103
Patna	56	92
Purba Champaran	77	102
Purnia	84	100
Ranchi	10	12
Rohtas	174	238
Saharsa	60	86
Samastipur	50	62
Saran	82	151
Singhbhum	2	1
Sitamarhi	33	43
Siwan	80	121
Vaishali	43	55
Gujarat		
Ahmadabad	60	84
Amreli	33	114
Banas Kantha	41	103
Bharuch	26	19
Bhavnagar	57	118
Gandhinagar	6	14
Jamnagar	32	76
Junagadh	46	141
Kachchh	21	41
Kheda	68	101
Mahesana	86	197
Panch Mahals	32	56
Rajkot	42	103
Sabar Kantha	38	71
Surat	12	26
Surendranagar	25	35
Vadodara	18	48
Valsad	3	6
Haryana		
Ambala	118	255
Bhiwandi	63	139
Faridabad	113	304
Gurgaon	109	239
Hisar	203	549
Jind	152	335
Karnal	225	601
Kurukshetra	248	682
Mahendragarh	82	204
Rohtak	153	374
Sirsa	126	348
Sonipat	128	315
Himachal Pradesh		
Bilaspur	26	34
Chamba	18	26
Hamirpur	37	34
Kangra	77	84
Kullu	19	25
Mandi	68	87
Simla	33	34
Sirmaur	30	45
Solan	24	24
Una	34	40
Jammu & Kashmir		
Badgam	2	1
Doda	8	6
Jammu	85	100
Kargil	2	2
Kathua	43	42
Leh	3	2
Punch	22	59
Rajauri	28	28
Udhampur	33	35
Karnataka		
Belgaum	54	28
Bellary	3	4
Bidar	15	16
Bijapur	106	79
Dharwad	95	43
Gulbarga	32	21
Raichur	28	9

Crop Production (Contd.)

STATE/DISTRICT	AREA IN '000 ha	PRODUCTION IN '000 tns.	STATE/DISTRICT	AREA IN '000 ha.	PRODUCTION IN '000 tns.	STATE/DISTRICT	AREA IN '000 ha.	PRODUCTION IN '000 tns.
WHEAT (Contd.)			**WHEAT (Contd.)**			**WHEAT (Contd.)**		
Madhya Pradesh			Solapur	46	35	Banda	183	239
Balaghat	18	12	Wardha	37	44	Bara Banki	152	286
Bastar	3	4	Yavatmal	22	19	Bareilly	164	307
Betul	62	45	**Orissa**			Basti	271	372
Bhind	82	119	Balangir	9	15	Bijnor	122	215
Bhopal	69	81	Baleshwar	5	10	Budaun	214	425
Bilaspur	26	34	Cuttack	9	17	Bulandshahr	221	573
Chhatarpur	101	133	Dhenkanal	4	8	Chamoli	28	24
Chhindwara	72	90	Kalahandi	4	6	Dehra Dun	27	29
Damoh	98	85	Koraput	5	10	Deoria	233	379
Datia	41	45	Mayurbhanj	5	12	Etah	166	371
Dewas	57	72	Sambalpur	8	14	Etawah	117	311
Dhar	71	26	Sundargarh	6	15	Faizabad	161	313
Durg	21	13	**Punjab**			Farrukhabad	136	281
East Nimar	24	35	Amritsar	313	930	Fatehpur	118	221
Guna	179	152	Bathinda	295	825	Garhwal	46	46
Gwalior	90	152	Faridkot	367	1,138	Ghaziabad	109	259
Hoshangabad	147	114	Firozpur	372	1,242	Ghazipur	120	208
Indore	73	116	Gurdaspur	190	493	Gonda	235	333
Jabalpur	140	100	Hoshiarpur	170	340	Gorakhpur	271	491
Jhabua	13	19	Jalandhar	222	655	Hamirpur	155	208
Mandla	79	46	Kapurthala	103	269	Hardoi	216	385
Mandsaur	54	82	Ludhiana	267	940	Jalaun	100	170
Morena	97	174	Patiala	314	957	Jaunpur	148	237
Narsimhapur	36	54	Rupnagar	75	208	Jhansi	105	165
Panna	82	79	Sangrur	359	1,160	Kanpur	181	406
Raigarh	5	6	**Rajasthan**			Kheri	176	243
Rajnandgaon	20	11	Ajmer	63	69	Lalitpur	90	109
Ratlam	34	45	Alwar	141	343	Lucknow	77	134
Rewa	130	106	Banswara	40	32	Mainpuri	164	374
Sagar	250	253	Barmer	14	26	Mathura	179	452
Satna	166	144	Bharatpur	146	341	Meerut	170	441
Sehore	113	167	Bhilwara	94	176	Mirzapur	110	135
Seoni	78	35	Bikaner	12	22	Moradabad	260	547
Shahdol	47	26	Bundi	92	183	Muzaffarnagar	160	385
Shajapur	49	78	Chittaurgarh	92	171	Naini Tal	112	236
Sidhi	33	21	Churu	3	6	Pilibhit	124	241
Surguja	12	13	Dungarpur	35	32	Pithoragarh	51	68
Tikamgarh	73	190	Ganganagar	303	638	Pratapgarh	117	176
Ujjain	69	84	Jaipur	203	437	Rae Bareli	164	186
Vidisha	237	234	Jaisalmer	2	3	Rampur	99	246
West Nimar	28	26	Jalor	45	56	Saharanpur	178	388
Maharashtra			Jhalawar	51	77	Shahjahanpur	198	389
Ahmadnagar	64	73	Jhunjhunun	30	56	Sitapur	186	268
Akola	45	29	Jodhpur	32	31	Tehri Garhwal	38	36
Amravati	45	31	Kota	165	289	Unnao	152	264
Aurangabad	47	34	Nagaur	37	34	Uttarkashi	15	15
Bhandara	29	18	Pali	75	87	Varanasi	156	234
Bid	45	27	Sawai Madhopur	132	343	**West Bengal**		
Buldana	28	18	Sikar	40	58	Bankura	8	15
Chandrapur	41	13	Sirohi	22	28	Barddhaman	9	22
Dhule	41	30	Tonk	120	126	Birbhum	17	30
Jalgaon	43	45	Udaipur	78	114	Hugli	7	14
Kolhapur	16	16	**Uttar Pradesh**			Jalpaiguri	9	16
Nagpur	76	51	Agra	152	410	Koch Bihar	13	26
Nanded	40	29	Aligarh	216	576	Maldah	19	48
Nashik	92	63	Allahabad	197	343	Medinipur	7	14
Osmanabad	56	41	Almora	64	79	Murshidabad	82	184
Parbhani	58	47	Azamgarh	219	382	Nadia	42	112
Pune	44	47	Baharich	196	268	24 Parganas	11	24
Sangli	38	24	Ballia	108	208	West Dinajpur	39	91
Satara	26	31				**INDIA**	22,569	42,853

MINERAL PRODUCTION
(1982-83)

STATE/DISTRICT	QUANTITY Tonnes	VALUE Rs. '000
AGATE		
Gujarat		
Bharuch	1,480	354
INDIA	1,480	354
ANDALUSITE		
Uttar Pradesh		
Mirzapur	146	34
INDIA	146	34
APATITE		
Andhra Pradesh		
Vishakhapatnam	4,080	1,726
West Bengal		
Puruliya	12,782	1,847
INDIA	16,862	3,573
ASBESTOS		
Andhra Pradesh		
Cuddapah	2,025	11,843
Bihar		
Singhbhum	1,484	1,115
Karnataka		
Hassan	158	65
Madhya Pradesh		
Sidhi	7	2
Tikamgarh	351	28
Rajasthan		
Ajmer	2,731	162
Bhilwara	87	5
Dungapur	122	5
Pali	1,448	154
Udaipur	16,102	918
INDIA	24,515	14,297
BARYTES		
Andhra Pradesh		
Cuddapah	378,435	35,996
Khammam	3,928	306
Kurnool	459	39
Mahbubnagar	9,000	225
Prakasam	716	34
Bihar		
Singhbhum	16	3
Himachal Pradesh		
Sirmaur	39	6
Madhya Pradesh		
Dewas	164	6
Maharashtra		
Chandrapur	1,502	76
Rajasthan		
Alwar	2,266	664
Chittaurgarh	225	15
Sikar	33	3
Udaipur	2,230	493
INDIA	399,013	37,866
BAUXITE		
Bihar		
Ranchi	609,430	16,860
Goa		
Goa	30,000	334
Gujarat		
Amreli	20	N.A.
Bhavnagar	1,030	24
Jamnagar	280,180	12,910
Junagadh	11,830	156
Kachchh	30,590	916
Kheda	9,820	308
Sabar Kantha	8,050	198
Himachal Pradesh		
Bilaspur	64	73
Udhampur	1,640	178
Karnataka		
Belgaum	21,510	818
Dakshin Kannad	53,920	1,537

STATE/DISTRICT	QUANTITY Tonnes	VALUE Rs. '000
BAUXITE (Contd.)		
Madhya Pradesh		
Jabalpur	46,980	1,411
Mandla	246,460	25,227
Satna	21,580	797
Shahdol	64,980	2,896
Sidhi	600	15
Maharashtra		
Kolhapur	85,830	3,054
Ratnagiri	211,730	8,836
Tamil Nadu		
Nilgiri	25,580	639
Salem	96,530	3,379
INDIA	1,858,354	80,566
CALCAREOUS		
Gujarat		
Jamnagar	685,376	7,488
INDIA	685,376	7,488
CALCITE		
Andhra Pradesh		
Anantapur	680	34
Gujarat		
Bharuch	150	6
Madhya Pradesh		
Jhabua	210	15
West Nimar	6,056	375
Rajasthan		
Jhunjhunun	426	23
Sikar	471	19
Sirohi	16,649	1,844
Udaipur	2,035	119
INDIA	26,677	2,515
CHALK		
Gujarat		
Junagach	85,982	7,340
INDIA	85,982	7,340
CHROMITE		
Andhra Pradesh		
Krishna	193	39
West Godavari	60	9
Bihar		
Singhbhum	590	149
Karnataka		
Hassan	54,760	9,270
Maharashtra		
Ratnagiri	2,560	870
Manipur		
Tengnoupal	320	22
Orissa		
Cuttack	185,530	98,390
Dhenkanal	30,770	13,709
Kendujhar	61,320	24,283
INDIA	336,103	146,741
CLAY		
Andhra Pradesh		
Anantapur	11,975	213
Kurnool	17,180	255
Jammu & Kashmir		
Pulwama	1,690	26
Karnataka		
Chikmagalur	18,641	141
Tamil Nadu		
South Arcot	2,602	42
Tiruchchirappalli	3,684	37
West Bengal		
Bankura	601	9
INDIA	56,373	723

STATE/DISTRICT	QUANTITY Tonnes	VALUE Rs. '000
COAL		
Andhra Pradesh		
Adilabad	4,018,620	543,518
Karimnagar	4,278,401	578,654
Khammam	2,890,258	390,907
Assam		
Dibrugarh	646,262	97,159
Bihar		
Dhanbad	30,747,168	4,726,762
Ranchi	12,073,217	1,856,016
Himachal Pradesh		
Bilaspur	4,072	5,007
Udhampur	31,624	6,214
Madhya Pradesh		
Betul	1,819,599	223,774
Chhindwara	3,497,121	430,076
Raigarh	5,441	669
Shahdol	6,420,079	789,541
Sidhi	4,467,630	549,429
Surguja	7,155,656	880,003
Maharashtra		
Chandrapur	3,669,557	451,209
Nagpur	2,850,664	350,518
Yavatmal	255,158	31,374
Meghalaya		
East Khasi Hills	110	16
Orissa		
Sambalpur	3,155,572	460,871
Uttar Pradesh		
Mirzapur	2,076,553	249,892
West Bengal		
Bardhamam	203	303
INDIA	90,062,965	12,621,912
COPPER ORE		
Andhra Pradesh		
Khammam	1,970	197
Bihar		
Singhbhum	1,245,320	215,768
Karnataka		
Chitradurga	32,291	5,488
Hassan	19,370	1,975
Rajasthan		
Alwar	12	169
Jhunjhunun	795,170	119,714
Sikkim		
East Sikkim	592	2,437
INDIA	2,094,725	345,748
CORUNDUM		
Andhra Pradesh		
Anantapur	N.A.	13
Khammam	3	41
Karnataka		
Bellary	1	16
Hassan	62	41
Madhya Pradesh		
Bastar	1	1
Sidhi	78	94
Maharashtra		
Bhandara	1,145	688
Rajasthan		
Tonk	2	1
INDIA	1,292	895
CRUDE OIL		
Assam		
Dibrugarh	2,808,384	2,218,455
Sibsagar	1,486,839	1,174,514
Gujarat		
Ahmadabad	634,058	461,911
Braruch	2,287,744	1,662,622
Mahesana	703,265	512,329
Maharashtra		
Greater Bombay	7,006,000	5,916,707
INDIA	14,926,290	11,946,538

Mineral Production (Contd.)

STATE/DISTRICT	QUANTITY Tonnes	VALUE Rs. '000
DIAMOND		
Madhya Pradesh		
Panna	14,542	17,588
INDIA	14,542	17,588
DIASPORE		
Madhya Pradesh		
Chhatarpur	610	108
Shivpuri	250	18
Tikamgarh	2,220	642
Uttar Pradesh		
Jhansi	440	91
Lalitpur	1,640	437
INDIA	5,160	1,296
DOLOMITE		
Andhra Pradesh		
Anantapur	148	3
Bihar		
Palamu	171,554	9,995
Gujarat		
Bhavnagar	6,630	194
Vadodara	242,398	3,337
Haryana		
Mahendragarh	50	1
Himachal Pradesh		
Bilaspur	824	375
Karnataka		
Bijapur	6,896	173
Tumkur	173	13
Madhya Pradesh		
Balaghat	8,705	161
Chhindwara	8,955	179
Durg	28,141	704
Jabalpur	14,004	234
Jhabua	20,824	277
Mandla	2,330	67
Satna	72	1
Seoni	1,358	13
Maharashtra		
Nagpur	25,094	812
Orissa		
Sundargarh	604,273	39,958
Rajasthan		
Jaipur	2,600	31
Jaisalmer	2,940	118
Jhunjhunun	3,520	53
Nagaur	10	N.A.
Sikar	1,715	34
Udaipur	2,859	71
Uttar Pradesh		
Dehra Dun	3,838	51
Mirzapur	24,176	798
West Bengal		
Jalpaiguri	25,998	573
INDIA	1,210,085	58,226
EMERALD		
Rajasthan		
Udaipur	1,000	N.A.
INDIA	1,000	N.A.
FELSPAR		
Andhra Pradesh		
Hyderabad	6,580	175
Mahbubnagar	320	7
Nellore	373	7
West Godavari	32	1
Bihar		
Palamu	199	4
Haryana		
Mahendragarh	204	9
Karnataka		
Bangalore	5,505	329
Madhya Pradesh		
Chhatarpur	20	N.A.
Jabalpur	438	9
Shahdol	844	17

STATE/DISTRICT	QUANTITY Tonnes	VALUE Rs. '000
FELSPAR (Contd.)		
Rajasthan		
Ajmer	34,547	1,338
Alwar	1,182	56
Jaipur	2,822	132
Pali	1,377	57
Tonk	345	11
Udaipur	1,170	53
Tamil Nadu		
Coimbatore	1,838	37
Tiruchchirappalli	2,956	142
INDIA	60,752	2,384
FIRE CLAY		
Andhra Pradesh		
Adilabad	1,700	26
East Godavari	7,048	286
West Godavari	8,045	202
Bihar		
Bhagalpur	4,700	71
Dhanbad	47,334	719
Giridih	18,548	260
Hazaribag	14,691	318
Palamu	136,443	3,384
Ranchi	11,935	299
Santhal Pargana	23,681	535
Singhbhum	10,540	197
Gujarat		
Kachchh	1,648	30
Mahesana	2,024	9
Panch Mahals	600	6
Rajkot	36,778	254
Sabar Kantha	1,860	19
Surat	4,848	47
Surendranagar	165,789	1,525
Valsad	16,781	163
Karnataka		
Bangalore	648	8
Tumkur	3,502	53
Kerala		
Quilon	205	5
Madhya Pradesh		
Jabalpur	70,442	1,181
Narsimhapur	1,275	18
Panna	834	11
Rajgarh	1,099	44
Shahdol	28,536	628
Orissa		
Cuttack	45,742	1,594
Puri	568	15
Sambalpur	47,110	3,259
Sundargarh	22,106	898
Rajasthan		
Alwar	4,409	133
Bharatpur	305	8
Bikaner	34,569	1,376
Chittaurgarh	1,960	29
Tamil Nadu		
Chengalpattu	4,170	6,567
South Arcot	16,068	213
Tiruchchirappalli	18,673	350
West Bengal		
Birbhum	13,098	171
INDIA	830,312	24,911
FLUORITE (Graded)		
Gujarat		
Vadodara	486	709
Rajasthan		
Dungarpur	2,678	2,607
Jalor	1,021	1,402
INDIA	4,185	4,718
GARNET (Abrasive)		
Andhra Pradesh		
Nellore	35	8
Rajasthan		
Ajmer	150	9
Bhilwara	74	81

STATE/DISTRICT	QUANTITY Tonnes	VALUE Rs. '000
GARNET (Abrasive) (Contd.)		
Sikar	1	N.A.
Tonk	16	5
Tamil Nadu		
Kanniyakumari	2,738	548
INDIA	3,014	651
GARNET (Gem)		
Rajasthan		
Bhilwara	1,033	10
Tonk	216	3
INDIA	1,249	13
GOLD		
Andhra Pradesh		
Anantapur	6	657
Chittoor	8	998
Bihar		
Sinhbhum	64	9,489
Karnataka		
Gulbarga	8	667
Kolar	1,396	166,917
Raichur	1,013	87,628
INDIA	2,495	266,356
GRAPHITE		
Andhra Pradesh		
East Godavari	465	93
Khammam	216	34
Vishakhapatnam	108	32
Bihar		
Palamu	17,268	1,754
Gujarat		
Panch Mahals	296	13
Orissa		
Balangir	21,361	2,263
Kalahandi	196	30
Phulabani	1,995	299
Sambalpur	16,390	1,111
Rajasthan		
Banswara	66	6
Tamil Nadu		
Madurai	660	39
INDIA	59,021	5,674
GYPSUM		
Gujarat		
Kachchh	8,340	62
Himachal Pradesh		
Sirmaur	76,000	33
Rajasthan		
Barmer	69,400	1,151
Bikaner	136,170	4,637
Jaisalmer	53,400	85
Nagaur	157,490	2,827
Pali	17,250	493
Tamil Nadu		
Coimbatore	33,540	1,202
Tiruchchirappalli	16,150	671
Uttar Pradesh		
Dehra Dun	400	12
Tehri Garhwal	4,230	167
INDIA	572,370	11,340
IRON ORE		
Andhra Pradesh		
Anantapur	77,838	1,262
Khammam	7,005	350
Nellore	227	9
Bihar		
Palamu	41,064	2,352
Singhbhum	6,809,876	248,186
Goa		
Goa	12,501,022	252,066
Karnataka		
Bellary	3,746,712	84,123
Bijapur	28,525	569
Chikmagalur	52,492	1,347

Mineral Production (Contd.)

STATE/DISTRICT	QUANTITY Tonnes	VALUE Rs. '000
IRON ORE (Contd.)		
Karnataka (Contd.)		
Chitradurga	120,310	1,689
Dharwad	20,100	342
Tumkur	18,387	297
Uttar Kannad	12,539	244
Madhya Pradesh		
Bastar	64,500	252,000
Durg	3,762,362	1,610,019
Orissa		
Cuttack	124,457	3,273
Kendujhar	4,209,108	141,032
Mayurbhanj	142,001	1,875
Sundargarh	1,970,923	90,585
Rajasthan		
Jaipur	2,183	48
Jhunjhunun	3,454	48
INDIA	33,715,085	1,242,718

JASPER

STATE/DISTRICT	QUANTITY Tonnes	VALUE Rs. '000
Rajasthan		
Jodhpur	3,356	245
INDIA	3,356	245

KAOLIN

STATE/DISTRICT	QUANTITY Tonnes	VALUE Rs. '000
Andhra Pradesh		
Adilabad	12,097	207
Cuddapah	1,814	63
East Godavari	135	54
Kurnool	15,759	290
West Godavari	19,775	838
Bihar		
Bhagalpur	75	26
Ranchi	239	49
Santhal Pargana	11,914	1,419
Singhbhum	32,518	6,025
Gujarat		
Banas Kantha	3,965	159
Bhavnagar	1,305	65
Kachchh	3,469	202
Mahesana	6,628	2,041
Sabar Kantha	8,943	2,969
Haryana		
Faridabad	7,736	103
Gurgaon	1,999	67
Jammu & Kashmir		
Udhampur	64	17
Karnataka		
Belgaum	647	26
Bidar	454	18
Dharwad	2,439	76
Hassan	5,229	594
Kolar	1,550	62
Shimoga	1,830	439
Kerala		
Cannanore	11,666	3,793
Kozhikode	14,326	8,929
Quilon	9,846	8,905
Madhya Pradesh		
Betul	9,330	168
Chhatarpur	1,231	15
Datia	38	1
Durg	75	1
Gwalior	72	2
Jabalpur	2,461	55
Satna	2,566	78
Shahdol	2,338	41
Sidhi	36	1
Maharashtra		
Chandrapur	4,568	114
Orissa		
Mayurbhanj	4,923	955
Sambalpur	7,537	268
Rajasthan		
Bhilwara	10,642	423
Bikaner	288	15
Chittaurgarh	110,731	2,077
Jaipur	25,751	579
Sawai Madhopur	93	3
Udaipur	231	3

STATE/DISTRICT	QUANTITY Tonnes	VALUE Rs. '000
KAOLIN (Contd.)		
Tamil Nadu		
South Arcot	2,694	997
West Bengal		
Birbhum	99,970	2,413
Delhi		
Delhi	45,916	922
INDIA	507,838	46,466

KYANITE

STATE/DISTRICT	QUANTITY Tonnes	VALUE Rs. '000
Andhra Pradesh		
Prakasam	38	17
Bihar		
Singhbhum	18,545	8,385
Karnataka		
Chikmagalur	223	44
Mysore	633	64
Maharashtra		
Bhandara	19,429	5,606
INDIA	38,868	14,116

LEAD CONCENTRATES

STATE/DISTRICT	QUANTITY Tonnes	VALUE Rs. '000
Andhra Pradesh		
Guntur	4,546	20,679
Rajasthan		
Udaipur	15,238	36,782
Sikkim		
East Sikkim	233	N.A.
INDIA	20,017	57,461

LIGNITE

STATE/DISTRICT	QUANTITY Tonnes	VALUE Rs. '000
Gujarat		
Kachchh	396	39,593
Tamil Nadu		
South Arcot	5,569,572	529,165
INDIA	5,569,968	568,758

LIME KANKAR

STATE/DISTRICT	QUANTITY Tonnes	VALUE Rs. '000
Andhra Pradesh		
Krishna	16,326	48
INDIA	16,326	48

LIME SHELL

STATE/DISTRICT	QUANTITY Tonnes	VALUE Rs. '000
Andhra Pradesh		
Nellore	4,296	187
Karnataka		
Uttar Kannad	22	1
Kerala		
Alleppey	42,485	4,304
Cannanore	2,628	145
Kottayam	49,423	2,020
Kozhikode	2,117	114
Mallappuram	4,100	554
INDIA	105,071	7,325

LIMESTONE

STATE/DISTRICT	QUANTITY Tonnes	VALUE Rs. '000
Andhra Pradesh		
Adilabad	500,065	10,692
Anantapur	25,181	524
Cuddapah	271,738	2,717
Guntur	904,394	10,809
Karimnagar	1,124,340	14,167
Krishna	100,669	2,930
Kurnool	639,229	12,119
Nalgonda	305,240	7,021
Vizianagaram	35,018	413
Assam		
Karbi Anglong	270,272	12,266
Bihar		
Palamu	552,759	31,925
Ranchi	193,296	8,289
Rohtas	1,006,321	37,145
Singhbhum	571,686	19,315
Gujarat		
Banas Kantha	12,112	369
Bharuch	86	3
Jamnagar	557,013	4,658

STATE/DISTRICT	QUANTITY Tonnes	VALUE Rs. '000
LIMESTONE (Contd.)		
Junagadh	1,720,455	33,326
Kheda	201,840	7,137
Vadodara	6,820	112
Haryana		
Ambala	515,148	74,131
Himachal Pradesh		
Bilaspur	411,000	10,240
Shimla	1,345	40
Sirmaur	163,597	3,507
Solan	2,670	50
Jammu & Kashmir		
Pulwama	19,000	418
Karnataka		
Belgaum	8,221	163
Bijapur	435,965	8,810
Chitradurga	3,605	54
Dharwad	477	17
Gulbarga	1,238,565	17,402
Shimoga	101,477	3,065
Tumkur	486,200	7,731
Uttar Kannad	20,507	656
Madhya Pradesh		
Bilaspur	27,458	474
Dhar	2,422	40
Durg	1,846,345	71,503
Jabalpur	1,444,584	24,826
Mandsaur	346,448	7,795
Narsimhapur	65	1
Raipur	1,039,862	20,216
Sehore	469	28
Maharashtra		
Chandrapur	31,097	777
Yavatmal	720,130	16,869
Meghalaya		
East Khasi Hills	233,873	5,402
Orissa		
Sambalpur	599,718	7,887
Sundargarh	2,113,758	126,604
Rajasthan		
Ajmer	320	3
Bundi	243,410	7,524
Chittaurgarh	1,262,207	16,050
Jaipur	26,000	390
Kota	359,273	5,641
Sawai Madhopur	649,524	20,421
Sikar	43,520	1,438
Sirohi	1,705	26
Udaipur	369,207	7,654
Tamil Nadu		
Coimbatore	448,673	13,909
Madurai	513,593	8,731
Ramanathapuram	880,552	23,228
Salem	805,929	27,074
Tiruchchirappalli	1,079,439	18,774
Tirunelveli	732,120	20,635
Uttar Pradesh		
Dehra Dun	395,868	9,451
Mirzapur	825,573	20,645
Naini Tal	5,841	44
Tehri Garhwal	55,376	1,499
West Bengal		
Jalpaiguri	987	19
INDIA	29,511,657	837,799

MAGNESITE

STATE/DISTRICT	QUANTITY Tonnes	VALUE Rs. '000
Karnataka		
Mysore	9,009	3,358
Rajasthan		
Ajmer	1,385	17
Dungarpur	48	2
Udaipur	22	1
Tamil Nadu		
Salem	354,199	96,718
Uttar Pradesh		
Almora	48,722	8,789
Pithoragarh	49,733	5,799
INDIA	463,118	114,684

Mineral Production (Contd.)

STATE/DISTRICT	QUANTITY Tonnes	VALUE Rs. '000
MANGANESE ORE		
Andhra Pradesh		
Adilabad	20,033	4,138
Vizianagaram	54,355	2,643
Bihar		
Singhbhum	1,345	61
Goa		
Goa	86,038	4,639
Gujarat		
Panch Mahals	330	39
Vadodara	1,056	74
Karnataka		
Belgaum	6,700	271
Bellary	195,148	21,857
Chitradurga	26,695	1,819
Shimoga	51,023	5,604
Tumkur	6,791	458
Uttar Kannad	158,033	15,379
Madhya Pradesh		
Balaghat	243,860	72,673
Maharashtra		
Bhandara	112,879	27,963
Nagpur	108,797	23,399
Orissa		
Kendujhar	337,025	44,411
Koraput	11,563	826
Sundargarh	103,888	13,766
INDIA	1,525,559	240,020
MICA (Waste & Crude)		
Andhra Pradesh		
Nellore	3,751	6,502
Vishakhapatnam	36	2
Bihar		
Giridih	3,443	5,415
Hazaribag	2,422	8,693
Munger	2	9
Nawada	1,296	5,310
Orissa		
Dhenkanal	10	15
Rajasthan		
Ajmer	82	210
Bhilwara	1,436	3,924
Udaipur	43	36
Tamil Nadu		
Nilgiri	104	295
INDIA	12,625	30,411
MOULDING SAND		
Andhra Pradesh		
Nellore	521	10
Goa		
Goa	45,998	296
Gujarat		
Bhavnagar	5,680	142
Jamnagar	1,216	4
Karnataka		
Belgaum	12,550	92
Dharwad	581	3
Gulbarga	12,184	122
Uttar Kannad	2,880	22
Madhya Pradesh		
Durg	22,173	122
INDIA	103,783	813
OCHRE		
Andhra Pradesh		
Kurnool	4,554	89
Bihar		
Sinhbhum	388	10
Gujarat		
Banas Kantha	343	16
Bhavnagar	15	1
Kachchh	132	8
Karnataka		
Bellary	12,130	474
Chikmagalur	60	7
Dharwad	723	21

STATE/DISTRICT	QUANTITY Tonnes	VALUE Rs. '000
OCHRE (Contd.)		
Madhya Pradesh		
Jabalpur	10	N.A.
Mandla	1,410	85
Rajnandgaon	100	6
Rewa	340	14
Satna	18,828	857
Shahdol	75	2
Maharashtra		
Chandrapur	1,416	24
Rajasthan		
Bharatpur	67	1
Bikaner	426	10
Chittaurgarh	25,908	232
Jaisalmer	14	N.A.
Sawai Madhopur	177	3
Sikar	102	1
Udaipur	19,209	250
Uttar Pradesh		
Banda	2,972	178
INDIA	89,399	2,289
PHOSPHORITE		
Madhya Pradesh		
Chhatarpur	16,258	3,577
Jhabua	67,258	18,883
Sagar	1,159	255
Rajasthan		
Udaipur	398,127	201,599
Uttar Pradesh		
Dehra Dun	62,032	9,392
Lalitpur	1,871	544
Tehri Garhwal	2,335	107
INDIA	549,040	234,357
PYRITES		
Bihar		
Ranchi	1	N.A.
INDIA	57,599	20,832
PYROPHYILLITE		
Madhya Pradesh		
Chhatarpur	96	7
Shivpuri	9,711	262
Tikamgarh	4,287	243
Rajasthan		
Udaipur	8,221	440
Uttar Pradesh		
Hamirpur	111	6
Jhansi	7,551	330
Lalitpur	10,087	243
INDIA	40,064	1,531
QUARTIZTE		
Bihar		
Munger	10,066	152
Madhya Pradesh		
Durg	20,388	245
Raigarh	9,106	2,430
Orissa		
Balangir	37,303	1,034
Dhenkanal	2,407	122
Kalahandi	6,725	231
Kendujhar	3,149	102
Mayurbhanj	13,082	142
Sambalpur	8900	400
Sundargarh	478	12
INDIA	103,594	4,780
QUARTZ		
Andhra Pradesh		
Guntur	629	20
Hyderabad	2,983	84
Khammam	41,307	1,399
Medak	8,560	204
Nellore	3,279	123
Prakasam	1,206	42
Rangareddi	3,359	74
Vishakhapatnam	656	16

STATE/DISTRICT	QUANTITY Tonnes	VALUE Rs. '000
QUARTZ (Contd.)		
Gujarat		
Panch Mahals	14,169	137
Haryana		
Mahendragarh	136	4
Karnataka		
Bangalore	963	21
Bellary	28,721	735
Hassan	3,055	184
Mandya	1,500	90
Shimoga	21,985	1,429
Tumkur	1,420	81
Madhya Pradesh		
Chhatarpur	20	N.A.
Shahdol	200	4
Rajasthan		
Ajmer	27,192	790
Alwar	1,874	73
Jaipur	684	16
Jhunjhunun	381	8
Pali	5,392	133
Sikar	2,086	42
Tonk	1,082	15
Udaipur	104	4
Tamil Nadu		
Coimbatore	33,540	1,202
Tiruchchirappalli	10,760	704
INDIA	217,243	7,634
SALT		
Himachal Pradesh		
Mandi	4,326	1,082
INDIA	4,326	1,082
SAND		
Andhra Pradesh		
Khammam	343,579	2,020
Bihar		
Giridih	84,606	328
Himachal Pradesh		
Bilaspur	171,000	950
Kerala		
Alleppey	37,072	253
Maharashtra		
Chandrapur	652,013	652
INDIA	1,288,270	4,203
SILICA SAND		
Andhra Pradesh		
Nellore	18,306	381
Bihar		
Santhal Pargana	32,352	1,483
Singhbhum	500	30
Gujarat		
Bharuch	24,243	9600
Kachchh	17,864	297
Sabar Kantha	5,618	97
Surendranagar	45,499	780
Karnataka		
Gulbarga	2,303	76
Kerala		
Alleppey	32,787	492
Madhya Pradesh		
Balaghat	706	14
Dewas	4,244	106
Morena	4,430	46
Maharashtra		
Kolhapur	3,061	94
Ratnagiri	103,340	2,691
Rajasthan		
Alwar	163	5
Barmer	414	21
Bharatpur	24,997	921
Bikaner	61	3
Bundi	18,633	838
Jaipur	65,278	1,822
Sawai Madhopur	6,114	240
Sikar	623	7
Tonk	8,492	317

Mineral Production (Contd.)

SILICA SAND (Contd.)

STATE/DISTRICT	QUANTITY Tonnes	VALUE Rs. '000
Uttar Pradesh		
Allahabad	169,617	3,719
Banda	28,022	656
INDIA	617,667	16,096

SILLIMONITE

STATE/DISTRICT	QUANTITY Tonnes	VALUE Rs. '000
Kerala		
Quilon	1,460	1,084
Maharashtra		
Bhandara	5,949	2,211
Meghalaya		
East Kasi Hills	4,457	2,370
West Kasi Hills	4,457	2,370
Rajasthan		
Udaipur	236	37
Tamil Nadu		
Kanniyakumari	233	166
INDIA	12,335	5,868

SILVER

STATE/DISTRICT	QUANTITY Tonnes	VALUE Rs. '000
Andhra Pradesh		
Vishakhapatnam	3,783	8,323
Bihar		
Dhanbad	12,914	28,411
Singhbhum	424	1,162
Kolar	98	252
Raichur	79	160
INDIA	17,298	38,218

SLATE

STATE/DISTRICT	QUANTITY Tonnes	VALUE Rs. '000
Andhra Pradesh		
Guntur	819	15
Prakasam	1,991	48
Haryana		
Mahendragarh	275	105
Madhya Pradesh		
Mandsaur	6,509	841
Rajasthan		
Alwar	345	200
INDIA	9,939	1,209

STAUROLITE

STATE/DISTRICT	QUANTITY Tonnes	VALUE Rs. '000
Karnataka		
Hassan	15	9
INDIA	15	9

STEATITE

STATE/DISTRICT	QUANTITY Tonnes	VALUE Rs. '000
Andhra Pradesh		
Anantapur	2,483	318
Chittoor	2,838	61
Cuddapah	755	39
Khammam	642	13
Kurnool	5,466	124
Warangal	78	5
Bihar		
Giridih	738	15
Singhbhum	1,007	71
Karnataka		
Chikmagalur	2,152	151
Mysore	268	38
Raichur	3,800	72
Tumkur	1,110	56
Madhya Pradesh		
Durg	119	3
Jabalpur	4,659	217
Narsimhapur	635	28
Maharashtra		
Bhandara	196	2
Orissa		
Kendujhar	231	7
Mayurbhanj	698	24
Rajasthan		
Ajmer	3,710	70
Alwar	14,335	637
Banswara	4,701	232
Bharatpur	24	1
Bhilwara	35,882	5,044
Dungarpur	33,320	1,955
Jaipur	26,709	4,480
Jhunjhunun	987	20
Pali	531	11
Sawai Madhopur	17,472	662
Tonk	430	43
Udaipur	130,901	9,603
Tamil Nadu		
Salem	28,420	611
Uttar Pradesh		
Almora	6,471	327
Lalitpur	300	8
Pithoragarh	3,693	202
INDIA	335,761	25,150

SULPHUR

STATE/DISTRICT	QUANTITY Tonnes	VALUE Rs. '000
Tamil Nadu		
Chengalpattu	4,170	6,567
INDIA	4,170	6,587

TUNGSTEN (Concentrates)

STATE/DISTRICT	QUANTITY Tonnes	VALUE Rs. '000
Rajasthan		
Nagaur	30,371	2,429
West Bengal		
Bankura	8,466	887
INDIA	38,837	3,316

VERMICULITE

STATE/DISTRICT	QUANTITY Tonnes	VALUE Rs. '000
Andhra Pradesh		
Nellore	1,847	311
Gujarat		
Vadodara	75	15
Karnataka		
Hassan	250	21
Rajasthan		
Ajmer	69	15
Tamil Nadu		
North Arcot	1,197	527
INDIA	2,778	889

WOLLASTOMITE

STATE/DISTRICT	QUANTITY Tonnes	VALUE Rs. '000
Rajasthan		
Sirohi	15,895	2,109
Udaipur	20	2
INDIA	15,915	2,111

ZINC CONCENTRATE

STATE/DISTRICT	QUANTITY Tonnes	VALUE Rs. '000
Rajasthan		
Udaipur	52,468	126,854
Sikkim		
East Sikkim	408	566
INDIA	52,876	127,420

INDUSTRIAL LOCATIONS
(1984)

STATE/DISTRICT	MANUFACTURING TOWNS
BEVERAGE & TOBACCO PRODUCTS	
Andhra Pradesh	
East Godavari	Bikkavoli
Guntur	Guntur
Hyderabad	Azamabad, Ghatkesar, Hyderabad, Secunderabad
Karimnagar	Sirsilla
Assam	
Cachar	Cachar, Dallabcherra, Serispore, Silchar
Darrang	Arun, Bhipoorkli & Orang, Bishnanth, Borai, Darrang, Deokaijuli, Dherai, Dibru & Dbullib, Dooars & Dalgaon, Kopati, Mangaldai, Tezpur, Tinkharia
Dibrugarh	Bagapani, Choonabhatti, Dehing, Dibrugarh, Dirok, Doom Dooma, Margherita, Moran, Namdang, Partol, Tingri
Lakhimpur	Dejoo/Harmutty, Lakhimpur, Seajuli
Nowgong	Nowgong
Sibsagar	Borputra, Duklingia, Golaghat, Gorunga, Jorhat, Sibsagar
Bihar	
Munger	Munger
Patna	Hathidah
Goa	
Goa	Goa, Marma Goa Harbour
Gujarat	
Bharuch	Ankleshwar
Jamnagar	Jamnagar
Junagadh	Kodinar, Mandali
Vadodara	Harni
Himachal Pradesh	
Solan	Solan
Karnataka	
Bangalore	Bangalore, Khushnagar
Belgaum	Ugar — Khurd
Bellary	Hospet
Chikmagalur	Chikmagalur
Dakshin Kannad	Mangalore
Hassan	Hassan
Kodagu	Kodagu, South Coorg.
Kerala	
Alleppey	Shertally
Ernakulam	Anamalai
Idukki	Cholai, Devikulam, Peermade, Udamapan
Quilon	Pathanamthitta Pathanapura
Trivandrum	Trivandrum
Maharashtra	
Ahmadnagar	Ahmednagar, Harigaon, Rahuri, Shivajinagar
Greater Bombay	Chakala, Parel, Vile Parle
Kolhapur	Sahunagar
Nagpur	Nagpur
Nashik	Nashik
Pune	Junner
Raigarh	Uran
Sangli	Sakharale
Solapur	Solapur
Thane	Thane, Panch Pakhadi
Rajasthan	
Ganganagar	Sriganganagar
Jaipur	Jaipur
Udaipur	Udaipur
Sikkim	
South Sikkim	Melli Bazar
Tamil Nadu	
Chengalpattu	Ambattur
Coimbatore	Appakudal, Coimbatore, Injipara, Karamalai, Monica, Mudis, Stanmore
Dharmapuri	Hosur
Madurai	High Ways

STATE/DISTRICT	MANUFACTURING TOWNS
BEVERAGE & TOBACCO (Contd.)	
Nilgiri	Cheranbadi, Kotagiri, Nilgiris, Punsandle, Thaishola, Wellington
South Arcot	Anamalai
Tirunelveli	Maryolai, Tirunelveli
Uttar Pradesh	
Agra	Tundla
Basti	Basti
Deoria	Captainganj, Gangeshwar, Ramkota
Ghaziabad	Modinagar
Lucknow	Lucknow
Meerut	Daurala, Mawana
Saharanpur	Saharanpur
West Bengal	
Calcutta	Calcutta, Kidderpore
Darjiling	Bagdogra, Darjeeling, Glenburun & Lingia, Tumsong Estate
Jalpaiguri	Banarhat, Baradighi, Beech, Bharnobari & Satali, Dooars, Jalpaiguri, Karballa, Kartick, Rheabari, Rydak, Twelve Estate
Twenty Four Parganas	Agarpara
Delhi	
Delhi	Delhi
CHEMICAL & PHARMACEUTICALS	
Andhra Pradesh	
Adilabad	Adilabad
Anantapur	Anantapur
Chittoor	Tirupati
East Godavari	Kakinada
Guntur	Tadepalli
Hyderabad	Azamabad, Balanagar, Hyderabad, Secunderabad, Uppal, Warangal Road
Karimnagar	Karimnagar, Ramgundam
Khammam	Paloncha
Krishna	Machilipatnam, Vuyyuru
Kurnool	Kurnool
Medak	Medak, Pattancheru, Pattancheruvu, Pathanchuru, Sangareddy
Nalgonda	Bibinagar, Nalgonda
Srikakulam	Srikakulam
Vishakhapatnam	Malkapuram, Venketapuram, Vishakhapatnam
Assam	
Dibrugarh	Namrup, Tinsukia
Goalpara	Bongaigaon, Jogighopa
Kamrup	Gawahati, Noonmati
Lakhimpur	Namrup
Bihar	
Begusarai	Barauni
Dhanbad	Sindri
Giridih	Gomia
Munger	Barauni
Muzaffarpur	Muzaffarpur
Palamu	Barwadih
Patna	Patna
Ranchi	Ranchi
Rohtas	Amjhere, Dalmianagar
Santhal Pargana	Dabrugarh, Jasidih
Singhbhum	Jaduguda, Jamshedpur
Goa	
Goa	Santa Monica, St. Jose de Areal Margao
Gujarat	
Ahmadabad	Ahmadabad, Dascroi, Kankaria, Kathwada, Khokhra, Mehmedabad, Maninagar, Rakhial
Amreli	Amreli
Banas Kantha	
Bharuch	Ankleshwar, Bharuch

STATE/DISTRICT	MANUFACTURING TOWNS
CHEMICAL & PHARMACEUTICALS (Contd.)	
Bhavnagar	Bhavnagar
Gandhinagar	Kalol
Jamnagar	Mithapur, Okha, Okhamandal, Sikka, Singhach
Junagadh	Porbandar, Verawal
Kachchh	Kandla
Panch Mahals	Kalol, Nurpura
Rajkot	Rajkot
Surat	Bhestan, Hazira, Olpad, Pandesara, Udhna, Walod
Surendranagar	Bamanbera, Dhrangadhra
Vadodara	Fertilisernagar, Jawaharnagar, Nandesari, Ranol, Sakarda, Vadodara, Wadi Wadi
Valsad	Atul, Billimora, Valsad, Vapi
Haryana	
Ambala	Yamunanagar
Faridabad	Faridabad
Gurgaon	Gurgaon
Karnal	Karnal, Panipat
Sonipat	Murthal
Jammu & Kashmir	
Jammu	Baki Brahmana, Bari Brahmana, Jammu
Karnataka	
Bangalore	Bangalore, Hoskote, Kottigenanalli, Yeshwantpur
Bellary	Hagari
Chikmagalur	Kudremukh
Dakshin Kannad	Mangalore, Penambur
Dharwad	Hubli
Hassan	Hospet
Mandya	Belagula, Mandya
Mysore	Mysore
Raichur	Chicksugar
Shimoga	Bhadravati, Shimoga
Uttar Kannad	Karwar & Gokarna
Kerala	
Ernakulam	Ambalamugal, Binanipuram, Edapally, Travancore Eloor & Cochin
Kottayam	Chingavanam
Kozhikode	Calicut, Kozhikode
Malappuram	Malappuram
Palghat	Palghat
Quilon	Kottakkal, Quilon
Trichur	Irinjalakude
Trivandrum	Kazakuttam, Trivandrum
Madhya Pradesh	
Bhopal	Bhopal
Dewas	Dewas, Nagda
Durg	Bhilai, Kumhari
East Nimar	Khandwa, Nepa Nagar
Guna	Vijaypur
Indore	Indore
Jabalpur	Behraghat, Bhitoni
Raipur	Raipur
Ratlam	Ratlam
Shahdol	Amlai
Ujjain	Birlagram
Maharashtra	
Ahmadnagar	Ahmadnagar, Belapur, Pravaranagar, Rahuri, Shivajinagar
Akola	Akola
Aurangabad	Aurangabad, Paithan
Dhule	Purshottam
Greater Bombay	Amboli, Andheri, Bhandup, Bombay, Chandivli, Chembur, Dadar, Deonar, Ghatkopar, Goregaon, Jogeshwari, Kandivali, Kurla, Lalbaug, Lawer Parel, Powai & Madh Island, Mahim, Mulund, Sakinaka, Sewri, Sion, Tardeo, Trombay, Vakola, Vikhroli, Vile Parle, Wadala, Worli

Industrial Locations (Contd.)

STATE/DISTRICT	MANUFACTURING TOWNS	STATE/DISTRICT	MANUFACTURING TOWNS	STATE/DISTRICT	MANUFACTURING TOWNS

CHEMICAL & PHARMACEUTICALS (Contd.)

STATE/DISTRICT	MANUFACTURING TOWNS
Jalgaon	Jalgaon, Panchor
Kolhapur	Sahunagar
Nagpur	Hingna Industrial Estate, Nagpur
Nanded	Nanded
Nashik	Bhausahebnagar, Nashik
Pune	Akurdi, Chinchwad, Nira, Pimpri, Pune
Raigarh	Khopoli, Nagothane, Panvel, Patalganga, Raigad, Rasayani, Roha, Taloja, Thal Vaishet
Ratnagiri	Lote Parshuram (Chiplun)
Thane	Ambernath, Balkum, Bhiwandi, Boisar, Dombivali, Ghansoli, Hingni, Kalwe, Kalyan, Kavesar, Kolshet, Majiwade, New Bombay, Patlipada, Saravali, Shahapur, Tarapur, Thane, Wado, Wardha
Orissa	
Cuttack	Cuttack, Paradeep
Dhenkanal	Talcher
Ganjam	Ganjam
Koraput	Thirumbadi
Mayurbhanj	Rairangpur
Puri	Near Bhubaneshwar
Sundargarh	Birmitrupar, Rourkela
Punjab	
Amritsar	Khasa
Bathinda	Bathinda
Faridkot	Jaitu
Hoshiarpur	Balachur, Hoshiarpur
Jalandhar	Jalandhar
Patiala	Dera Bassi, Rajpura
Rupnagar	Mohali, Nangal
Sangrur	Sanghera
Rajasthan	
Alwar	Alwar, Bhiwadi
Jaipur	Jaipur
Jodhpur	Jodhpur
Kota	Kota, Ladpura
Sawai Madhopur	Billopa, Sawai Madhopur
Tonk	Tonk
Udaipur	Debak, Debari, Udaipur
Tamil Nadu	
Chengalpattu	Ambattur, Avadi & Kaduvatti, Chromepet, Ennore, Manali
Coimbatore	Coimbatore, Sirumugal
Dharmapuri	Krishnagiri
Madras	Madras
Madurai	Madurai, Sholavandan
Nilgiri	Ootacamund
North Arcot	Arkonam, Katpadi, Ranipet, Vanthrangal
Ramanathapuram	Koviloor
Salem	Mettur, Mettur Dam
South Arcot	Cuddalore, Tindivanam
Tiruchchirappalli	Tiruchchirappalli
Tirunelveli	Sahupuram, Sankarnagar, Tuticorin
Uttar Pradesh	
Aligarh	Salempur
Allahabad	Jhansi & Naini, Phulpur
Bara Banki	Bara Banki
Bareilly	Bareilly, Clutterbuckgunj
Bijnor	Seabara
Budaun	Babrala
Bulandshahr	Bulandshahr
Dehra Dun	Dehra Dun, Mussoorie, Rishikesh
Deoria	Deoria
Etah	Etah
Etawah	Auraiya
Ghaziabad	Ghaziabad, Modinagar, Shahibabad
Gorakhpur	Gorakhpur
Jhansi	Jhansi
Kanpur	Kanpur, Panki

CHEMICAL & PHARMACEUTICALS (Contd.)

STATE/DISTRICT	MANUFACTURING TOWNS
Lucknow	Lucknow
Mathura	Mathura
Mirzapur	Renukot
Moradabad	Gajaraula
Rampur	Bilaspur
Shahjahanpur	Rosa, Shahjahanpur
Sitapur	Hargaon
Sultanpur	Jagdishpur
Unnao	Magarwara
Varanasi	Sahapuri
West Bengal	
Barddhaman	Asansol, Durgapur
Calcutta	Behala, Calcutta, Panihati & Maniktala
Haora	Haora, Goabaria
Hugli	Konnagar, Rishra, Serampore, Sibpur, Tribeni & Bansberia
Medinipur	Haldia
Twentyfour Parganas	Birlapur, Budge Budge, Garia, Khardah, Kossipur, Naihati
West Dinajpur	Baidynath
Chandigarh	
Chandigarh	Chandigarh
Delhi	
Delhi	Delhi, Okhla, Palam Airport
Pondicherry	
Pondicherry	Kalapet, Mettapalyam

FOOD PRODUCTS

STATE/DISTRICT	MANUFACTURING TOWNS
Andhra Pradesh	
Anantapur	Sudhanagar
Chittoor	Chittoor
East Godavari	Chelluru, Rajahmundry, Samalkot
Guntur	Chagalla, Ganapavaram, Guntur, Vadlamudi
Hyderabad	Hyderabad, Secunderabad
Karimnagar	Muthyampet, Vuyyuru
Medak	Madhunagar
Nalgonda	Kodad
Nizamabad	Shakarnagar
Prakasam	Chirala
Vishakhapatnam	Vishakhapatnam
West Godavari	Chagallu, Tadepalligudem, Tanuku & Kovvur, Venkatarayapuram
Assam	
Lakhimpur	Aira Estate
Bihar	
Bhagalpur	Bhagalpur
Madhubani	Pandaul
Pashchim Champaran	Bettiah, Charpatia, Harinagar, Jogapatti, Mayhaula, Narkatiaganj, Ramnagar
Purba Champaran	Motihar
Ranchi	Ranchi
Rohtas	Dalmianagar
Saharsa	Supaul
Samastipur	Barrah
Saran	Hathua, Marhowrah
Siwan	Siwan
Goa	
Goa	Margao, Vasco-De-Gama
Gujarat	
Ahmadabad	Ahmadabad, Naroda, Rakhial
Amreli	Amreli
Bharuch	Palej
Bhavnagar	Bhavnagar, Mahuva
Jamnagar	Jamnagar, Singhach
Junagadh	Junagadh, Kodinar, Manvardar, Porbandar
Kachchh	Ramania
Kheda	Kheda
Panch Mahals	Dahod, Panch Mahals
Rajkot	Dhoraji, Khandheri, Rajkot
Surat	Bardoli, Chaltan, Madhi, Surat, Udhna, Vyaro-Paniari, Walod
Valsad	Gamdevi, Vapi

FOOD PRODUCTS (Contd.)

STATE/DISTRICT	MANUFACTURING TOWNS
Haryana	
Ambala	Yamunanagar
Hisar	Fatehabad
Jind	Jind
Karnal	Panipat
Kurukshetra	Shahbad-Thaneswar
Rohtak	Rohtak
Sonipat	Bahalgarh, Kandli, Rasoi
Himachal Pradesh	
Mandi	Sainj
Jammu & Kashmir	
Baramula	Baramula
Srinagar	Srinagar
Karnataka	
Bangalore	Bangalore
Belgaum	Belgaum, Chikodi, Hubli, Sankeshwar, Ugar-Khurd
Bellary	Hospet
Bidar	Hallikhed
Bijapur	Saidpur
Dakshin Kannad	Mangalore, Puttur
Dharwad	Hubli
Hassan	Hassan
Kodagu	Kodagu
Kolar	Kolar
Mandya	Mandya
Mysore	Mysore
Raichur	Munirabad, Pragati Nagar
Shimoga	Harige
Tumkur	Tumkur
Kerala	
Cannanore	Tellichery
Ernakulam	Angamally
Quilon	Kundara
Trichur	Chalakudy
Madhya Pradesh	
Bhopal	Bhopal
East Nimar	Khandwa
Gwalior	Gwalior
Indore	Indore
Jabalpur	Jabalpur
Sehore	Sehore
Ujjain	Ujjain
Vidisha	Hoshangabad
Maharashtra	
Ahmadnagar	Ahmadnagar, Bhende Burruk, Gautamnagar, Harigaon, Koregaon, Lakshmiwadi, Pravaranagar, Rangangaon, Shivajinagar, Shrigonda, Tilaknagar
Akola	Akola, Risod
Amravati	Achalpur, Amravati, Chandur
Aurangabad	Aurangabad, Raghunathpur, Sillod
Bid	Bid
Buldana	Malkpur
Dhule	Bhadne, Purshottam, Shahada, Shirpur
Greater Bombay	Bhardup, Bombay, Goregaon, Mazgaon, Mulund, Tardeo, Wadala, Vile Parle
Jalgaon	Amalner, Bhoras, Jalgaon, Kasoda, Panchor, Yawal
Kolhapur	Birdri, Ganganagar, Ichalkaranji, Kaditre, Kamathe, Kasba, Kolhapur, Shiroli, Warananagar
Nagpur	Monda, Nagpur
Nashik	Bhausahebnagar, Kakasahebnagar, Malegaon, Ravalgaon, Pimpalas
Osmanabad	Dheki, Dhoki, Osmanabad
Parbhani	Vadakalas
Pune	Bhawani Nagar, Chinchwad, Induri, Malegaon, Pune, Someshwar Nagar, Theor
Raigarh	New Bombay
Sangli	Atpadi, Sakharale, Sangli
Satara	Marali, Phaltan, Shivnagar, Wai

Industrial Locations (Contd.)

FOOD PRODUCTS (Contd.)

STATE/DISTRICT	MANUFACTURING TOWNS
Solapur	Akaluj, Dhamangaon, Malkapur, Pancharpur, Shreepur, Solapur
Thane	Ghansoli, Pakhadi, Panch, Thane
Yavatmal	Pusad
Orissa	
Cuttack	Badamba
Ganjam	Aska
Koraput	Rayagada
Puri	Kurda, Nayagarh
Sambalpur	Tharsuguda
Punjab	
Amritsar	Taran
Faridkot	Moga
Gurdaspur	Pandar
Hoshiarpur	Hoshiarpur
Jalandhar	Nakoda
Ludhiana	Ludhiana, Samrala
Patiala	Bahadurgarh, Nabha, Rajpura
Rajasthan	
Bharatpur	Bharatpur
Bhilwara	Bhilwara
Ganganagar	Sriganganagar, Sangaria, Kasrisinghpur
Jaipur	Jaipur
Tamil Nadu	
Chengalpattu	Ennore, Paladam
Coimbatore	Appakudal, Coimbatore, Tudiyalu
Dharmapuri	Dharmapuri, Palacode
Madras	Madras
Madurai	Nilakottai, Sholavandan
Periyar	Satyamangalam
Salem	Mahanur
South Arcot	Mundiyampakkam, Nellikuppam, Villupuram, Vridachalam
Thanjavur	Vedaraniyam
Tiruchchirappali	Kattur, Pugalur, Tiruchchirappalli, Trichy
Uttar Pradesh	
Aligarh	Aligarh, Satha
Azamgarh	Sathiaon
Bahraich	Bahriach, Nandara
Ballia	Rasra
Bara Banki	Burhwal
Bareilly	Baheri
Basti	Amroha, Basti
Bijnor	Bijnor, Dhampur, Hempur, Seabara
Budaun	Sheikhupur
Bulandshahr	Jahangirabad
Deoria	Gangeshwar, Gauri Bana, Kathkuiyam, Padrapuna, Pratappur, Ramkota, Seorahi
Etah	Etah
Farrukhabad	Kaimganj
Ghaziabad	Ghaziabad, Modinagar, Simbhaoli
Gonda	Balrampur, Tulsipur
Gorakhpur	Pipraich, Sardarnagar
Kanpur	Kanpur
Kheri	Golagokarannath, Paliakalan
Mathura	Chhata
Meerut	Mawana, Meerut, Mohiuddinpur
Moradabad	Amroha, Chandausi
Muzaffarnagar	Khotauli, Mansurpur, Shamli
Naini Tal	Bazpur, Kashipur, Kichha, Nadehi, Sifarganj
Pilibhit	Majhola
Rampur	Rampur
Saharanpur	Deoband, Dikhanj, Gangeshwar, Jabalpur, Sarsawa
Shahjahanpur	Rosa, Tilhar
Sitapur	Hargaon
Varanasi	Aurai, Varanasi
West Bengal	
Calcutta	Calcutta

FOOD PRODUCTS (Contd.)

STATE/DISTRICT	MANUFACTURING TOWNS
Haora	Howrah, Sibpur
Hugli	Rishra
Jalpaiguri	Jalpaiguri
Twentyfour Parganas	Belgharia, Budge Budge, Gosaha
Chandigarh	
Chandigarh	Chandigarh
Delhi	
Delhi	Delhi
Pondicherry	
Pondicherry	Ariyur, Oreleans

LEATHER, RUBBER & PLASTIC PRODUCTS

STATE/DISTRICT	MANUFACTURING TOWNS
Andhra Pradesh	
Guntur	Mangalagiri
Hyderabad	Hyderabad
Mahbubnagar	Thimmapur
Rangareddi	Vallabhnagar
Vishakhapatnam	Venketapura, Vishakhapatnam
Assam	
Goalpara	Goalpara
Bihar	
Patna	Bataganj & Mokamehghat
Goa	
Goa	Goa
Gujarat	
Bharuch	Ankleshwar, Palej, Panoli
Kheda	Kheda, Mehmedabad
Mahesana	Kadi
Panch Mahals	Halol, Kalol, Panchmahals
Rajkot	Rajkot
Vadodara	Nandesari
Valsad	Billimora
Haryana	
Faridabad	Faridabad
Gurgaon	Ballabgarh, Dharuhera
Hisar	Mayur
Sonipat	Rai
Karnataka	
Bangalore	Bangalore Nagasandra, Peenya, Yeshwantpur
Belgaum	Hukkari
Mysore	Metagalli, Mysore
Kerala	
Cannanore	Baliapatam
Idukki	Cholai, Devikulam, Peermade, Udampan,
Kottayam	Kottayam
Malappuram	Malappuram
Quilon	Pathanamthitta, Pathanapura
Trichur	Perambra
Trivandrum	Peroorkada, Trivandrum
Madhya Pradesh	
Dewas	Dewas
Maharashtra	
Aurangabad	Aurangabad, Chikhalthana, Gevral
Chandrapur	Chandrapur
Greater Bombay	Bhandup, Bombay, Goregaon, Jogeshwari, Kandivli, Kurla, Sewri, Vile Parle,
Jalgaon	Jalgaon
Nagpur	Nagpur
Nashik	Nashik, Satpur
Pune	Pune, Telegaon
Raigarh	Taloja
Satara	Satara
Thane	MIDC Estate & Kolshet, Tarapur, Thane
Orissa	
Baleshwar	Baleshwar
Puri	Bhubaneshwar
Punjab	
Amritsar	Chhehorta
Hoshiarpur	Hoshiarpur
Ludhiana	Ludhiana
Rajasthan	
Sirohi	Mount Abu
Udaipur	Kankorli

LEATHER, RUBBER & PLASTIC (Contd.)

STATE/DISTRICT	MANUFACTURING TOWNS
Tamil Nadu	
Chengalpattu	Ambattur, Manali, Pallavaram Tiruvottiyur
Madras	Madras
Madurai	Kochadai, Kodaikanal, Madurai
North Arcot	Arkonam, Ranipet, Wallajahpet
Salem	Mettur Dam
Uttar Pradesh	
Bareilly	Bareilly, Bhitaura
Bulandshahr	Salempur, Sikandrabad
Fatehpur	Fatehpur
Ghaziabad	Sahibabad
Kanpur	Kanpur
Naini Tal	Kumon Hills
Sultanpur	Amethi, Jagdishpur
West Bengal	
Calcutta	Calcutta, Kharagpur
Hugli	Chakundi, Rishra, Sahaganj
Medinipur	Haldia
Twenty-four Parganas	Batanagar, Panihati, Kankinara
Pondicherry	
Pondicherry	Pondicherry
Dadra & Nagar Haveli	Selwas

TEXTILE INDUSTRY

STATE/DISTRICT	MANUFACTURING TOWNS
Andhra Pradesh	
Adilabad	Adilabad, Sirpur
Anantapur	Anantapur, Kirikera, Kodigenaballi, Tadepah
Chittoor	Chittoor
East Godavari	Kakinada
Guntur	Guntur
Hyderabad	Hyderabad, Secunderabad
Kurnool	Adoni
Mahbubnagar	Kalwakuthy
Medak	Medak
Vishakhapatnam	Chitavalsah, Vishakhapatnam
Vizianagaram	Nellimara, Vizianagaram
Warangal	Kamlapuram, Warangal
Assam	
Cachar	Chandrapore
Darrang	Chariduwar
Kamrup	Kamrup, Natkuchi
Nowgong	Jogi Road, Nowgong
Bihar	
Gaya	Gaya
Katihar	Katihar
Patna	Mokameh
Purnia	Forbesganj, Kishanganj, Purnia
Samastipur	Muktapur
Siwan	Siwan
Goa	
Goa	Goa
Gujarat	
Ahmadabad	Ahmadabad, Gomtipur, Kankaria, Maninagar, Mehmedabad, Mithipur, Naroca Road, Raikhad, Rajpur Road, Rakhial Road, Shahibag, Saraspur, Viramgam
Bharuch	Ankleshwar, Bharuch
Bhavnagar	Bhavnagar
Jamnagar	Jamnagar
Junagadh	Porbandar, Veraval
Kachchh	Kandla, Nadiad, Petlad
Kheda	Nadiad, Petlad,
Mahesana	Laxmipura, Kalol, Mahesana
Rajkot	Morbi, Rajkot
Surat	Surat, Udhna
Vadodara	Petrofils, Vadodara
Haryana	
Bhiwani	Bamla
Faridabad	Faridabad
Hisar	Bhiwani, Fatehabad, Hissar

Industrial Locations (Contd.)

TEXTILE INDUSTRY (Contd.)

STATE/DISTRICT	MANUFACTURING TOWNS
Jind	Jind
Mahendragarh	Dharuher, Rewari
Himachal Pradesh	
Hamirpur	Barotiwala, Malagarh, Solan
Jammu & Kashmir	
Jammu	Jammu
Kathua	Kathua
Srinagar	Rambagh
Karnataka	
Bangalore	Bangalore
Belgaum	Gokak Falls
Bidar	Bidar
Dharwad	Gadag, Harihar, Hubli
Gulbarga	Gulbarga
Hassan	Belur
Mysore	Nanjangud
Kerala	
Alleppey	Komalapuram
Cannanore	Cannanore
Eranakulam	Ermathala
Kozhikode	Mavoor
Palghat	Kanikoda
Trichur	Koratti, Trichur
Trivandrum	Nedumangal, Trivandrum
Madhya Pradesh	
Bhopal	Balgarh, Bhopal
East Nimar	Burhanpur
Gwalior	Birlanagar, Gwalior
Indore	Indore
Jhabua	Meghnagar
Raisen	Raisen
Rajnandgaon	Rajnandgaon
Ratlam	Ratlam
Shahjapur	Shajapur
Ujjain	Nagda, Ujjain
Maharashtra	
Ahmadnagar	Ahmadnagar
Akola	Akola
Aurangabad	Aurangabad, Khamgaon
Dhule	Navapur
Greater Bombay	Bombay, Borivli, Byculla, Cinchpokli, Chembur, Dadar, Ghatkopar, Goregaon, Kurla, Lalbaug, Lower Parel, Mazagaum, Naigaum, Parel, Prabhadevi, Sakinaka, Vikhroli, Worli,
Jalgaon	Jalgaon, Kolhapur
Kolhapur	Kolhapur
Nagpur	Nagpur
Nanded	Nanded
Osmanabad	Latur
Pune	Bhosari, Chinchwad, Haveli, Pimpri, Pune
Raigarh	Khopoli Karsundi, Mahad, Patalganga, Roha
Ratnagiri	Ratnagiri
Sangli	Madhavnagar
Solapur	Barsi, Sangola, Solapur
Thane	Ambernath, Jaykaygram, Kalyan, Tarapur, Thane
Wardha	Hinganghat, Wardha
Yavatmal	Yavatmal
Orissa	
Cuttack	Chowdar, Cuttack
Dhenkanal	Baulpur
Punjab	
Amritsar	Amritsar, Chhehoria,
Bathinda	Doomwali
Faridkot	Abohar, Malout
Gurdaspur	Dhariwal
Hoshiarpur	Chohal, Hoshiarpur
Ludhiana	Ludhiana
Patiala	Kharar, Rajpura
Rupnagar	Kharar
Sangrur	Barnala, Malerkotla
Rajasthan	
Ajmer	Ajmer
Alwar	Alwar, Bhiwadi, Desula
Banswara	Banswara

TEXTILE INDUSTRY (Contd.)

STATE/DISTRICT	MANUFACTURING TOWNS
Bhilwara	Bhilwara, KhARIgram, Raila
Jaipur	Jaipur, Podarpuri
Jhalawar	Bhiwani, Mandi, Pachpahar
Kota	Kota, Rangpur
Pali	Pali
Udaipur	Udaipur
Tamil Nadu	
Changalpattu	Ambattur, Kathirvedu, Manali
Coimbatore	Appakudal, Coimbatore, Ganapathy, Kaniyamuthur, Karamadai, Madathukulam, Palladam, Peelamedu, Pulankinar, Singanallur, Sirumugal, Sowri Palayam, Tiruppur, Udamalpet, Upplipalayam, Vadamathurai
Dharmapuri	Belathur, Dharmapuri
Madras	Madras, Perambur
Madurai	Dindigul, Kalyampathur, Kappallur, Madurai
Pudukkottai	Pudukkottai, Viralimalai
Ramanathapuram	Chettinad, Kamudakudi, Rajapalayam
Salem	Gandhinagar, Mettur Dam, Suramangalam
Thanjavur	Thanjavur
Tiruchchirappalli	Kathirvedu, Nanoparai
Tirunelveli	Ambasamudram, Ettayapuram, Kovilpatty, Shencottah, Tuticorin
Uttar Pradesh	
Aligarh	Hathras
Almora	Lal Kuan
Azamgarh	Maunath Bhanjan
Banda	Banda
Bara Banki	Barabanki
Bulandshahr	Bulandshahr, Sikandrabad
Faziabad	Akbarpur
Fatehpur	Fatehpur
Ghaziabad	Dasna, Modinagar, Ghaziabad, Sahibadad
Ghazipur	Ghazipur
Kanpur	Kanpur, Panki
Lucknow	Lucknow
Mathura	Mathura
Meerut	Modinagar
Moradabad	Gajaraula, Thakurdwara
Naini Tal	Lalkha, Naini Tal
Rae Bareli	Modigram, Rae Bareli
Rampur	Jwalanagar
Saharanpur	Saharanpur
Varanasi	Varanasi
West Bengal	
Bankura	Bankura, Barsul
Barddhaman	Barsul
Calcutta	Calcutta, Kanknaarrah, Kantalpara, Karmarhatty, Sarampur, Taratolla Road
Haora	Bally, Barnagore, Bauria, Chengail, Ghusuri, Howrah, Rajgunge Awdir, Sijberia, Sibpore
Hugli	Bansberia, Bhadreswar, Champdany, Gondalpara, Hoogly, Rishra, Samnuggur, Sibpore, Serampur
Medinipur	Tantigaria
Twentyfour Parganas	Birati, Birlapur, Bishnupur, Bhalpara, Budge, Jagatlal, Jegatdal, Harinagar, Hazinagar, Kamarhatty, Naihati, Panihati, Piragachha, Shamnagar, Talpukua, Titaghur, Twentyfour Parganas
Chandigarh	
Chandigarh	Chandigarh
Delhi	
Delhi	Delhi
Pondicherry	
Mahe	Mahe
Pondicherry	Modaliarpet, Nedungadu, Pondicherry

ELECTRICAL MACHINERY

STATE/DISTRICT	MANUFACTURING TOWNS
Andhra Pradesh	
Hyderabad	Cherlapalli, Hyderabad Kukatpali
Krishna	Machilipatnam
Medak	Medak
Nellore	Andhra Kesarinagar
Vishakhapatnam	Vishakhapatnam
Bihar	
Ranchi	Ranchi
Singhbhum	Ghatsila, Jamshedpur
Goa	
Goa	Panji, Zuarinagar
Gujarat	
Ahmadabad	Ahmadabad, Maninagar, Naroda, Vatva
Kheda	Kheda, Mogar
Panch Mahals	Halol
Rajkot	Morvi, Wankaner
Vadodara	Samiala, Vadodara
Haryana	
Ambala	Panchkula
Faridabad	Faridabad
Gurgaon	Ballabhgarh, Dharuhera
Hisar	Hisar
Jind	Jind
Mahendragarh	Mahendragarh, Malpur, Rewari
Sonipat	Sonipat
Jammu & Kashmir	
Badgam	Badgam, Srinagar
Srinagar	Srinagar
Karnataka	
Bangalore	Bangalore, Byappanahalli, Islahalli, Modival Port, Rajajinagar, Settingere Village, Yeshwantpur
Dharwad	Hubli
Gulbarga	Gulbarga
Mysore	Mysore
Kerala	
Alleppey	Mannar
Palghat	Palghat
Quilon	Pathanamthitta
Trichur	Trichur
Trivandrum	Trivandrum
Madhya Pradesh	
Bhopal	Bhopal
Jabalpur	Deori
Raipur	Raipur
Raisen	Mandideep
Rewa	Udyog
Satna	Satna
Maharashtra	
Aurangabad	Aurangabad
Greater Bombay	Andheri, Bhandup, Bombay, Borivli, Jogeshwari, Kandivli, Kanjur, Madh Island, Mulund, Parel, Powai, Santacruz, Seepz, Wadala, Worli
Nagpur	Nagpur, Warora
Nashik	Ambad, Satpur
Pune	Chinchwad, Loni, Pimpri, Pune
Raigarh	Patalganga, Taloja
Satara	Khandala, Shirwal
Thane	Dombivali, Kolshet, Tarapur, Thane
Meghalaya	
East Khasi Hills	Barapani
Orissa	
Puri	Bhubaneshwar
Sambalpur	Bargarh, Hirakud
Punjab	
Amritsar	Goindwal Sahib
Hoshiarpur	Hoshiarpur
Patiala	Patiala, Rajpura
Rupnagar	Karar, Mohali
Rajasthan	
Alwar	Alwar
Jaipur	Jhotwara

Industrial Locations (Contd.)

ELECTRICAL MACHINERY (Contd.)

STATE/DISTRICT	MANUFACTURING TOWNS
Kota	Cablenagar, Kota
Sirohi	Mount Abu
Tamil Nadu	
Chengalpattu	Ambattur, Porur, Pozhal
Dharmapuri	Hosur
Madras	Guindy, Madras
Pudukkottai	Kalathur, Mathur
South Arcot	Sevakuppam, Vadalur
Tiruchchirappalli	Tiruchchirappalli
Uttar Pradesh	
Agra	Agra
Allahabad	Naini
Bulandshahr	Sikandrabad
Dehra Dun	Dehra Dun
Garhwal	Garhwal
Ghaziabad	Dadri, Ghaziabad, Mankapur, Modinagar, Sahibabad
Jhansi	Jhansi
Kanpur	Kanpur
Lucknow	Lucknow
Mainpuri	Shikohabad
Mathura	Mathura
Meerut	Partapur
Naini Tal	Naini, Naini Tal
Rai Bareli	Rai Bareli
Saharanpur	Hardwar
West Bengal	
Barddhaman	Rupnarainpur
Calcutta	Calcutta, Cossipur, Dum Dum, Garden Reach, Narkeldanga, Salt Lake, Taratalla Road
Haora	Bauria
Medinipur	Haldia
Twentyfour Parganas	Budge Budge, Joka, Khardah, Shamnagar
Chandigarh	
Chandigarh	Chandigarh, Mohali
Delhi	
Delhi	Delhi, Kazkaji Industrial Estate

METAL PRODUCTS

STATE/DISTRICT	MANUFACTURING TOWNS
Andhra Pradesh	
Adilabad	Adilabad
Cuddapah	Cuddapah
Hyderabad	Hyderabad, Kukatpally, Lingampally, Secunderabad, Uppal Road
Khammam	Kojhagudam, Manuguru, Palonda
Krishna	Vijaywada
Medak	Medak, Pattancheru, Rudrarum, Zaheerabad
Nalgonda	Nagarjunasagar
Nellore	Sulurpet
Rangareddi	Medchal
Srikakulam	Srikakulam
Vishakhapatnam	Malkapuram, Vishakhapatnam
Vizianagaram	Shreeramnagar
Warangal	Warangal
Bihar	
Dhanbad	Bokaro, Dhanbad, Kumardhubi, Tundoo
Hazaribag	Bokaro, Patratu
Muzaffarpur	Muzaffarpur
Patna	Mokameh
Ranchi	Ranchi
Samastipur	Muktapur
Singhbhum	Adityapur, Chandil, Ghatsila, Golmuri, Indranagar, Jamshedpur
Goa	
Goa	Shiroda
Gujarat	
Ahmadabad	Ahmadabad, Dehagam, Kali, Kankaria, Lambha, Sardarnagar, Vatawa, Vatva
Bharuch	Ankleshwar, Bharuch, Dahej, Narmadanagar, Panoli
Bhavnagar	Bhavnagar
Kheda	Kanjeri, Karamsad

METAL PRODUCTS (Contd.)

STATE/DISTRICT	MANUFACTURING TOWNS
Mahesana	Mahesana
Panch Mahals	Kalol
Surat	Hazira, Mora Choryasi, Surat, Udhna
Surendranagar	Limbdi
The Dangs	The Dangs
Vadodara	Baroda, Chhani, Vadodara
Haryana	
Ambala	Pinjore, Sahabadpur, Yamunanagar
Bhiwani	Bhiwani
Faridabad	Faridabad
Gurgaon	Amarnagar, Ballabgarh, Mujesar
Hisar	Hissar
Jind	Jind, Zafarbagh
Mahendragarh	Dharuhera, Mahendragarh, Rewari
Rohtak	Bahadurgarh
Sonipat	Kundli, Ganaur, Murthal
Himachal Pradesh	
Kullu	Kullu
Sirmaur	Rampur
Solan	Parwanoo
Karnataka	
Bangalore	Bangalore, Krishnarajapuram, Mahadevapupra, Rajajinagar, Yelahanka
Belgaum	Belgaum
Bellary	Kariganur, Hospet, Vijaynagar
Bidar	Bidar
Dakshin Kannad	Baikampady, Mangalore, New Mangalore
Dharwad	Dharwad, Karur, Hubli, Hurli, Sattar
Gulbarga	Gulbarga, Shahabad
Kolar	Bangarpeth, Kolar
Mysore	Belwadi, Mysore
Raichur	Taloja
Shimoga	Bhadravati, Shimoga
Kerala	
Ernakulam	Binanipuram, Ernakulam
Kozhikode	Nellalam Feroke
Quilon	Kundara
Trichur	Athani
Madhya Pradesh	
Bhopal	Bhopal
Dewas	Dewas
Dhar	Badnawar, Pithampur
Durg	Bhilai
Gwalior	Birlanagar
Indore	Indore
Morena	Morena
Raipur	Chalisgarh, Raipur
Ratlam	Dosigaon, Ratlam
Ujjain	Ujjain
Maharashtra	
Ahmadnagar	Ahmadnagar
Aurangabad	Aurangabad, Chikhalthana, Gangapur, Paithan
Bhandara	Tumsur
Chandrapur	Chanda Hill Road, Chandrapur
Greater Bombay	Andheri, Bhandup, Bombay, Borivli, Byculla, Chandivli, Chembur, Goregaon, Kalina, Kandivli Kanjur, Kurla, Lalbaug, Malad, Madh Island, Mahalaxmi, Mulund, Powai, Reay Road, Sion, Sewree, Trombay, Vidya Vihar, Vikhroli
Jalgaon	Jalgaon
Kolhapur	Kolhapur, Shiroli
Nagpur	Bagadganj, Hingna, Khandelwalnagar, Nagpur
Nanded	Nanded
Nashik	Nashik, Satpur
Pune	Akurdi, Chinchwad, Dapodi, Lonawala, Khadki, Kothrud, Mundhwa, Pimprai, Pune, Saswad, Walchandnagar

METAL PRODUCTS (Contd.)

STATE/DISTRICT	MANUFACTURING TOWNS
Raigarh	Alibagh, Khopoli, Kolhapur, Raigad, Roha, Taloja
Ratnagiri	Kasarde, Kudal Pinghli, Ratnagiri
Sangli	Kirleskarwadi
Satara	karad, Satara
Solapur	Tikekarwadi.
Thane	Ambernath, Boisar, Dhige, Dombivali, Kalwe, Kolshet, Majiwada, Manpada, New Bombay, Thane, Tarapur, Ulhasnagar
Orissa	
Baleshwar	Randia
Cuttack	Chowdwar, Jaipur Road, Kyjang
Dhenkanal	Angul, Dhenkanal, Talcher
Kendujhar	Bamaripali, Bandnipal, Bilaspada, Barbil, Daitari, Keonjhar
Koraput	Therubali
Mayurbhanj	Rairangpur
Puri	Bhubaneshwar
Sambalpur	Hirakud
Sundargarh	Kalinga
Punjab	
Amritsar	Amritsar
Firozpur	Abohar
Hoshiarpur	Hoshiarpur, Chohal
Ludhiana	Kalan, Ludhiana
Patiala	Patiala, Bahadurgarh
Rupnagar	Asron, Mohali, Ropar
Sangrur	Jitwal Kalan, Malerkotla, Sangrur
Rajasthan	
Ajmer	Ajmer
Alwar	Alwar
Bharatpur	Bharatpur
Chittaurgarh	Parsoli
Jaipur	Jaipur
Jalor	Jolar
Jhunjhunun	Khetrinagar
Jodhpur	Jodhpur
Kota	Kota
Sikar	Sikar
Tonk	Rewari
Udaipur	Debari
Tamil Nadu	
Chengalpattu	Ambattur, Elavur, Ennore, Gummidipoondi, Padi, Pallavaram, Thoraipakkam, Tiruvottiyur, Trininravur, Trivellore
Coimbatore	Coimbatore, Ganapathy, Kaniyur, Singanallur, Tiruchi, Perianaikenpalayan
Dharmapuri	Hosur
Madras	Guindy, Madras, Nandambakkam, Perambur, Perangudi, Saidapet
Madurai	Madurai
North Arcot	Arkonam, Ranipet
Pudukkottai	Kalathur
Ramanathapuram	Krishnapuram, Sivakasi, Valinokkam
Salem	Mettur Dam, Salem
Thanjavur	Nagapatinam
Tiruchchirappalli	Tiruchchirappalli
Uttar Pradesh	
Allahabad	Allahabad, Amethi, Naini
Bulandshahr	Sikandrabad, Surajpur
Dehra Dun	Dehra Dun
Etah	Etah
Fatehpur	Fatehpur
Ghaziabad	Dadri, Ghaziabad, Meerut Road, Sahibabad, Modinagar
Kaunpur	Kanpur
Mathura	Mathura
Mirzapur	Renukoot
Rampur	Rampur
Unnao	Sonik

Industrial Locations (Contd.)

METAL PRODUCTS (Contd.)

STATE/DISTRICT	MANUFACTURING TOWNS
West Bengal	
Barddhaman	Burnpur, Durgapur, Kulti
Birbhum	Suri
Calcutta	Ballygunge, Calcutta, Kamarhatti, Kidderpore, Woodland Road
Haora	Andal Road, Belpur, Belur, Dankuni, Howrah, Liluah,
Hugli	Angus & Adisaptagram, Bamunari, Dhankuni, Hugli, Konnagar, Rajyadharpur, Rishra, Sahaganj, Serampore, Sibpore
Medinipur	Haldia
Puruliya	Tamna
Twenty Four Parganas	Agarpara, Belgharia, Doharia, Dum Dum, Khardah, Panihati, Salt Lake, Sakchal, Samnagar, Sodepore Sukchur, Talkhara
Chandigarh	
Chandigarh	Chandigarh
Delhi	
Delhi	Delhi, Shahdra

NON-ELECTRICAL MACHINERY

STATE/DISTRICT	MANUFACTURING TOWNS
Andhra Pradesh	
Anantapur	Anantapur
Medak	Pattancheru
Goa	
Goa	Panaji
Gujarat	
Ahmadabad	Naroda
Kheda	Vidyanagar
Haryana	
Gurgaon	Dharuhera
Jammu & Kashmir	
Badgam	Badgam
Srinagar	Srinagar
Karnataka	
Bangalore	Bangalore, Rajajinagar
Dharwad	Attikola
Tumkur	Tumkur
Kerala	
Cannanore	Cannanore
Maharashtra	
Greater Bombay	Lalbaug
Pune	Pune
Thane	Mumbra, Thane
Punjab	
Rupnagar	Rupnagar
Rajasthan	
Ajmer	Ajmer
Tamil Nadu	
Madras	Perungudi
Madurai	Kodaikanal
Uttar Pradesh	
Dehra Dun	Dehra Dun
Ghaziabad	Sahibabad
Naini Tal	Ranibagh
West Bengal	
Haora	Haora
Andaman & Nicobar	
Andaman	Andaman

NON-METALLIC PRODUCTS

STATE/DISTRICT	MANUFACTURING TOWNS
Andhra Pradesh	
Adilabad	Adilabad, Asifabad, Devapur, Mancherial
Anantapur	Rayalacheruvu
Chittoor	Tirupati
Cuddapah	Kalamalla, Yerraguntala
Guntur	Durgapuram, Kistna, Macherla, Petasannigamdle
Hyderabad	Hyderabad
Karimnagar	Basantnagar
Krishna	Jaggayapeta, Vijaywada
Kurnool	Ankereddipelli, Bugganipalli
Medak	Medak

NON-METALLIC PRODUCTS (Contd.)

STATE/DISTRICT	MANUFACTURING TOWNS
Nalgonda	Bibinagar, Huzurnagar, Mahankaligudem, Mathampally, Mellachauvu, Miryalaguda, Rampurna, Wadapalli
Rangareddi	Tandur
Srikakulam	Takkali
Vishakhapatnam	Agnampudi, Porlupalem, Vishakhapatnam
Assam	
Goalpara	Golapara
Karbi Anglong	Bokajan
North Cachar Hills	Umrangshu
Bihar	
Bhojpur	Dumraon
Dhanbad	Dhanbad, Jharia, Kumardhubi, Sindri
Giridih	Bhandaridah
Hazaribag	Berdi-Begda, Bhurukunda, Jhumritalaiya, Marar
Palamu	Jagannathpur, Japle
Ranchi	Khalari
Rohtas	Banjari, Dalmianagar, Muri
Santhal Pargana	Jasidih
Singhbhum	Chaibasa
Gujarat	
Ahmadabad	Digvijaynagar, Rakhial
Amreli	Jafarabad
Banas Kantha	Banas Kantha
Bharuch	Ankleshwar
Bhavnagar	Mahuva
Jamnagar	Digvijaygram, Dwarka, Jamnagar, Sikka
Junagadh	Junagadh, Porbandar, Ranavar, Veraval
Kheda	Mehmedabad, Nadiad, Vallabh Vidyanagar
Rajkot	Morvi, Junagadh Distt., Parapipaliya, Wankaner
Sabar Kantha	Poshina
Surat	Magdalla
Surendranagar	Dhrangadhra, Surendranagar
Vadodara	Nandesari
Valsad	Dharmapur
Haryana	
Ambala	Bhupendra
Bhiwani	Charki Dadri
Gurgaon	Ballabgarh
Jind	Jind
Mahendragarh	Rewari
Rohtak	Bahadurgarh, Kassar
Sonipat	Selvi
Himachal Pradesh	
Chamba	Bina
Sirmaur	Rajban, Rajgarh, Paonte
Jammu & Kashmir	
Anantnag	Anantnag
Baramula	Baramula
Karnataka	
Bangalore	Bangalore, Malleswaram, Whitefield
Belgaum	Gokak
Bellary	Hospet
Bijapur	Bagalkot, Bijapur
Dharwad	Karur
Gulbarga	Chittapur, Kurkunta, Gulbarga, Malkhed, Shahabad, Wadi
Hassan	Hospet
Tumkur	Adiyapatna, Gubbi
Kerala	
Kottayam	Nattakam
Madhya Pradesh	
Damoh	Amlai, Narsingarh
Dewas	Amona, Nagda
Durg	Bhilai, Durg, Jamul
Jabalpur	Jabalpur, Kymore, Mehgaon
Mandsaur	Jawad, Neemuch
Raipur	Baikunth, Baloda, Chandi, Mandhar

NON-METALLIC PRODUCTS (Contd.)

STATE/DISTRICT	MANUFACTURING TOWNS
Rewa	Naubasta
Satna	Maihar, Raghurajnagar, Satna
Sidhi	Gopal Banas, Sidhi
Maharashtra	
Aurangabad	Chikhalthana
Chandrapur	Awarpur, Chanda, Chandrapur, Manikgarh, Rajura
Greater Bombay	Andheri, Bhandup, Bombay, Borivli, Haybunder, Kandivali, Kurla & Mazagaum, Marol Naka, Parel, Wadala
Nagpur	Nagpur
Pune	Pune
Raigarh	Mora, Taloja
Ratnagiri	Ratnagiri
Thane	Kalwe, Thane, Tarapur
Yavatmal	Chanaka (NU)
Orissa	
Cuttack	Barang, Cuttuck
Dhenkanal	Dhankanal, Talcher
Koraput	Malkangiri, Sunk
Sambalpur	Bargarh, Belpahar, Hirakud
Sundargarh	Latkota, Rajgangpur, Rourkela, Sundargarh
Punjab	
Bathinda	Bhatinda
Hoshiarpur	Hoshiarpur
Rajasthan	
Ajmer	Beawar
Alwar	Behilol
Bharatpur	Dholpur
Bhilwara	Bhilwara
Bundi	Lakheri
Chittaurgarh	Chittaugarh, Nimbahera, Parsoli, Shambhupura
Kota	Adityapur, (Kotah), Ladoura, Ramgandmani
Nagaur	Gotan
Sawai Madhopur	Sawai Madhopur
Sikar	Neem-ka-Thana
Sirohi	Basantgarh, Pindwara, Sirohi
Udaipur	Udaipur
Tamil Nadu	
Chengalpattu	Ambattur, Tiruvottiyur
Coimbatore	Coimbatore, Madukkarai, Mettupalayam
Dharmapuri	Hosur
Madras	Madras
Madurai	Vedasandur
North Arcot	Arkonam, Ranipet
Ramanathapuram	Alangulam, Ramasamyarjaenagar
Salem	Salem, Sankari
South Arcot	Vadalur
Tiruchchirappalli	Ariyalur, Dalmiapuram, Mayanur, Paliyur Village
Tirunelveli	Sankarnagar
Uttar Pradesh	
Allahabad	Iradatganj, Naini
Banda	Banda
Bareilly	Bareilly
Bijnor	Dhampur
Dehra Dun	Kalsi & Bharuwala, Rishikesh
Ghaziabad	Kanpur Dehat, Shahibabad
Mathura	Koshi
Pithoragarh	Devalthal
West Bengal	
Barddhaman	Burnpur, Durgapur, Kulti
Calcutta	Alambazar, Ariada, Calcutta, Salt Lake, Taratala
Hugli	Bansberia, Dhenkuni, Konnagar, Rishra
Puruliya	Madhukunda
Twenty four Parganas	Budge Budge, Sodepur
Delhi	
Delhi	Delhi
Pondicherry	
Pondicherry	Pondicherry
Yanam	Yanam

Industrial Locations (Contd.)

STATE/DISTRICT	MANUFACTURING TOWNS
PAPER PRODUCTS	
Andhra Pradesh	
Adilabad	Adilabad
East Godavari	Kakinada, Rajahmundry
Hyderabad	Hyderabad
Khammam	Sarapaka
Kurnool	Vasant Nagar
Medak	Pattancheru
Rangareddi	
Warangal	Kamalapuram
West Godavari	Vendra
Assam	
Cachar	Panchgram
Goalpara	Jogighopa
Nowgong	Jogi Road
Bihar	
Darbhanga	Rameshwarnagar
Pashchim Champaran	Betiah, Kumarbagh
Rohtas	Dalmianagar
Saharsa	Baijnathpur
Goa	
Goa	Goa
Gujarat	
Ahmadabad	Ahmedabad
Bharuch	Ankleshwar, Palej
Surat	Gunasada, Utran
Valsad	Billimora, Khadki, Vapi
Haryana	
Ambala	Jagadhri, Yamunanagar
Faridabad	Faridabad
Mahendragarh	Sehgalnagar
Sonipat	Kundli
Karnataka	
Bangalore	Bangalore, Nelamangale
Mandya	Belagula
Mysore	Nanjangud
Shimoga	Bhadravati
Uttar Kannad	Dandeli
Kerala	
Kottayam	Kottayam, Velloor
Quilon	Punalur
Madhya Pradesh	
Bhopal	Bhopal
East Nimar	Burhanpur, Nepa Nagar
Shahdol	Amlai
Maharashtra	
Aurangabad	Paithan
Bhandara	Gondia
Chandrapur	Ballarpur
Dhule	Shahada
Greater Bombay	Bhandup, Bombay, Powai & Madh Island, Prabhadevi
Jalgaon	Bhusawal
Nagpur	Nagpur
Nanded	Nanded
Nashik	Nashik
Raigarh	Khopoli, Roha
Satara	Khandala
Thane	Thane
Nagaland	
Mokokchung	Tuli
Orissa	
Cuttack	Chowdar
Kalahandi	Kesinga
Koraput	Jaykaypur
Sambalpur	Brajrajnagar
Punjab	
Hoshiarpur	Banah, Hoshiarpur
Sangrur	Sangrur
Tamil Nadu	
Chengalpattu	Tiruvottiyur
Coimbatore	Coimbatore, Sirumugal
Madras	Madras
Madurai	Madurai

STATE/DISTRICT	MANUFACTURING TOWNS
PAPER PRODUCTS (Contd.)	
Ramanathapuram	Virudhunagar
Salem	Pallipalayam
Uttar Pradesh	
Almora	Lal Kuan
Bijnor	Dhampur
Ghaziabad	Ghaziabad
Moradabad	Gajaraula
Naini Tal	Naini Tal
Saharanpur	Saharanpur
West Bengal	
Barddhaman	Raniganj
Calcutta	Alambazar, Calcutta
Hugli	Chandrahati
Twenty Four Parganas	Kankinara, Kossipur, Titaghur
Delhi	
Delhi	Delhi
Pondicherry	
Pondicherry	Pillarkuppam
TRANSPORT EQUIPMENT	
Andhra Pradesh	
Anantapur	Anantapur
Hyderabad	Hyderabad
Medak	Pattancheru, Zaheerabad
Vishakhapatnam	Gandhigram, Vishakhapatnam
Bihar	
Munger	Jamalpur
Muzaffarpur	Muzaffarpur
Patna	Mokameh
Ranchi	Ranchi
Singhbhum	Jamshedpur
Goa	
Goa	Goa, Sannoale, Tolani, Vasco-da-Gama
Gujarat	
Ahmadabad	Maninagar, Odhav
Panch Mahals	Kalol
Vadodara	Vadodara
Haryana	
Ambala	Pinjore
Gurgaon	Ballabgarh, Dharuhera, Gurgaon, Seekri
Hisar	Hissar
Jind	Jind
Mahendragarh	Rewari
Sonipat	Sonipat
Himachal Pradesh	
Solan	Parwanoo, Solan
Karnataka	
Bangalore	Adugodi, Bangalore, Doddaballapur
Bidar	Bidar
Mandya	Maddur
Mysore	Belawadi, Mysore
Kerala	
Ernakulam	Cochin
Trichur	Perambra
Madhya Pradesh	
Dewas	Dewas
Dhar	Dhar, Pithampur
Indore	Pithampur
Maharashtra	
Ahmadnagar	Ahmadnagar, Kedgaon.
Aurangabad	Aurangabad, Chikhalthana, Valuj
Bhandara	Bhandara, Sakoli.
Greater Bombay	Andheri, Bhandup, Bombay, Ghatkopar, Kurla, Mazagon, Wadala
Nashik	Madhasanghvi, Nashik, Satpur
Pune	Akurdi, Haveli, Pune, Urwade.
Raigarh	Karjat, Nhava-Sheva
Satara	Satara, Satara Road
Thane	Dombivali, Kausa, Manpada, Tarapur

STATE/DISTRICT	MANUFACTURING TOWNS
TRANSPORT EQUIPMENT (Contd.)	
Orissa	
Koraput	Koraput.
Punjab	
Hoshiarpur	Hoshiarpur
Ludhiana	Dhanderi, Ludhiana
Rupnagar	Mohali, Ropar
Rajasthan	
Alwar	Alwar
Bharatpur	Bharatpur.
Tamil Nadu	
Chengalpattu	Ambattur & Avadi, Ennore, Padi, Perunguluthur
Dharmapuri	Hosur
Madras	Madras, Perambur, Thiriperumbuddur
North Arcot	Ranipet.
Pudukkottai	Pudukkottai.
Tiruchchirappalli	Ponamali, Tiruchchirappalli
Uttar Pradesh	
Bulandshahr	Sikandrabad.
Deoria	Deoria.
Etah	Etah
Ghaziabad	Dadri, Ghaziabad, Sahibabad, Surajpur
Kanpur	Kanpur
Lucknow	Goela, Lucknow, Sarojininagar
Meerut	Meerut, Modipuram
Moradabad	Gajaraula
Pratapgarh	Pratapgarh
Rae Bareli	Salsa
Sitapur	Sitapur.
Sultanpur	Jagdishpur, Korwa
Varanasi	Varanasi.
West Bengal	
Barddhaman	Asansol, Burnpur, Chittaranjan, Durgapur, Raniganj
Calcutta	Calcutta, Kamarhutty
Darjiling	Siliguri
Haora	Angus, Sahaganj, Uttarpara
Medinipur	Kharagpur
Twentyfour Parganas	Garden Reach, Kanyanagar
WOOD PRODUCTS	
Arunachal Pradesh	
Dibang Valley	Bizari
Tirap	Jairampur, Namchik
West Kameng	Bhalak Pong
Assam	
Sibsagar	Sibsagar
Gujarat	
Ahmadabad	Ahmadabad
Panch Mahals	Kalol
Kerala	
Cannanore	Baliapatam
Madhya Pradesh	
Betul	Betul
Bhopal	Bhopal
Maharashtra	
Greater Bombay	Andheri Kurla Road, Bhandup
Nashik	Ambad
Thane	Tarapur
Orissa	
Koraput	Koraput
Phulabani	Phulabani
Tamil Nadu	
Madras	Madras
West Bengal	
Calcutta	Calcutta
Twenty-four Parganas	Joka
Andaman & Nicobar	
Andaman	Andaman, Port blair

INDEX

Dalu

Gunupur

Dalu (Meg) 94 B2
Daman (D&D) 149 C2, 149 D2
Daman-i-koh Plain (J&K) 58 B3
Damanganga (Guj) 47 C3
Dambal (Kar) 65 B3
Dambo-Rongjeng (Meg) 94 B2
Dambuk (Ar.P) 24 C1
Dambuk-aga (Meg) 94 B2
Damchera (Tri) 128 B1
Damdama Lake (Har) 52 D4
Damnagar (Guj) 45 B3
Damodar (W.B) 139 B3
Damodar (Ori) 100 DE1
Damodar (Bih) 31 B3, 37 B4
Damodar Valley (W.B) 139 B3
Damodar Valley (Bih) 31 BC3
Damoh (M.P) 77-79 E3
Damoh Plateau (M.P) 76 E3
Danda (U.P) 135 D4
Dandakaranya (M.P) 76 F5
Dandarmal (Guj) 43 C1
Dandavathi (Kar) 67 B3
Dandeli (Kar) 65, 68 B3
Dandeli W.L.S. (Kar) 66 B3
Dangri (Har) 50, 52 C2
Dankhar Gompa (H.P) 55 D2
Dankuni (W.B) 145 B2
Danta (Raj) 113 D2
Danta (Guj) 44 C1
Danta Hills (Guj) 43 C1
Dantan (W.B) 140 C7
Dantewara (M.P) 77, 80 F5
Dantewara Plateau (M.P) 76 F5
Danwan (Bih) 34 B4
Dapodi (Mah) 89 B2
Dapoli (Mah) 85 B3
Daporijo (Ar.P) 24 C1
Darauli (Bih) 32 B3
Daraundha (Bih) 32 B3
Darbandore (Goa) 41 B1
Darbhanga (Bih) 32, 34 C3, 36 B2, 40 B1
Dared (Guj) 47 B2
Dari (Guj) 48 C3
Dariapur (Bih) 32 C4
Daring (Ar.P) 25 C2
Darjeeling Hills (W.B) 139 C1
Darjiling (W.B) 140 D1, 142, 143, 145, 146 C1
Darlawn (Miz) 96 A1
Darna (Mah) 88 C2
Daroli (M.P) 80 E3
Darrah W.L.S. (Raj) 114 D3
Darranga (Ass) 26 BC2, 28 B2, 29 B1
Darsi (A.P) 17 C3
Darugiri (Meg) 94 B2
Darwa (Mah) 87 D1
Darwha (Mah) 85, 86 D1
Daryabad (U.P) 132 C4
Daryapur (Mah) 85, 86 D1
Dasada (Guj) 44, 46 B2
Dashapalla (Ori) 101 C2
Dasiari (Bih) 34 D3
Dasna (U.P) 132, 136 A3
Daspalla (Ori) 103 C2
Daspur (W.B) 140 C6
Dasua (Pun) 107 C2, 108 C3
Dasuya (Pun) 109 C2
Dataganj (U.P) 131, 132 B3
Datia (M.P) 77-79 E2
Dattapukar (W.B) 142 C3
Dattapur (Mah) 86 D1
Dattor Branch (Raj) 111 C1
Daudnagar (Bih) 32, 34 B4, 36 B3
Dauki (Meg) 94 D2
Daulatabad (Mah) 87 C2
Daund (Mah) 85, 86 C2
Dausa (Raj) 112-114 E2
Davangere (Kar) 64-66, 68-70 B3
Dawath (Bih) 32 B4
Daya (Ori) 100 C2, 104 C3-2
Dayapar (Guj) 44 A2
Debagram (W.B) 142 C3
Debari (U.P) 132 B3
Debari (Raj) 116 C3
Debi (Ori) 100, 104 D2
Debra (W.B) 140 C6
Dechua (Ori) 103 B2
Dedhapada (Guj) 44 C3
Dedipada (Guj) 44, 46 C3
Degana (Raj) 112, 114 D2
Deganga (W.B) 140 D6
Deh (Raj) 114 C2
Deh (Miz) 96 A2
Dehra Dun (U.P) 131, 132, 134 B2, 136, 137 A1
Dehri (Bih) 32, 34 B5, 36 B3
Dehri on Son (Bih) 39 B3
Dehti (Bih) 34 E3
Dehu (Mah) 87 B2
Deihi (Mah) 88 BC1
Dekadong (Ass) 28 B2
Delhi (Del) 150 C3
Delhi Cantonment (Del) 150 C3
Delhi Branch (Har) 50 CD4-3
Delhi Ridge (Har) 50 CD4
Delvada (Guj) 45 B3
Demi I (Guj) 47 B2
Demi II (Guj) 47 B2
Denkada (A.P.) 20 E1
Denkanikota (T.N) 121, 122 A1
Deo (Tri) 128 B1, 129 B2
Deo (Bih) 32, 34 B5
Deo Tibba (H.P) 54 C2
Deoband (U.P) 131, 132, 136 A3
Deogad (Mah) 88 B3
Deogaon (Mah) 88 D1
Deogarh (Ori) 101-103 C2
Deogarh Hills (M.P) 76 G3
Deoghar (Bih) 34 D5, 36 C3
Deoghar Upland (Bih) 31 C2
Deola (W.B) 139 C1
Deolali (Mah) 87 B2
Deolali Pravra (Mah) 86 C2
Deoli (Mah) 86 E1
Deorajan Nala (Mah) 88 D2
Deorha (H.P.) 54 C3
Deori (M.P) 78 E3, 81 F3
Deori (Bih) 32 D5
Deori (Mah) 87 D1
Deoria (Raj) 115 D3
Deoria (U.P) 131, 132, 134, 136-138 D4
Deoria (Bih) 34 C3
Deosai Basin (J&K) 58 B2
Depalpur (M.P) 77 C3
Depsang Basin (J&K) 58 C2
Dera Baba Jaimal Singh (Pun) 109 C2
Dera Baba Nanak (Pun) 109 C1
Dera Bassi (Pun) 109 D3, 110 D4
Dera Gopipur (H.P) 54, 55 B3
Derapur (U.P) 131 B4
Dergaon (Ass) 26 C2, 27 E2
Deshbandhu (Bih) 37 C4
Deshgaon (M.P) 79 D4
Deshnok (Raj) 113, 114 C2
Desuri (Raj) 112, 114 C3
Deulgaon Raja (Mah) 79 D1
Deulpur (W.B) 142 C3
Devadurga (Kar) 64, 65 C2
Devakottai (T.N) 121-123 B3
Devalmarhi Hills (Mah) 84 EF2
Devanhalli (Kar) 64, 65 C4
Devaprayag (U.P) 131, 134 B2
Devapur (A.P) 21 C1
Devar Hippargi (Kar) 66 C2
Devar Malai 1992 (Ker) 71 B3
Devarayadurga (Kar) 66 C4
Devarkonda (A.P) 17-19 C2
Devarkonda (Kar) 63 B4
Devarshola (T.N) 122 A2
Devendranagar (M.P) 80 E2
Devgad Baria (Guj) 48 D2
Devgadh Bariya (Guj) 44-46 C2
Devgarh (Raj) 112 C3
Devgarh (Mah) 85 B3
Devghar (Bih) 32 D5, 39 C3
Devikolam (Ker) 71, 74 B2
Devikot (Raj) 114 B2
Devikut (W.B) 143, C2
Devli (Raj) 112-114 D3
Dewal Thal (U.P) 136 C3
Dewas (M.P) 77-79, 81 D3
Dewas (Raj) 115 C3
Dhab (Bih) 39 B3
Dhabauli (Bih) 34 D4
Dhadhar (Guj) 43, 47 C2
Dhaka (Bih) 32, 34 C3, 37 B2
Dhakiar (J&K) 58 B3
Dhakuakhana (Ass) 26 D2
Dhalai (Ass) 26 C3
Dhalbhumgarh (Bih) 39 C4
Dhalbhumgarh (Bih) 32 D7
Dhaleswari (Miz) 96 A2
Dhaleswari (Ass) 26, 28 C3
Dhamangaon (Mah) 86 E1
Dhamauti Ramnath (Bih) 34 C3
Dhamdaha (Bih) 32, 34 E4
Dhamnagar (Ori) 101 D2
Dhamnod (M.P) 78 C4
Dhampur (U.P) 131, 132, 136 B3
Dhamra (Ori) 100 D2
Dhamtari (M.P) 77, 78 F4
Dhana (Har) 51 C3
Dhanai (Ori) 100 C3
Dhanana (Har) 51 C4
Dhanapur (U.P) 132 D5
Dhanarua (Bih) 32 C4
Dhanaula (Pun) 108 C4
Dhanaur (Bih) 34 C3
Dhanaura (U.P) 132 B3
Dhanauti Ramnath (Bih) 34 C3
Dhanbad (Bih) 32, 34 D6, 36, 38 C4, 40 C3
Dhanderi (Pun) 110 C4
Dhandhuka (Guj) 44-46 B2
Dhane (Ori) 104 C3
Dhanera (Guj) 44 C1
Dhaniakhali (W.B) 140 D6
Dhansiri (Ass) 26, 28 C2
Dhanwar (Bih) 32 C5
Dhaola Dhar (H.P) 54 D3
Dhar (H.P) 55 C3
Dhar (M.P) 77-79, 81 C3
Dharam Kot (Pun) 108 C4
Dharampur (Guj) 44, 45, 47 C3, 48 D3
Dharampur Plain (Bih) 31 C2
Dharangaon (Mah) 86 C1
Dharapuram (T.N) 121-123 A2
Dharchula (U.P) 131 C3
Dhargal (Goa) 41 A1
Dharhara (Bih) 32 D4
Dharhara (U.P) 132 D5
Dhari (Guj) 44, 45 B3, 47 B2
Dhariwal (Pun) 108, 109 C2, 110 C3
Dharma (Kar) 63, 67 B3
Dharma (Kar) 67 B3
Dharmabad (Mah) 86, 87 D2
Dharmanagar (Ori) 101 B3
Dharmanagar (Tri) 128 B1
Dharmapuri (T.N) 121-123, 125, 126 B1
Dharmastala (Kar) 66 B4
Dharmavaram (A.P) 17-20 B3
Dharmgarh (Ori) 101 B3
Dharmjaygarh (M.P) 77, 79 G3
Dharmpuri (M.P) 78 C3
Dharmsala (H.P) 54, 55 B2
Dharsul (Har) 52 B3
Dharuhera (Har) 52, 53 C4
Dharur (Mah) 86 D2
Dharwad (Kar) 64-66, 68, 70 B3
Dharwar Plateau (Kar) 63 B3
Dhasa (Guj) 45 B3
Dhasan (M.P) 76, 80 E2
Dhasan (U.P) 130 B5
Dhatarwadi (Guj) 47 B3
Dhauliganga (U.P) 135 B2
Dhaulpur (Raj) 112-114, 116, 117 E2
Dhaura Tanda (U.P) 132 B3
Dhaurahra (U.P) 131, 132 C4
Dhawa (Bih) 37 B3
Dhebar (Jai Samand) (Guj) 43 C1
Dhebar Lake (Jai Samand) (Raj) 111 C3
Dhekaha (Bih) 34 BC3
Dhekiajuli (Ass) 26 C2, 27 D2, 29 C1
Dheku (Mah) 88 C1
Dhemaji (Ass) 26 D2
Dhenkanal (Ori) 101-103, 105-106 C2
Dhige (Mah) 89 B2
Dhil Tank (Raj) 115 E3
Dhing (Ass) 26 C2, 27 D2
Dhinoj (Guj) 45 C2
Dhoki (Mah) 89 D2
Dhola (Guj) 46 B3
Dhola Range (Guj) 43 AB2
Dholbaha (Pun) 109 C1
Dholi (Bih) 32 C3-4
Dholka (Guj) 44, 45 C2
Dhom (Mah) 88 BC3
Dhone (A.P) 17, 18 B3
Dhoraji (Guj) 44, 45 B3, 48 C3
Dhorimanna (Raj) 114 B3
Dhrangadhra (Guj) 44-46 B2, 48 C2
Dhrol (Guj) 44, 45 B2
Dhuan (M.P) 80 F3
Dhuandhar (M.P) 80 F3
Dhuburi (Ass) 26-28 A2
Dhulagiri (W.B) 142 C3
Dhule (Mah) 85-87, 89, 91 C1
Dhulian (W.B) 142 B2
Dhulijan (Ass) 27 G1
Dhund (Raj) 111 E2
Dhupgari (W.B) 140 DE2, 142 C1
Dhuraiya (Bih) 32 E5
Dhuri (Pun) 108 C4, 109, C3
Dhurki (Bih) 32 A5
Diamond Harbour (W.B) 140 D6, 142 B3, 143 C3
Dibang or Sikang (Ar.P) 24 C1
Dibru (Ass) 26, 28 D2
Dibrugarh (Ass) 26-30 D2
Dibrugarh Plain (Ass) 26 D2
Didihat (U.P) 131 C3
Didwan Basin (Raj) 111 D2
Didwana (Raj) 112-114, 116 D2
Dig (Raj) 112-114 E2
Digapahandi (Ori) 101, 102 C3
Digboi (Ass) 26, 28 D2, 27 G1, 29 D1
Digha (W.B) 140 C7, 143 B4
Dighal (Har) 51 C4
Dighalbank (Bih) 32 E3
Dighi Kalan (Bih) 34 C4
Dighwa (Bih) 34 B3
Dighwara (Bih) 32, 34 C4
Diglipor (A&N) 148 B1
Diglur (Mah) 85-87 D2
Dignala (W.B) 142 B3
Digod (Raj) 112 E3
Digras (Mah) 86 D1
Dihang or Siang (Ar.P) 24 C1
Dikhu (Nag) 98, 99 B2
Diksanga (Kar) 67 C2
Dimapur (Nag) 98, 99 A3
Dimau (Ass) 26 D2
Dimbhe Kukadi (Mah) 88 B2
Dimjor la (J&K) 58 D2
Dina Nadi (Mah) 88 EF2
Dina Nagar (Pun) 108 C2
Dinara (Bih) 32 B4
Dindi (A.P) 16 C2
Dindigul (T.N) 121, 122, 125 A2
Dindori (M.P) 77-79 F3
Dindori (Mah) 85, 87 B1
Dinhata (W.B) 140 E2, 142 C1
Dip (Bih) 34 D3
Diphu (Ass) 26, 28, 30 C3, 27 E3
Dirang (Ar.P) 25 B2
Disa (Guj) 44-46 C1
Dispur (Ass) 26, 28 B2
Disteghil Sar (J&K) 58 B1
Ditari (Ori) 102 C2
Diu (D&D) 149 C2
Diu I (Guj) 43 B3
Divi (A.P) 17 D2
Diyodar (Guj) 44 B1
Diyung (Nag) 98, 99 B2
Dobbi (Bih) 36 B3
Dod Ballapur (Kar) 64-66, 68 C4
Doda (J&K) 59, 60 B3
Doda Betta (T.N) 120 A2
Dogachhi (W.B) 142 C2-3
Dogal (Pun) 109 D4
Doharia (W.B) 145 C3
Dohrighat (U.P) 134 D4
Dohrighat Pumped Canal (U.P) 135 D4
Dolungmukh (Ar.P) 25 C2
Domakonda (A.P) 17 C1
Domariaganj (U.P) 131, 134 D4
Domchanch (Bih) 34 C5, 39 B3
Domkal (W.B) 140 D4
Don (Guj) 47 A2
Don (Kar) 63, 67 C2, 67 B2
Don Paol Beach (Goa) 41 A1
Dondaicha (Mah) 86, 87 C1
Dongargaon (Mah) 88 D2
Dongargarh (M.P) 78 F4
Doranala (A.P) 19 C3
Dornakal (A P) 18, 19 D2
Doru (J&K) 59 B3
Doswada (Guj) 47 C3
Dowlaiswaram (A.P) 18 D2
Draksharamam (A.P) 19 E2
Dras (J&K) 60, 62 B2
Dras Basin (J&K) 58 B2
Dronachellam (A.P) 19 B3
Dubak (A.P) 17 C1
Dubrajpur (W.B) 140 C5, 142, 143 B3
Dudada (A.P) 19 C2
Dudhani (Mah) 86 D3
Dudhawa (M.P) 76 FG4, 80 F4
Dudhawa (Ori) 100 A2
Dudhganga (Mah) 88 BC3
Dudhinagar (U.P) 131 D5
Dudhnai (Ass) 26 B3
Dudhnai (Meg) 94 B2, 95 B2
Dudhsagar Falls (Goa) 41 B2
Dudhwa (U.P) 134 C3
Dudhwa (Mah) 84, 88 F1
Dudna (Mah) 84, 88 D2
Dudu (Raj) 112, 114 D2
Dukhri Kheda Tank (M.P) 80 E3
Dullabchara (Ass) 29 C2
Dum Dum (W.B) 140 D6, 143 C3, 145 C3
Dum Duma (Ass) 26 D2, 27 G1, 29 D1
Dumar Bahal (Ori) 104 B2
Dumaria (Bih) 32 B5, 32 D7
Dumas (Guj) 46 C3
Dumbur (Tri) 128 A2
Dumbur Hydel Project (Tri) 128 A2
Dumjor (W.B) 140 D6, 142 C3
Dumka (Bih) 32, 34 E5, 36 C3
Dumka Upland (Bih) 31 C2
Dumraon (Bih) 32, 34 B4, 38 B3
Dumri (Bih) 32 B6, 32 D6, 34 C4, 34 B2, 36 C4
Dunagiri (U.P) 130 B2
Duncan Passage (A&N) 148 B2
Dunda (U.P) 131 B2
Dungargarh (Raj) 113 D1
Dungla (Raj) 112 D3
Duogarh (M.P) 76 G3
Durg (M.P) 77, 79, 81, 82 F4
Durg-Bhilai Nagar (M.P) 78 F4
Durgachak (W.B) 140 D6
Durgapur (W.B) 140 C5, 142 B3
Durgauti (Bih) 32 A4
Dutang (H.P) 54 D2
Dwarka (Guj) 49 B2
Dwaraka Coast (Guj) 43 A2-3
Dwarband (Ass) 28 C3
Dwarka (Guj) 44-46 A2, 48 B2
Dwarkeswar (W.B) 139 B3
Dzongri (Sik) 119 A2
East Bhagirathi Plain (W.B) 139 C2-3
East Garo Hills (Meg) 94 B2
East Godavari (A.P) 17 DE2
East Hope Town (U.P) 132 A2
East Kameng (Ar.P) 24 B2
East Kathiawar Plains (Guj) 43 BC2
East Khasi Hills (Meg) 94 CD2
East Ladak Plateau (J&K) 58 D2
East Manipur Hills (Man) 92 A2-3
East Nimar (M.P) 77 D4
East Siang (Ar.P) 24 C1
East Sikkim (Sik) 118 B2
Eastern Ghats (A.P) 16 DE2
Eastern Ghats (Mahendragiri South) (A.P) 16 C3-4
Eastern Ghats (North) (Ori) 100 BC3
Eastern Himalaya (Sik) 118 AB2-3
Eastern Ramganga Dam (U.P) 135 C3
Eastern Siwalik (W.B) 139 C1
Edakkad (Ker) 72 A2
Edalabad (Mah) 85-87 D1
Edalwada (Guj) 47 D2
Edapally (Ker) 74 B2
Egra (W.B) 140 C7, 143 B4
Eka (U.P) 132 B4
Ekangarsarai (Bih) 32 C4
Ekburjee (Mah) 88 D1
Eklingji (Raj) 114 C3
Ekma (Bih) 32 B4
Ekrambe (Kar) 65 B2
Elagiri (T.N) 120 B1
Elamalai (Ker) 71 B2-3
Elamanchili (A.P) 17, 18 E2
Elavur (T.N) 125 C1
Elayavoor (Ker) 72 A2
Elephanta (Mah) 87 B2
Ellenabad (Har) 51, 52 A3
Ellora (Mah) 87 C1
Ellora Hills (Mah) 84 C1
Eloor (Ker) 74 B2
Eluru (A.P) 17-19 D2
Emani (A.P) 17 D2
Emmiganur (Kar) 65 C3
Emmiganuru (A.P) 17, 18 B3
Ennore (T.N) 125 C1
Eral (T.N) 122 B3
Erandol (Mah) 85-87 C1
Erangal (Mah) 87 B2
Eranholi (Ker) 72 A2
Eravipuram (Ker) 72 B3
Erinpura Hills (Raj) 111 C3
Erki (Tamar II) (Bih) 32 C7
Ernad (Manjeri) 71 B2
Ernakulam (Ker) 71, 73, 74 B3
Erode (T.N) 121-123 A2
Erragondapalem (A.P) 17 C2
Erraguntla (A.P) 18, 21 C3
Erramala (A.P) 16 BC3
Eruvadi (T.N) 122 A3
Etah (U.P) 131, 132, 136 B4
Etawah (U.P) 131, 132, 134, 138 B4
Etawah Branch (U.P) 130 B4
Ethipothala (A.P) 19 C2
Etmadpur (U.P) 131, 132 B4
Ettaiyapuram (T.N) 122 A3, 123 B3
Ettumanur (Ker) 73 B3
Eturnagaram (A.P) 17, 19 C1
Fagu (H.P) 55 C3
Faizabad (U.P) 131, 132, 134 D4, 137, 138 D3
Faizpur (Mah) 86 C1
Falakata (W.B) 140 E2, 142 C1
Falka (Bih) 32 E4
False Point (Ori) 100 D2
Fankusi (Tri) 128 B1
Farakka (W.B) 140 C4, 142 B2, 144 BC2
Fardapur (Mah) 87 C1
Faridabad (Har) 50-53 D4
Faridkot (Pun) 107, 109 B3, 108 B4
Faridnagar (U.P) 132 A3
Faridpur (U.P) 131, 132 B3
Faridpur (W.B) 140 C5
Farooqnagar (A.P) 18 C2
Farrukhabad (U.P) 131, 138 B4
Farrukhabad-cum-Fatehgarh (U.P) 132 B4
Farukhnagar (A.P) 19 C2
Fatahpur (U.P) 134 A2
Fatarpa (Goa) 41 A2
Fategadh (Guj) 47 B2
Fateh Nehar (A.P) 20 C2
Fatehabad (Har) 50-53 B3
Fatehabad (U.P) 131, 132 B4
Fatehgarh (U.P) 131, 134 B4
Fatehgarh (Pun) 107, 109 D3
Fatehgarh Churian (Pun) 108 B3
Fatehnagar (Raj) 113 D3
Fatehpur (U.P) 131, 132 C4
Fatehpur (Bih) 32 C5
Fatehpur (Raj) 112 D2, 113 D1-2
Fatehpur Sikri (U.P) 132, 134 A4
Fatwah (U.P) 32, 34 C4
Fazilka (Pun) 107, 109 B3, 108 B4
Fatehpur (U.P) 132 C5
Ferrargunj (A&N) 148 B2
Fertilisernagar (Guj) 48 D2
Firozabad (U.P) 131, 132 B4
Firozpur (Pun) 107, 109, B3, 108 B4
Firozpur Head Works (Pun) 109 B2
Firozpur Jhirka (Har) 50, 51 C5
Forbesganj (Bih) 32, 34 E3
Fulzar II (Guj) 47 A2
Gadag (Kar) 64-66, 68, 69 B3
Gadagad (Mah) 88 C1
Gadai Maharajpur (W.B) 142 B2, 142 D2
Gadarwara (M.P) 77, 78 E3
Gadhada (Guj) 44-46 B3
Gadhinglaj (Mah) 85, 86 C3
Gadhra (Bih) 34 D7
Gadilam (T.N) 120 B2
Gadola (Raj) 115 D3
Gadwal (A.P) 17, 18 B2
Gagan (M.P) 80 C3
Gahmar (U.P) 132 D5
Gaighata (W.B) 140 D6
Gaighatti (Bih) 32 C3
Gaikhuri Hills (Mah) 84 EF1
Gairatganj (M.P) 77 E3
Gajansar (Guj) 47 A2
Gajansar Nara (Guj) 47 A2
Gajapatinagaram (A.P) 17 E1
Gajendragarh (Kar) 65 B3
Gajner (Raj) 114 C2
Gajod (Guj) 47 A2
Gajol (W.B) 140 D3
Gajraula (U.P) 134, 136 B3
Gajuladinne (A.P) 20 B3
Gajwel (A.P) 17 C2
Galatga (Kar) 65 B2
Galibaga (Goa) 41 B2
Galhati (Mah) 88 C2
Gali Konda (A.P) 16 E1
Galna Hills (Mah) 84 C1
Galsi (W.B) 140 C5
Galwa (Raj) 115 E3
Gambhiramgadda (A.P) 20 E2
Gambheri (Raj) 115 D3
Gamdi (Guj) 45 C2
Gamharia (Bih) 32 D7
Ganapavaram (A.P) 17 D2
Ganaur (Har) 51, 53 CD3
Gandak (U.P) 135 D4
Gandak (Bih) 31, 37 B2
Gande (Bih) 32 D5
Ganderbal (J&K) 59, 60 B2
Gandevi (Guj) 44-46 C3, 47 D3
Gandhamardan Parbat (Ori) 100 B2
Gandhi Sagar (M.P) 76 C2
Gandhi Sagar (Raj) 111 D3
Gandhidam (Guj) 45, 46 B2
Gandhinagar (Guj) 44-46, 48 C2
Gandipalam (A.P) 20 C3
Ganeshpur (W.B) 142 C4
Gang Plain (Raj) 111 C1
Ganga (W.B) 139 C2, 144 C2
Ganga (M.P) 76 F2
Ganga (U.P) 130, 135 B3
Ganga (Bih) 31 B2, 37 C3
Ganga Delta (W.B) 139 C3
Ganga Ghat (Bih) 31 B3
Ganga Ghat (U.P) 132 C4
Ganga Parshad (Bih) 34 E4
Ganga Yamuna Doab (U.P) 130 B3-4
Gangabal (J&K) 60 B2
Gangadhara (A.P) 17 C1
Gangajalghati (W.B) 140 C5
Gangakher (Mah) 85-87 D2
Ganganagar (Raj) 112-114, 116 C1
Gangang (Mah) 88 D1
Gangapar Plain (U.P) 130 D5
Gangapur (Bih) 32 B4
Gangapur (Raj) 112-114 E2
Gangapur (Raj) 113 D3
Gangapur (Kar) 66 C2
Gangapur (Mah) 85-87, 89 C2
Gangarampur (W.B) 140 D3, 142 C2
Gangau (M.P) 79, 80 E2
Gangavali (Kar) 67 B3
Gangawathi (Kar) 64-66 C3
Gangdhar (Raj) 112 D4
Gangetic Plain (U.P) 130 CD3-4
Ganghara (Bih) 34 B4
Gangoh (U.P) 132 A3
Gangoli (Kar) 65 B4
Gangotri (U.P) 134 B2
Gangpur Basin (Ori) 100 C1
Gangrar (Raj) 112 D3
Gangtok (Sik) 118, 119 B2
Gangulpara (M.P) 80 F4
Ganj Dundwara (U.P) 132 B4
Ganjam (Ori) 101, 102, 105 C3
Ganjam Coast (Ori) 100 C3
Ganmo (H.P) 54 C2
Gannavaram (A.P) 17, 18 D2
Ganpatipule (Mah) 87 B3
Garalbari (W.B) 142 C1
Garautha (U.P) 131 B5
Garbara (Guj) 45 D2
Garchiroli (Mah) 85, 91 F1
Garda Hills (Guj) 43 A2
Garden Reach (W.B) 140 D6, 145 A2
Gardoani (W.B) 142 C3
Garhakota (M.P) 78, 79 E3
Garhbeta (W.B) 142 B3, 140 C6
Garhchiroli (Mah) 87 D1
Garhi (Raj) 112 D4
Garhjai Hills (Ori) 100 C2
Garhmuktesar (U.P) 131, 132, 134 B3
Garhshankar (Pun) 107 D2, 108 D3
Garhwal (U.P) 131 B3, 137 B2
Gariaband (M.P) 79 G4
Gariyadhar (Guj) 44, 45 B3
Garkha (Bih) 32 B4
Garo Hills (Meg) 94 B2
Garobadha (Meg) 94 A2
Garot (M.P) 77, 78 C2
Garu (Bih) 32 B6
Garwa (Bih) 32, 34 A5, 36 A3
Gasherbrum (J&K) 58 C2
Gaspar Dias Beach (Goa) 41 A1
Gaumukh (U.P) 134 B2
Gaunaha (Bih) 32 B2
Gaundongrem (Goa) 41 B2
Gaur (W.B) 143 C2
Gaur Plain (W.B) 139 BC2
Gaurela (M.P) 78 F3
Gaurhpuri (Mah) 87 B2
Gauri Bazar (W.B) 136 D4
Gauribidanur (Kar) 64-66 C4
Gauripur (Ass) 27 A2
Gawan (Bih) 32 C5, 39 B3
Gawilgarh Hills (Mah) 84 D1
Gaya (Bih) 32, 34, 36, 38 B5, 39 B3, 40 B2
Gayatri (Ker) 71, 74 B2
Gayatri Jalashaya (T.N) 120 A1
Gayatri Jalashaya (Kar) 63 C4
Gayatri Jellashaya (Kar) 67 C4
Gersoppaghat (Kar) 63 B3
Gevrai (Mah) 85, 86 C2
Ghaggar (Pun) 107 D3
Ghaggar Branch (Pun) 107 C3-4, 109 CD3
Ghaggar Nali (Har) 50 C2
Ghagghar (Har) 50, 52 C2, 52 B3
Ghaghar (U.P) 130 D5, 135 D5
Ghaghara (U.P) 130 D4, 130 C4, 135 C4, CD4
Ghaghara (Bih) 31 B1, 37 B2
Ghaghara Barrage (U.P) 135 C3
Ghaghra (Bih) 32 B6, 36, 37 B4
Ghaghri (Bih) 37 B4
Ghamarwin (H.P) 54 B3
Ghanpur (A.P) 17 C2
Ghanshyampur (Bih) 32 D3
Ghansoli (Mah) 86 B2
Gharaunda (Har) 51, 52 C3
Gharghoda (M.P) 77 G3
Gharprabha (Kar) 66, 67 B2
Ghaspani (Nag) 99 B3
Ghatal (W.B) 140 C6, 142, 143 B3
Ghatampur (U.P) 131, 132, 134, 136 C4
Ghatanji (Mah) 86 E1
Ghataro Chaturbhuj (Bih) 34 C4
Ghatisubramanya (Kar) 66 C4
Ghatol (Raj) 112 D4
Ghatprabha (Mah) 84 C3
Ghatshila (Bih) 32, 34 D7, 36, 38 C4
Ghaziabad (U.P) 131, 132, 134, 136 A3
Ghazipur (U.P) 131, 132, 134, 136 D5
Ghee (Guj) 47 A2
Ghelo (Guj) 47 B3
Gheora (Del) 150 B3
Ghirni (Mah) 88 D2
Ghod (Mah) 88 C2
Ghodadhro (Guj) 47 B2
Ghodavadi (Guj) 47 B3
Ghodegaon (Mah) 85 B2
Ghogha (Guj) 44, 45 C3
Ghogha Mahal (Guj) 44 C3
Ghontval (Guj) 43 C3
Ghorajhari (Mah) 88 E1
Ghorasahan (Bih) 32 C3
Ghori (U.P) 135 D5
Ghosai (Bih) 34 DE4
Ghoshpur (W.B) 142 B3
Ghosi (U.P) 131, 132 D4
Ghosi (Bih) 32 C4
Ghotaru (Raj) 114 B2
Ghoti (Mah) 86 B2
Ghughumari (W.B) 142 C1
Ghugri (W.B) 139 B2
Ghugri (Bih) 31 C2, 37 C3
Ghugus (Mah) 86 E2
Ghujerab Range (J&K) 58 B1
Ghum (W.B) 143 C1
Gidam (M.P) 79 F5
Gidan Mata (Raj) 111 D3
Giddalur (A.P) 17-19 C3
Giddarbaba (Pun) 108 B4
Gidheswari (Bih) 37 C3
Gidi (Bih) 34 C6
Gidni (W.B) 146 B3
Gilgit (J&K) 58 A1, 59-61 B2
Gilgit Mountains (J&K) 58 A1
Gilgit Wazarat (J&K) 59 B2
Gilhuria (Bih) 39 C3
Gingee (T.N) 121, 123 B1
Gingee Hills (T.N) 120 B1
Ginnuragarh (M.P) 79 D3
Gir Range (Guj) 43 B3
Gird Gwalior (M.P) 77 DE1
Girgithi (U.P) 135 D4
Giriak (Bih) 32 C4
Giridih (Bih) 32, 34 D5, 36, 39 C3
Giridih Upland (Bih) 31 C2
Girinanda (Raj) 115 CD2
Girna (Mah) 84, 88 C1
Girnar (Guj) 46 B3
Girnar (Guj) 43 B3
Girnar Hills (Guj) 43 B3
Giroli (Mah) 88 D1
Goa Vela (Goa) 41 A1
Goalpara (Ass) 26-29 B2
Goalpokhai (W.B) 140 D2
Goaltor (W.B) 140 C6
Gobardanga (W.B) 142 C3
Gobindganj (Bih) 36 B2
Gobindgarh (Pun) 108 D4
Gobindpur (Bih) 32 C5, 32 D6, 34 C4
Gocham (Ar.P) 25 C2
Godahado (Ori) 104 C3
Godavari (A.P) 16, 20 D2
Godavari (Mah) 84, 88 C2
Godavari (Mah) 76 F5
Godavari (Upper) (Mah) 84 C2
Godavari Canal (Mah) 88 C2
Godavari Delta (A.P) 16 DE2
Godavari Delta (M.P) 20 D2
Godda (Bih) 32, 34 E5, 36 C3
Godhatad (Guj) 47 A2
Godhra (Guj) 44-46 C2
Godri Nadi (M.P) 80 F5
Godwar Plain (Raj) 111 C3
Goghat (W.B) 140 C6
Gogri (Bih) 32 D4
Gogunda (Raj) 112 C3
Goh (Bih) 32 B5
Gohad (M.P) 77, 78 E1
Gohana (Har) 50-52 C3
Goharganj (M.P) 77 D3
Gohelwar Coast (Guj) 43 B3
Gohpur (Ass) 26 C2
Goiiavagu (A.P) 20 C1
Goilkera (Bih) 32 C7
Gojharia (Guj) 45 C2
Gokak (Kar) 64-66, 68 B2
Gokaran (Kar) 66 B3
Gokarn (Kar) 65, 68 B3
Goki (Mah) 88 E1
Gola (Bih) 32 C6, 36, 39 B4
Gola Gokarannath (U.P) 132, 136 C3
Golaghat (Ass) 26, 28 C2, 27 E2, 29 C1
Golai (Bih) 37 B3
Golakganj (Ass) 26 A2
Golconda (A.P) 17, 19 BC2
Golconda Plateau (A.P) 16 C2
Golianahati (W.B) 142 C1
Gollapalli Plateau (M.P) 76 F5
Golmuri-cum-Jugsalai (Bih) 32 D7
Goma (Guj) 47 B2
Gomati (Guj) 43 D1
Gomati (U.P) 130, 135 C4
Gomati Plain (U.P) 130 CD4
Gomoh (Bih) 34 D6
Gomti Creek (Guj) 43 A2
Gomukhi (T.N) 124 B2
Gona (W.B) 139, 144 C4
Gonda (U.P) 131, 132, 134, 137 C4
Gondal (Guj) 44-46 B3
Gondar (Har) 51 C3
Gondia (Mah) 85, 86, 89, 91 F1
Gondli (Guj) 47 B2
Gondli Tank (M.P) 80 F4
Gondwara Hills (M.P) 76 EF3
Goniana (Pun) 108 B4
Gooty (A.P) 17-19 B3
Gopalapuram (A.P) 17 D2
Gopalganj (Bih) 32, 34 B3, 36 B2
Gopalnagar (W.B) 142 C3
Gopalpur (Ori) 102, 103, C3
Gopalpur (Bih) 32 E4
Gopalpur (W.B) 142 B2
Gopalpura (Raj) 115 E3
Gopeshwar (U.P) 131, 134 B2
Gopiballabhpur (W.B) 140 B6
Gopichettipalaiyam (T.N) 121, 122, 126 A2
Gopiganj (U.P) 132 D5
Gopikandar (Bih) 32 E5
Gorakhpur (U.P) 131-134, 136-138 D4
Gorakhpur (Har) 51 B3
Goras (M.P) 79 D2
Goraul (Bih) 32 C4
Goraya (Pun) 108 C3
Goreakothi (Bih) 32 B3
Gorebal (Kar) 65 C3
Gorhia (U.P) 132 C4
Goriabahar (M.P) 80 FG5
Gormi (M.P) 78 E1
Gorsoppa (Kar) 66 B3
Gorubathan (W.B) 140 D2
Gorumahisani (Ori) 105 D1
Gorumara W.L.S. (W.B) 143 C1
Gosaba (W.B) 140 D6
Gosainganj (U.P) 132 D4
Goshikhurd (Mah) 88 E1
Gossaigaon (Ass) 26 A2
Gosthani Reservoir (A.P) 16 E1
Gotan (Raj) 116 C2
Gotehole (Kar) 67 B4
Gothera (Raj) 113 D2
Gothini (Bih) 32 B3
Gotlia Creek (Guj) 43 A2
Govardhan (U.P) 132 A4
Govind Ballabh Pant Sagar (M.P) 76 G2
Govind Ballabah Pant Sagar (U.P) 130 D5
Govind Palli (Ori) 104 B3
Govind Sagar (H.P) 54, 55, 56 B3
Govind Sagar (U.P) 130, 135 B5
Govindgarh (M.P) 79 F2
Govindgarh Pisangan (Raj) 115 D2
Govindghat (U.P) 134 B2
Govindwal (Pun) 109 C2
Govindwal Sahib (Pun) 110 C3
Gowai (Bih) 37 C4
Grand Anicut (T.N) 124 B2
Grandi I (Goa) 41 A1
Great Gandak (U.P) 130 E4
Great Gandak (Bih) 31 B1, 37 B2
Great Himalaya (J&K) 58 BC2-3
Great Himalaya (H.P) 54 CD2-3
Great Himalaya (Sik) 118 AB1
Great Himalaya (Ar.P) 24 BC1
Great Himalaya (U.P) 130 BC2
Great Indian Desert (Raj) 111 BD1-2
Great Karakoram (J&K) 58 BD1-2
Great Nicobar (A&N) 148 B3
Greater Bombay (Mah) 90 B2
Gua (Bih) 34 C7
Guasuba (W.B) 139, 144 C4
Gubbi (Kar) 64, 65, 68 C4
Gubra (M.P) 79 E3
Gudalur (T.N) 121-123, 126 A2, 122 A3
Gudalur Pass (Ker) 71 B3
Gudari (Ori) 102 B3
Gudem (A.P) 18 D2, 19 E2
Gudgeri (Kar) 65 B3
Gudha (Raj) 115 D3
Gudibanda (Kar) 64 C4
Gudivada (A.P) 17, 18 D2
Gudiyatham (T.N) 121-123 B1
Gudur (A.P) 17-19 C3
Guhagar (Mah) 85 B3, 87 B3
Guhal (Guj) 47 C2
Guhla (Har) 50 C2
Guindy N.P (T.N) 123 C1
Gujarat Coast (Guj) 43 C3
Gujarat Plains (Guj) 43 C2-3
Gula (Har) 52 C2
Gulam (Bih) 34 D4
Gulavthi (U.P) 132 A3
Gulbarga (Kar) 64-66, 68, 70 C2
Guledagudda (Kar) 65 B2
Gulf of Kachchh (Guj) 43 A2
Gulf of Khambhat (Guj) 43 C3
Gulf of Manar (T.N) 120 B3
Gulganj (M.P) 79 E2
Gulgulpat (M.P) 76 H3
Gulmarg (J&K) 60 B2
Gumani (Bih) 37 C3
Gumia (Bih) 32, 34 C6, 38 B4
Gumla (Bih) 32, 34 B6, 36, 39 B4
Gummidipundi (T.N) 121, 125 C1
Gumti (Tri) 128 A2
Guna (M.P) 77-79 D2
Gunamanipara (Tri) 128 A2
Gunasada (Guj) 48 D3
Gunda (Guj) 47 B2
Gunda Road (Kar) 66 C3
Gundalamma Konda (A.P) 16 E2
Gundar (Ker) 74 B3
Gundlakamma (A.P) 16 C3
Gundlavagu (A.P) 20 D1
Gundlupet (Kar) 64-66 C5
Gundrampalli (A.P) 18 C2
Gunnar (T.N) 120, 124 B3
Gunnaur (U.P) 131, 132 B3
Gunri (Bih) 34 B4
Guntakal (A.P) 18, 19 B3
Guntupalle (A.P) 19 D2
Guntur (A.P) 17-19, 21 D2
Guntur Plains (A.P) 16 CD2
Gunupur (Ori) 101, 102 B3

179

184

ABBREVIATIONS

States: A.P —Andhra Pradesh; Ar.P—Arunachal Pradesh; Ass—Assam; Bih—Bihar; Goa—Goa; Guj—Gujarat; Har—Haryana; H.P—Himachal Pradesh; J&K—Jammu & Kashmir; Kar—Karnataka; Ker—Kerala; M.P—Madhya Pradesh; Mah—Maharashtra; Man—Manipur; Meg—Meghalaya; Miz—Mizoram; Nag—Nagaland; Ori—Orissa; Pun—Punjab; Raj—Rajasthan; Sik—Sikkim; T.N—Tamil Nadu; Tri—Tripura; U.P—Uttar Pradesh; W.B-West Bengal.
Union Territories: A&N—Andaman and Nicobar Islands; Chan—Chandigarh; DNH—Dadra & Nagar Haveli; D&D—Daman & Diu; Del—Delhi; Lak—Lakshadweep; Pon—Pondicherry.